THE ETHICS OF
CLIMATE CH

M000204942

Global climate change is one of the most daunting ethical and political challenges confronting humanity in the twenty-first century. The intergenerational and transnational ethical issues raised by climate change have been the focus of a significant body of scholarship. In this new collection of essays, leading scholars engage and respond to first-generation scholarship and argue for new ways of thinking about our ethical obligations to present and future generations. Topics addressed in these essays include moral accountability for energy consumption and emissions, egalitarian and libertarian perspectives on mitigation, justice in relation to cap-and-trade schemes, the ethics of adaptation, and the ethical dimensions of the impact of climate change on nature.

DENIS G. ARNOLD is the Jule and Marguerite Surtman Distinguished Scholar in Business Ethics at the University of North Carolina at Charlotte. He is the author of *The Ethics of Global Business* (2011) and the editor of *Ethics and the Business of Biomedicine* (Cambridge, 2009).

THE ETHICS OF GLOBAL CLIMATE CHANGE

EDITED BY

DENIS G. ARNOLD

CAMBRIDGE
UNIVERSITY PRESS

CAMBRIDGE
UNIVERSITY PRESS

University Printing House, Cambridge CB2 8BS, United Kingdom

Published in the United States of America by Cambridge University Press, New York

Cambridge University Press is part of the University of Cambridge.

It furthers the University's mission by disseminating knowledge in the pursuit of
education, learning and research at the highest international levels of excellence.

www.cambridge.org
Information on this title: www.cambridge.org/9781107666016

© Cambridge University Press 2011

First published 2011
First paperback edition 2014

A catalogue record for this publication is available from the British Library

Library of Congress Cataloguing in Publication data
The ethics of global climate change / edited by Denis G. Arnold.
p.   cm.
Includes bibliographical references and index.
ISBN 978-1-107-00069-8
1. Environmental ethics.   2. Environmental responsibility.   3. Climatic changes – Moral
and ethical aspects.   4. Global warming – Moral and ethical aspects.   I. Arnold, Denis Gordon.
GE42.E844   2011
179′.1–dc22
2010031762

ISBN 978-1-107-00069-8 Hardback
ISBN 978-1-107-66601-6 Paperback

Contents

v

Illustrations

vii

CHAPTER 9

Contributors

W. NEIL ADGER is Professor of Environmental Economics, School of Environmental Sciences, University of East Anglia, and Programme Manager, Tyndall Centre for Climate Change Research.

DENIS G. ARNOLD is the Jule and Marguerite Surtman Distinguished Scholar in Business Ethics at the University of North Carolina at Charlotte.

LUC BOVENS is Professor and Head of the Department of Philosophy, Logic and Scientific Method, London School of Economics and Political Science.

PHILIP CAFARO is Associate Professor of Philosophy at Colorado State University.

SIMON CANEY is Professor in Political Theory, University Lecturer, and Fellow and Tutor in Politics, Magdalen College, University of Oxford.

MARY R. ENGLISH is Research Leader, Institute for a Secure and Sustainable Environment, University of Tennessee, Knoxville.

STEPHEN GARDINER is Associate Professor of Philosophy at the University of Washington.

DALE JAMIESON is Director of Environmental Studies, Professor of Environmental Studies and Philosophy, and Affiliated Professor of Law at New York University.

ALICE KASWAN is Professor of Law at the University of San Francisco School of Law.

SARAH KRAKOFF is Professor of Law at the University of Colorado Law School.

DARREL MOELLENDORF is Professor of Philosophy and Director of the Institute for Ethics and Public Affairs at San Diego State University.

RICHARD D. MORGENSTERN is Senior Research Fellow, Resources for the Future.

SOPHIE NICHOLSON-COLE is Senior Research Associate at the Tyndall Centre for Climate Change Research.

JOHN NOLT is Professor of Philosophy at the University of Tennessee, Knoxville.

CLARE PALMER is Professor of Philosophy, Texas A&M University.

HENRY SHUE is Senior Research Fellow at Merton College, University of Oxford.

ROBERT H. SOCOLOW is Professor of Mechanical and Aerospace Engineering and Co-Director of the Carbon Mitigation Initiative at Princeton University.

Acknowledgements

This volume originated with a conference at the University of Tennessee in Knoxville on "Energy and Responsibility" in April 2008. Early versions of half of the papers collected here were presented at that conference, while the remainder of the papers were commissioned specifically for this volume. The conference was co-sponsored by a variety of organizations at the University of Tennessee including the Department of Philosophy, the Center for Applied and Professional Ethics, the Institute for a Secure and Sustainable Environment, the College of Architecture, the College of Business Administration, and the College of Law. The conference also benefited from the financial support of Alcoa, Oak Ridge National Laboratory, the Tennessee Valley Authority, and the Charter of Human Responsibilities. Special thanks are due to conference organizers John Nolt, Mary R. English, Heather Douglas, Nina Gregg, Becky Jacobs, and Corrine Martin, as well as conference supporters John Hardwig and Bruce Bursten.

My work on this project was supported by the Surtman Foundation and the Belk College of Business at the University of North Carolina, Charlotte. Jason Hubbard and Rebecca Glavin provided excellent research assistance. I am grateful for this support. My editor at Cambridge, Hilary Gaskin, once again provided wise counsel and timely support. Her dedication to the highest standards of publishing is much appreciated. For their love, support, and patience with this project I am especially grateful to Sara and Skye.

Introduction: climate change and ethics

Denis G. Arnold

This book is an interdisciplinary collection of mainly normative essays written by philosophers, scientists, legal scholars, and an economist. The complex intergenerational ethical issues that climate change raises have been the subjects of a significant body of recent scholarship.[1] The original scholarship collected in this volume is distinctive in that this is the first second-generation collection of essays to appear on the ethics of climate change. The contributors to this volume engage and respond to the first-generation literature, and because climate change is perhaps the largest collective action problem ever confronted by humanity, several contributors argue for new ways of conceptualizing our ethical obligations in order to address a problem of this scope and difficulty. This volume is also the first in which a group of scholars critically engages the outcomes of the Copenhagen climate conference (Conference of the Parties 15) in which the nations of the world once again tried to reach an agreement about how to slow or stop anthropogenic climate change.

It is well understood that the Earth's climate is changing as a result of human activity. More specifically, the climate is changing because of the inefficient consumption of fossil fuels and rapid deforestation. A changing climate will place present and future human populations in jeopardy and the poor will be most adversely impacted. By climatologists, geologists, oceanographers, and other scientists working on problems related to climate change this is well understood. To many in the business community these are facts that have been incorporated into current operations and long-term strategic plans. To many policymakers at the local, state or provincial, and federal levels, these are facts that demand sound public policy. But to a vocal minority such claims are no more than the rantings of muddle-headed environmentalists and wealthy liberals. Skepticism about harmful, anthropogenic climate change, and the need for mitigation and adaption,

---

[1] The bibliography at the end of this volume provides a select overview of that literature.

is expressed by small businessmen, physicians, graduate students, politicians, and even some research scientists. A recent Gallup Poll found that 40 percent of Americans were "only a little" or "not at all" concerned about global warming.[2] Approximately the same percentage (39 percent) believe that there is "a lot of disagreement" among climate scientists about whether the Earth has been warming in recent years and (42 percent) about whether human activity causes climate change.[3] Remarks by conservative commentators in the media imply that anthropogenic climate change is a myth perpetrated by liberals with guilty consciences and that redeploying resources to combat climate change will cause more harm than good.[4]

Each of the contributors to this book grounds his or her arguments on the premise that anthropogenic climate change has been occurring, continues to occur, and poses a significant threat to human populations. But such a premise is one that remains contentious outside the scientific community, academia, and the boardrooms of corporations. Readers of this book who are either skeptical of climate change, or unsure of what to think, may be unwilling to grant the premise. For this reason it is necessary to spend some time assessing current scientific opinion on climate change.

I IS THERE A SCIENTIFIC CONSENSUS REGARDING GLOBAL CLIMATE CHANGE?

Views expressed on science may be divided into three broad categories. First, there is the peer-reviewed research that appears in leading scientific journals. This work is typically vetted by editors and external peer reviewers who have expertise on the precise issues being addressed. The editorial boards of these journals are populated by senior academic and government scientists. Second, there are summaries of such research, concurring statements, and policy statements prepared for use by policymakers and the general public by teams of scientists. This includes the work of the Intergovernmental Panel on Climate Change (IPCC) as well as that of the American Association for the Advancement of Science, the National Research Council of the National Academy of Sciences, and the European Science Foundation. Third, there are opinion pieces in newspapers, blogs,

---

[2] Gallup Poll completed between March 5 and 8, 2009. N = 1,012 adults nationwide. MoE ± 3 (for all adults). Available at www.pollingreport.com/enviro.htm.

[3] *Newsweek* Poll conducted by Princeton Survey Research Associates International, August 1–2, 2007. N = 1,002 adults nationwide. MoE ± 4 (for all adults). Available at www.pollingreport.com/enviro.htm.

[4] For examples, consult the editorial pages of the *Wall Street Journal* (the *Journal*'s science reporters do not echo the stance of the editorial page writers) or *The Glenn Beck Program* on Fox News Channel.

industry-sponsored position papers, and even vanity journals published with the intention of advancing an ideological perspective rather than advancing science.[5]

Critics of the view that there is a consensus on climate change typically appeal to sources in the third category. Instead of advancing their position via credible scientific papers, critics typically broadcast their message through the pronouncements of think tanks and self-proclaimed experts. In fact, according to one review of this debate, nearly all climate change skeptics are "economists, business people or politicians, not scientists."[6] In its article, "Meet the Global Warming Skeptics," the magazine *New Scientist* examines the connections of many of the prominent climate change skeptics. The authors note that the Competitive Enterprise Institute, a free-market lobby organization, is made up of two lawyers, an economist, a political scientist, a graduate in business studies, and a mathematician. Similarly, the American Enterprise Institute, another free-market lobbying organization, has only one natural scientist, a chemist. Both of these lobby groups are funded by ExxonMobil, as are the George C. Marshall Institute and the International Policy Network, leading think tanks promoting global climate change skepticism.[7] The tobacco industry used similar techniques in an effort to promote its agenda and undermine public health efforts regarding the dangers of smoking.[8] Indeed, former US Senator and former US Undersecretary of State Timothy Wirth argues that climate change deniers "patterned what they did after the tobacco industry. Both figured, sow enough doubt, call the science uncertain and in dispute. That's had a huge impact on both the public and Congress."[9]

---

[5] For an overview of the climate change denial industry, see Sharon Begley, "The Truth About Denial," *Newsweek*, August 13, 2007. Available at www.newsweek.com/id/32482; and "Meet the Global Warming Skeptics," *New Scientist*, 2486 (February 12, 2005): 40. For an example of a vanity journal publication on climate change, see "Environmental Effects of Increased Atmospheric Carbon Dioxide," *Journal of American Physicians and Surgeons*, 12(3) (Fall 2007). Available at www.jpands.org/vol12no3/robinson.pdf. This journal is not listed in major scientific databases such as PubMed or ISI Web of Knowledge. However, the article is featured prominently on the pages of the Heartland Institute, a "free market" think tank and a center of climate change skepticism. See www.heartland.org/policybot/results/22434/Environmental_ Effects_of_Increased_Atmospheric_Carbon_Dioxide_updated.html.
[6] Fred Pearce, "Climate Change: Menace or Myth?," *New Scientist*, 2486 (February 12, 2005): 38.
[7] *New Scientist*, "Meet the Global Warming Skeptics," 40.
[8] For an overview, see, for example, Allan M. Brandt, *The Cigarette Century: The Rise, Fall, and Deadly Persistence of the Product That Defined America* (New York: Basic Books, 2009). Creationists use similar methods to cast doubt on Darwinian evolutionary theory. For an assessment of their arguments, see, for example, Philip Kitcher, *Abusing Science: The Case Against Creationism* (Cambridge, MA: MIT Press, 1983).
[9] Begley, "The Truth About Denial."

An analogy may help us to better understand the influence of these non-scientifically grounded lobbying efforts. Imagine for a moment that you had to make a life or death medical decision about a loved one. Imagine, for example, that you are contemplating approving a new procedure for the treatment of heart valve disease in your child. Although the procedure is new, there is a clear consensus in the peer-reviewed medical journals about the usefulness and safety of the procedure. However, the opposite view is represented in the various other sources including vanity journals, the opinion pages of some magazines, and some research bought and paid for by companies that stand to lose financially if the procedure is widely adopted. Upon which sources would you rely to make an informed judgment? Obviously most people would rely on the consensus opinion of research published in respected peer-reviewed medical journals. The analogy is not far-fetched since a radically altered climate is likely to adversely impact the welfare of future children.

But is there really a consensus in the scientific literature regarding climate change? Are climate scientists really in agreement on this question? What evidence is there for such conclusions? Before answering these questions, it will be helpful to briefly review some recent history. In the 1980s scientists noticed that the Earth's climate was changing. In the late 1980s the Intergovernmental Panel on Climate Change (IPCC) was formed in order to investigate these changes.[10] The IPCC quickly gained credibility by offering cautious conclusions concerning climate change that were grounded in rigorous scientific studies.[11] The third IPCC climate change report, released in 2001, confirmed that the majority of Earth scientists were convinced that climate change was happening and that the release of anthropogenic greenhouse gases (GHGs), such as carbon dioxide and methane, was the main cause. Does the peer-reviewed scientific literature support these conclusions? It does so unequivocally. In an important study of the scientific literature Naomi Oreskes examined 928 articles on climate change published in peer-reviewed journals between 1993 and 2003.[12] She found that *none* of these articles disagreed with the main conclusions of the IPCC. According to Oreskes, "there is a scientific consensus on the reality of anthropogenic climate change.

[10] Spencer R. Weart, *The Discovery of Global Warming* (Cambridge University Press, 2003), 160.
[11] Ibid., p. 162.
[12] Naomi Oreskes, "Beyond the Ivory Tower: The Scientific Consensus on Climate Change," *Science*, 306 (December 3, 2004): 1686.

Climate scientists have repeatedly tried to make this clear. It is time for the rest of us to listen."[13]

More recently Peter Doran and Maggie Kendall Zimmerman surveyed Earth scientists with academic affiliations along with those at state geologic surveys, at US federal research facilities, and at US Department of Energy national laboratories about their views on climate change.[14] Of the 3,146 Earth scientists who completed the survey, 90 percent believe that global temperature levels have risen in comparison to pre-1800s levels and 82 percent believe that "human activity is a significant contributing factor in changing mean global temperatures."[15] Among those surveyed, "the most specialized and knowledgeable respondents (with regard to climate change) are those who listed climate science as their area of expertise and who also have published more than 50% of their recent peer-reviewed papers on the subject of climate change."[16] Of these specialists, 96.2 percent (76 of 79) believe that global temperature levels have risen in comparison to pre-1800s levels" and 97.4 percent (75 of 77) believe that "human activity is a significant contributing factor in changing mean global temperatures."[17]

The two areas of expertise in the survey with the smallest percentage of participants indicating that they believed that climate change resulted from anthropogenic activity were those in economic geology, the study of the Earth for economic gain, with 47 percent (48 of 103), and meteorology, which tends to focus on short-term climate patterns, with 64 percent (23 of 36).[18] Doran and Kendall Zimmerman reach the following conclusion: "It seems that the debate on the authenticity of global warming and the role played by human activity is largely nonexistent among those who understand the nuances and scientific basis of long-term climate processes."[19]

To revisit our medical analogy, imagine that 97 percent of pediatric cardiothoracic surgeons share a judgment about the usefulness and safety of the heart procedure you are considering for your child, but only 47 percent of pharmacologists working for industry and 64 percent of dermatologists agree. Upon whose judgment will you rely? The consensus opinion of those physicians who perform the procedure and attend continuing medical education classes that review best practices in pediatric heart surgery, or

[13] Ibid. See, also, her recent book with Erik M. Conway, *Merchants of Doubt: How a Handful of Scientists Obscured the Truth on Issues from Tobacco Smoke to Global Warming* (New York: Bloomsbury Press, 2010).
[14] Peter Doran and Maggie Kendall Zimmerman, "Examining the Scientific Consensus on Climate Change," *EOS*, 20(3) (January 20, 2009): 21–22.
[15] Ibid., 21.   [16] Ibid.   [17] Ibid.   [18] Ibid., 22.   [19] Ibid.

the somewhat mixed opinions of non-specialists (some of whom may lose work if the procedure is widely adopted)?

It is not surprising then that the findings of the IPCC regarding global climate change have been endorsed by most major scientific organizations including the Academies of Science for the G8+5 in a joint statement.[20] This includes the National Science Academies of Brazil, Canada, China, France, Germany, India, Italy, Japan, Mexico, Russia, South Africa, the UK, and the USA. Additionally, the IPCC's findings have received concurring assessments from the American Association for the Advancement of Science, the National Research Council, the European Science Foundation, the American Geophysical Union, the European Federation of Geologists, the European Geosciences Union, the Australian Meteorological and Oceanographic Society, the American Meteorological Society, the Australian Meteorological and Oceanographic Society, the American Chemical Society, the American Physical Society, the American Statistical Association, and others.[21]

Leading global companies have also recognized that a scientific consensus exists regarding anthropogenic climate change and have taken proactive measures to address GHG emissions.[22] These include Alcan, Alcoa, BP, BHP Billiton, Dow Chemical, Iberdrola, Novo Nordisk, Scottish Power, Royal Dutch Shell, STMicroelectronics, and Weyerhauser, among others.[23] As early as 1997, then Alcoa Chairman and Chief Executive Officer Paul O'Neil recognized the scientific consensus on climate change and directed his team to reduce GHG emissions.[24] (O'Neil later served as US Treasury Secretary in 2001–2002.) During his tenure as president of Shell Oil Company, John Hofmeister criticized those who still argue that the science is unclear. "We have to deal with greenhouse gases," he said at a 2006 speech at the National Press Club. "From Shell's point of view, the debate is over. When 98 percent of scientists agree, who is Shell to say, 'Let's debate the science'?" On another occasion he stated, "It's a waste of time to debate it.

[20] "Joint Science Academies' Statement: Climate Change Adaptation and the Transition to a Low Carbon Society" (June 2008). Available at www.nationalacademies.org/includes/climatechangestatement.pdf.
[21] Links to the original documents may be found at Wikipedia contributors, "Scientific Opinion on Climate Change," *Wikipedia, The Free Encyclopedia*, en.wikipedia.org/w/index.php?title=Scientific_opinion_on_climate_change&oldid=278716034.
[22] For a discussion of the ethical obligations of businesses regarding climate change, see Denis G. Arnold and Keith Bustos, "Business, Ethics, and Global Climate Change," *Business and Professional Ethics Journal*, 22(2/3) (Summer/Fall 2005): 103–130.
[23] *Business Week*, "The Race Against Climate Change: How Top Companies Are Reducing Emissions of CO_2 and Other Greenhouse Gases," December 12, 2005.
[24] Alcoa, *1997 Annual Report*, 4.

Policymakers have a responsibility to address it. The nation needs public policy. We'll adjust."[25] John Chambers, chairman and chief executive officer of Cisco, has said, "It [climate change] is not a question of if. It is." He added, "There is no doubt in hardly any of the well-educated minds that if we don't act quickly, we are going to have a tremendous problem on our hands."[26] According to Chambers, "Mitigating the impacts of climate change is critical to the world's economic and social stability."[27]

We have seen that there is a clear consensus in the scientific community regarding climate change that has been endorsed not merely by environmental organizations but by leading corporations and governmental agencies. But what is the consensus view of climate change and its impact on the planet? These questions will be answered in the next section.

2 ENERGY CONSUMPTION AND CLIMATE CHANGE

The production and consumption of fossil fuels produces GHG emissions (e.g., carbon dioxide, methane, and nitrous oxide) that alter the energy balance of the Earth's climate and contribute to climate change. According to the US Energy Information Administration (EIA), if current laws and policies remain unchanged, global energy consumption is projected to increase by "50% from 462 quadrillion British thermal units (Btu) in 2005 to 563 Btu in 2030."[28] Demand is projected to rise by 85 percent in non-Organization for Economic Co-operation and Development (OECD) countries and by 19 percent in OECD countries, with the difference between the two groups primarily resulting from the projected economic growth in non-OECD countries.[29] While fossil fuels are expected to continue to supply much of the energy used worldwide, the rising costs of liquid fossil fuels are projected to drive their share from 37 percent in 2005 to 33 percent in 2030 due to projected increases in oil prices.[30] Still, demand for fossil fuel liquids is expected to increase from 84.3 million barrels per day in 2005 to 112.5 million barrels per day in 2030.[31]

[25] Associated Press, "Shell Oil Chief: U.S. Needs Global Warming Plan," September 8, 2006. Available at www.msnbc.msn.com/id/14733060/.
[26] Michael Kanellos, "Cisco CEO Takes Jab at Climate Change Deniers," CNET News, February 20, 2008.
[27] Antony Savvas, "NASA and Cisco Build Climate Change Reporting Platform," ComputerWeekly.com, March 4, 2009. Available at www.computerweekly.com/Articles/2009/03/04/235132/nasa-and-cisco-build-climate-change-reporting-platform.htm.
[28] Energy Information Administration (2008), "International Energy Outlook 2008: Highlights," www.eia.doe.gov/oiaf/ieo/pdf/highlights.pdf.
[29] Ibid., 1.   [30] Ibid.   [31] Ibid., 2.

According to the EIA, world electricity generation is expected to nearly double, increasing from about 17.3 trillion kilowatt hours (kW h) in 2005, to 24.4 trillion kW h in 2015, and finally 33.3 trillion kW h in 2030.[32] Sustained economic growth in non-OECD countries is expected to drive increased energy consumption in these nations by an average of 4.0 percent annually from 2005 to 2030. This is compared to a projected average increase of 1.3 percent annually for OECD countries over the same period.[33] Coal and natural gas are projected to account for the largest increases in fuel consumption for energy production, with a 3.1 percent projected annual growth rate for coal and a 3.75 percent annual growth rate for natural gas.[34]

The IPCC reports that between its Third Assessment Report (TAR) in 2001 and the Fourth Assessment Report (FAR) in 2007, new observations and modeling have "led to improvements in the quantitative estimate of radiative forcing."[35] (The impact of the warming or cooling properties of GHGs is measured in radiative forcing.) In particular, the IPCC's Fourth Assessment Report indicates with "very high confidence" that significant changes in the Earth's climate have occurred as a result of GHG emissions – as well as of deforestation and other anthropogenic factors – since 1750 and have resulted in warming.[36] Among these changes are an increase in the global atmospheric concentration of carbon dioxide from a pre-industrial value of about 280 ppm to 379 ppm in 2005, of methane from of about 715 ppb to 1,774 ppb in 2005, and of nitrous oxide from a preindustrial value of about 270 ppb to 319 ppb.[37] From 1995 to 2005 carbon dioxide radiative forcing increased by 20 percent, the largest change in at least the last 200 years.[38] "The combined radiative forcing due to increases in carbon dioxide, methane, and nitrous oxide is $+2.30$ W m^{-2}, and its rate of increase during the industrial era is *very likely* to have been unprecedented in more than 10,000 years."[39]

Given these findings the US Environmental Protection Agency (EPA) has stated that the "science clearly shows that concentrations of these gases are at unprecedented levels as a result of human emissions, and these high levels are very likely the cause of the increase in average temperatures and other changes in our climate."[40] The IPCC has also predicted future

---

[32] Ibid., 3.   [33] Ibid.   [34] Ibid., 4.

[35] "IPCC, 2007: Summary for Policymakers," in S. Solomon, D. Qin, M. Manning, Z. Chen, M. Marquis, K. B. Averyt, M. Tignor, and H. L. Miller (eds), *Climate Change 2007: The Physical Science Basis. Contribution of Working Group I to the Fourth Assessment Report of the Intergovernmental Panel on Climate Change* (New York: Cambridge University Press, 2007), 2.

[36] Ibid., 2–3 ff.   [37] Ibid., 2–3   [38] Ibid., 4.   [39] Ibid.

[40] US Environmental Protection Agency, "EPA Finds Greenhouse Gases Pose Threat to Public Health, Welfare/Proposed Finding Comes in Response to 2007 Supreme Court Ruling," April 17, 2009.

changes to the Earth's climate. These changes include "A warming of about 0.2°C per decade for the next two decades"[41] resulting in heatwaves,[42] heavy precipitation at high latitudes and decreases in precipitation in subtropical regions,[43] more intense typhoons and hurricanes,[44] and sea-level rises. It is well understood in the scientific community that global warming and sea-level rise will "continue for centuries due to the time scales associated with climate processes and feedbacks, even if greenhouse gas concentrations were to be stabilized."[45] It should also be pointed out that these predictions reflect conservative estimates based on consensus forecasting rather than the more pessimistic outcomes predicted by some scientists and recently reported to be occurring.[46]

The forecasted impact of climate change on ecosystems and human populations is substantial and largely negative. Negative forecasts include significant increases in droughts, floods, and coastal flooding; more severe weather events; loss of fisheries; widespread species extinctions; and widespread migration away from low-lying coastal regions and other high-risk areas.[47] The major risks to human health include the following:

- increases in malnutrition and consequent disorders, with implications for child growth and development;
- increased deaths, disease, and injury due to heatwaves, floods, storms, fires, and droughts;
- the increased burden of diarrheal disease;
- the increased frequency of cardiorespiratory diseases due to higher concentrations of ground-level ozone related to climate change;
- the altered spatial distribution of some infectious disease vectors.[48]

Given these scientific predictions one can understand why the EPA ruled that "greenhouse gases contribute to air pollution that may endanger public health or welfare."[49] In addition to adverse impacts on human populations, climate change will adversely impact other species. The IPCC estimates that 20–30 percent of plant and animal species assessed are very likely "to

---

[41] "IPCC, 2007: Summary for Policymakers," in M. L. Parry, O. F. Canziani, J. P. Palutikof, P. J. van der Linden, and C. E. Hanson (eds), *Climate Change 2007: Impacts, Adaptation and Vulnerability. Contribution of Working Group II to the Fourth Assessment Report of the Intergovernmental Panel on Climate Change* (Cambridge University Press, 2007), 12.

[42] Ibid., 15.   [43] Ibid., 16.   [44] Ibid.   [45] Ibid.

[46] Julienne Stroeve, Marika M. Holland, Walt Meier, Ted Scambos, and Mark Serreze, "Arctic Sea Ice Decline: Faster Than Forecast," *Geophyical Research Letters*, 34, L09501, doi:10.1029/2007GL029703. See also Sharon Begley, "Climate Pessimists Were Right," *Wall Street Journal*, February 9, 2007, B1.

[47] "IPCC, 2007: Summary for Policymakers," in M. L. Parry *et al.*, 11–12.   [48] Ibid., 12.

[49] US Environmental Protection Agency, "EPA Finds Greenhouse Gases Pose Threat to Public Health" (April 17, 2009).

be at increased risk of extinction if increases in global average temperature exceed 1.5–2.5°C."[50]

3 ONE PLANET: ETHICAL ISSUES

Energy consumption, and the impact of that consumption on climate change, raise a range of important ethical issues regarding responsibility and, historically, accountability for the causes of climate change, duties to future generations and to the millions of species that coinhabit Earth with humans, and the just distribution of the costs of mitigation and adaptation.

In Chapter 1, "Energy, ethics, and the transformation of nature," Dale Jamieson grapples with the question of what choices we ought to make in order to respond to the challenge of anthropogenic climate change. He begins by providing an overview of the history of energy usage and then asks which transformations of nature for energy production are morally acceptable. He argues that no matter which alternative energy choices we choose as a means of addressing climate change, we will be confronted by difficult choices that require incentives for adopting a coherent and consistent energy policy and personal sacrifices – we will need to "grasp the nettle." He sees humanity's best hope as a highly motivated, global citizens' movement that leads to effective political action. In Chapter 2, "Is no one responsible for global environmental tragedy? Climate change as a challenge to our ethical concepts," Stephen Gardiner engages Jamieson's view, defended in Chapter 1 and in numerous other essays, that a new value system grounded in a respect for nature is needed to adequately confront the global environmental crisis. Gardiner agrees that climate change presents a daunting challenge to conventional ethical thinking, but argues that the focus should not be on re-envisioning our ethical systems as much as it should be on delegating political responsibility for collective action and holding political actors accountable for their responses to climate change. As Gardiner acknowledges, his position is not so much an alternative to Jamieson's position as it is a complementary position.

In Chapter 3, "Greenhouse gas emission and the domination of posterity," John Nolt argues that our emissions of GHG constitute unjust domination of future generations, analogous in many morally significant respects to certain historical instances of domination of people, such as those based on race or gender, that are now almost universally condemned. Further, he argues that no benefits that we may bequeath to future

---

[50] "IPCC, 2007: Summary for Policymakers," in M. L. Parry et al., 11.

generations can compensate for that injustice. Nolt argues that anyone who consumes energy that is not necessary has a duty to reflect on his or her energy consumption and to reduce it relative to the amount of his or her unnecessarily consumed energy.

The next three chapters take up the right to produce GHG emissions. As Simon Caney notes in Chapter 4, "Climate change, energy rights, and equality," at least three types of entities may be said to have such rights: individuals, corporations, and states. Accepting that a serious program of mitigation is required, he raises the question of how the opportunity to use fossil fuel energy (and other energy sources that emit GHGs) should be distributed. Caney begins by considering and rejecting the pragmatic view that GHG emissions should be distributed on the basis of the distribution of emissions now or at some fixed point in the recent past. He also considers and rejects as ethically objectionable the egalitarian view that the distribution of emission rights should be distributed on an equal per capita basis. Caney instead defends a position whereby the right to emit GHGs is distributed in such a way that people can enjoy "a minimally decent standard of living." In Chapter 5, "Common atmospheric ownership and equal emissions entitlements," Darrel Moellendorf defends a similar position to that of Caney regarding equal emissions rights but with a surprising conclusion. Moellendorf argues that the principle of common atmospheric property rights, which libertarians defend, requires an approach to mitigating climate change that is more demanding than a principle of equalizing intergenerational proportional costs that egalitarian liberals defend. Hence, in his view, libertarianism is committed to even greater anthropogenic climate change mitigation than is egalitarian liberalism. In his contribution to this book, Luc Bovens defends a position distinctly at odds with both Caney and Moellendorf. In Chapter 6, "A Lockean defense of grandfathering emission rights," Bovens provides a sustained defense of grandfathering grounded in Lockean property rights and a concern for respecting the differential investments made by different actors prior to their knowledge of the impact of GHG emissions on climate change. His arguments present a direct challenge to the arguments defended by both Caney and Moellendorf.

In Chapter 7, Sarah Krakoff argues that temporal lags and spatial dispersion issues make climate change a truly global issue that requires us to rethink our attitudes toward the Earth. In "Parenting the planet," she argues that we confront difficult collective action problems across generations that require us to embrace an ethics for a potentially tragic age. She points to practical examples of states, cities, and communities that have embraced

ways of living that combat climate change and that are compatible with
sustaining the Earth. Drawing from the work of the psychologist Erik
Erikson, Krakoff observes that love, care, and wisdom are the central virtues
of parents and goes on to argue that we need to think of ourselves as parenting
the planet – of caretaking and reducing our demands on the planet – in order
to preserve the Earth for current and future human communities in company
with other species.

Next, in Chapter 8, "Living ethically in a greenhouse," Robert H. Socolow
and Mary R. English provide a state-of-the-art account of the ways in which
energy consumption is causally connected to climate change. They identify
four important ways that climate change intersects with ethical issues regarding the planet and five important ways that climate change intersects with
ethical issues regarding humans. They argue that considerations of justice
require that the 27 percent of the world's population that emits 79 percent
of carbon dioxide emissions must substantially reduce their emissions if we
are to achieve long-term stabilization of the Earth's climate. Further, they
argue that if the planet's climate is to be stabilized then the 73 percent of
the population that currently produces low emissions must not be allowed
to "catch-up" to the high emitters. In a climate-stabilized world, permissible
carbon dioxide emissions per capita would be well below the per capita
emissions of well-off people. They argue that human populations must utilize
the tools of moral imagination to forge a new planetary identity in order to
manage climate change over the coming decades within the parameters of
our present capacities. To accomplish the task before us, they argue that we
need to develop technological innovations that are within our reach, and
they advocate the technological "wedge" approach made famous by Stephen
Pacala and Socolow as a means of achieving this goal.[51]

In Chapter 9, "Beyond business as usual: alternative wedges to avoid
catastrophic climate change and create sustainable societies," Philip Cafaro
defends a radical rethinking of our attitudes toward consumption and
growth. He argues that because there is a strong likelihood that business
as usual with respect to consumption and growth cannot avoid catastrophic
climate change, we need to consider slowing or ending growth. Drawing
upon Pacala and Socolow's "wedge" approach, discussed at length by
Socolow and English in Chapter 8, Cafaro defends an alternative wedge
scheme focused on reductions in consumption, population growth, and
economic growth. Cafaro's wedge scheme is meant to complement, rather

---

[51] Stephen Pacala and Robert Socolow, "Stabilization Wedges: Solving the Climate Problem for the
Next 50 Years with Current Technologies," *Science*, 305 (2004): 968–972.

than supplant, the technological wedge scheme of Pacala and Socolow. However, in calling for a fundamental rethinking of our assumptions regarding the merits of luxury consumption, personal travel, meat consumption, and economic growth, Cafaro's alternative wedge scheme provides a significant challenge to existing and proposed mitigation strategies.

The range of ethical perspectives represented in this book are all in agreement that significant and rapid action is needed to protect the Earth for the benefit of human and other populations both in the present and in the future. To meet this challenge it is necessary to put in place public policies that help to solve our collective action problems regarding energy and climate. In Chapter 10, "Addressing competitiveness in US climate policy," Richard D. Morgenstern reviews the mandatory policies to reduce US GHG emissions – mainly cap-and-trade systems and carbon taxes – that are being seriously debated in the political arena. He points out that a frequent concern raised in these debates is the potential for adverse impacts on the competitiveness of US industry, particularly on firms or in sectors that face high energy costs and significant international competition. Morgenstern explains how production costs across individual manufacturing industries could be affected by a unilateral policy that establishes a price on carbon dioxide emissions. He then examines a range of possible policy responses to address these impacts including the use of standards instead of market-based policies for some sectors, different types of free allowance allocation under a cap-and-trade system, and trade-related policies. While no easy answers emerge, the pros and cons of each of the options are considered, with an emphasis on both the efficiency and equity issues involved.

Cap-and-trade systems are sometimes criticized by environmental justice advocates on the grounds that such systems fail to adequately take into account the interests of the individuals most adversely impacted by pollution. These are the individuals living in close proximity to coal-fired electric generation plants and other carbon-intensive industrial facilities. In the case of GHGs, co-pollutants often directly affect the health and welfare of communities that are in closest proximity to the sources of emissions. In Chapter 11, "Reconciling justice and efficiency: integrating environmental justice into domestic cap-and-trade programs for controlling greenhouse gases," Alice Kaswan argues that a just cap-and-trade system that focuses on efficiency in the interest of reduced GHG emissions will overlook the unequal distribution of burdens with respect to co-pollutants. Power plants and factories that are permitted to increase emissions will typically increase the amount of co-pollutants being emitted, thus adversely impacting the surrounding communities, which, in the USA, disproportionately comprise poor and minority populations. Further, these populations are not typically included in the process of determining emission distribution levels. Kaswan argues that these outcomes are unjust and, as an alternative, recommends direct regulations that require all facilities to adopt available cost-effective emission reduction mechanisms alongside a cap-and-trade system. In this way, she argues, traditional concerns of the environmental justice movement can be addressed along with climate change abatement.

W. Neil Adger (who worked on the IPPC's adaptation team) and Sophie Nicholson-Cole focus our attention on the need to address the ethical dimensions of adaptation in Chapter 12, "Ethical dimensions of adapting to climate change-imposed risks." They explain that adaptation to the impact of climate change is already with us and argue that in order for adaptation schemes to be implemented fairly we need a better understanding of the risks of climate change relative to other environmental stresses, wider consideration of the social justice dimensions of adaptation policy, support for those most adversely impacted by climate change, and an inclusive and participatory decision-making process.

All of the chapters in this book thus far have been concerned with the impact of climate change on human populations. In her contribution to this volume, Clare Palmer considers the impact of climate change on the *non-human* world – on nonhuman animals, on species and on ecosystems. In Chapter 12, "Does nature matter? The place of the nonhuman in the ethics of climate change," Palmer rightly points out that the existing literature on the ethics of climate change largely ignores the impact of climate change on the nonhuman world. She explores a series of subtle and often contentious arguments regarding the harm climate change might inflict on the nonhuman world. She concludes that we need much more specific knowledge of the potential impact of climate change on the nonhuman world and better accounts of why those impacts matter morally.

This book concludes with a contribution by Henry Shue, who first began publishing scholarly articles about the ethical dimensions of climate change in 1993.[52] Shue believes it has been shown that our inaction regarding climate change will infringe on the basic rights of individuals in the near and distant future. Despite this, he notes that most of the world's nation states have assumed a lackadaisical approach to climate change and have made little progress on the issue since the first climate conference in Rio de Janeiro in 1982. In Chapter 14, "Human rights, climate change, and the trillionth ton," Shue argues that rapid climate change calls for the

[52] Henry Shue, "Subsistence Emissions and Luxury Emissions," *Law* and *Policy*, 15(1) (1993): 39–59.

construction of rights-protecting institutions. These institutions should be international in composition and scope, intergenerational in their aims and ambitions, and immediately constituted. They should have as their goals, he argues, significantly limiting carbon emissions. For example, if we want to limit global warming to 2°C above pre-industrial levels, a figure endorsed by increasing numbers of scientists as the minimum amount we can reasonably permit given the likely harm to human populations, we must avoid emitting the trillionth metric ton (Tt C) of carbon to gain a 50 percent chance of meeting this goal. But we have already emitted 0.5 Tt C, which will likely result in 1°C of warming, so Shue argues we need to take immediate action. In this judgment Shue is in agreement with many of the other contributors to this volume.

Beyond this Shue argues that any carbon-trading scheme that is put in place internationally must protect the basic human rights of those who are presently fossil-fuel-market-dependent poor; that is, those who require fossil fuels to subsist and otherwise fulfill their basic rights. Shue argues that as a matter of fairness such individuals should be provided free permits for the use of fossil fuels while those more affluent should be required to pay for the limited remaining permits to use fossil fuels. Shue is pessimistic that his generation will have the fortitude to meet these basic duties to present and future generations.

The normative theorists who have contributed to this volume are in uniform agreement that mitigating and adapting to anthropogenic climate change presents a fundamental ethical challenge for humanity. It remains to be seen whether the generations succeeding Shue's are up to the ethical, political, economic, and technological challenges of protecting our planetary atmosphere so as to ensure a livable planet for future generations of humans and the multitudes of species that coinhabit the planet with us.

CHAPTER I

Energy, ethics, and the transformation of nature

Dale Jamieson

Climate change is a complex problem that can be approached from many different perspectives: atmospheric science, global change biology, environmental economics, international law, environmental philosophy, and so on. It involves almost every sector of society from land use planning to forest management. One particularly productive way of framing the problem of climate change is as a problem of energy policy. This is the perspective that I will take in this paper. I begin with some brief remarks about the role of different energy sources in human history. I go on to claim that every currently available energy policy entails difficult trade-offs and that technology will not deliver us from the agony of choice, at least on the time-scale on which we must act to avoid "dangerous anthropogenic interference with the climate system."[1] I then bring these observations to bear explicitly on the problem of climate change and discuss their implications for policies that are now under active consideration. Finally, I draw some conclusions.

I ENERGY'S HISTORY

Energy use has been central to the development of human civilization, society, and economy. As a first approximation, we can say that the story of human development has been the story of increased use of energy. Indeed, we can even think of human history as falling into epochs marked by the human ability to exploit various sources of energy. According to Vaclav Smil:

All preindustrial societies derived their energy from sources that were almost immediate transformations of solar radiation (flowing water and wind) or that took relatively short periods of time to become available in a convenient form; just a few months of photosynthetic conversion to produce food and feed crops, a few

---

[1] This is the obligation undertaken by the USA and the other 193 countries who are parties to the Framework Convention on Climate Change. For the text of the treaty, visit http://unfccc.int/essential_background/convention/background/items/2853.php.

years of metabolism before domestic animals and children reached working age; or a few decades to accumulate phytomass in mature trees to be harvested for fuel wood and charcoal.[2]

What is distinctive about the energy profile of modern societies is their use of fossilized stores of solar energy in the form of coals and hydrocarbons, and their use of electricity generated by burning these fuels, as well as by water, nuclear fission, wind, and the Earth's heat.

Coal was used for various purposes in antiquity, but it was a marginal energy source until the late seventeenth century when Great Britain became the first country to adopt coal as its primary energy source. The transition to coal did not occur in the USA until the 1880s. The demand for coal has continued to increase and the International Energy Agency predicts that global energy demand will grow by 40 percent between 2007 and 2030, and the demand for coal will grow more than the demand for any other energy source.[3] Petroleum, like coal, was known in antiquity but played almost no role as an energy source. Until 1870 global petroleum production was negligible, but by 1950 it had risen to 10.42 million barrels per day, and by 2005 production was more than 84 million barrels per day.[4] Today petroleum makes up 40 percent of total energy consumption in the United States, and fossil fuels make up 86 percent of the world's energy supply.

At some point energy sources became value-valenced, not just by their relative ability to do work, but also by other features with which they are associated. So, for example, in the 1950s nuclear energy was widely seen as good. It was viewed as a cheap, clean, unlimited source of power, and the US government actively promoted its adoption throughout the world.[5] Nuclear fission had previously been associated with the bombs that had been used to destroy Hiroshima and Nagasaki, and made possible the death grip of "mutually assured destruction" in which the USA and the Soviet Union were then locked. Nuclear energy symbolized the transformation of the power of the atom from a destructive force to one that could serve human development and progress (a word that was frequently invoked in that far-off time). Atoms for peace was an expression of the biblical injunction to "beat . . . swords into ploughshares."[6]

[2] Vaclav Smil, *Energy in Nature and Society* (Cambridge, MA: MIT Press, 2008), 203.
[3] www.iea.org/country/graphs/weo_2009/fig1-1.jpg.
[4] www.earth-policy.org/Updates/2007/Update67_data2.htm#table1.
[5] See, e.g., President Dwight Eisenhower's December 1953 "Atoms for Peace" speech. For an account, visit www.eisenhower.archives.gov/digital_documents/Atoms_For_Peace/Atoms_For_Peace.html.
[6] Isaiah 2:2–4.

Sometime in the 1960s nuclear power became bad, as it came to be seen as a high-risk technology that imposes harms on future generations. The value-valence of nuclear power turned for a number of reasons. First, it soon became clear that nuclear power could not live up to the grand claims that had been made on its behalf (e.g., as producing electricity that would be "too cheap to meter"[7]). Second, once nuclear tests in the atmosphere had been abolished, nuclear plants were the only remaining anthropogenic source of cancer-causing radiation. A third concern was the thermal pollution in waterways caused by nuclear plants, which ran head on into the ecological concerns of the nascent environmental movement. Fourth, it seemed completely unacceptable to many people that we would build nuclear plants that inevitably produce waste that would have to be managed for thousands of years, without any clear plan about how to manage it. Finally, although the Cold War continued into the 1980s, the specter of nuclear war began to diminish in the 1960s, and the fear of catastrophic nuclear disaster began to migrate from nuclear war to nuclear power. These fears seemed to be confirmed by the 1979 accident at the Three Mile Island nuclear plant in Pennsylvania, which occurred just weeks after the release of a popular film depicting nuclear catastrophe.[8] These fears were reinforced by the 1986 Chernobyl disaster.

Perhaps the most interesting charge against nuclear power was that it is "inappropriate technology." That nuclear reactions were being used to boil water to create steam to turn turbines seemed to many people like using a scalpel to butter toast or, more revealingly, blanketing entire neighborhoods with persistent pesticides (such as DDT) in order to control mosquitoes. Both seemed to be needlessly complex, high-risk approaches to solving fundamentally manageable problems. In addition, these approaches were seen as inelegant and out of tune with the idea of the simple life that is modest in means and rich in ends. The 1960s also saw a rise of libertarian ideals across the political spectrum that celebrated decentralized approaches to problem solving. Nuclear energy was seen by many as the ultimate expression of highly centralized, technological, "modernist" management of human life and society.

At the same time that nuclear power was being demonized, a premodern energy source, wood, was being valorized. Wood had largely been given up

---

[7] This sentiment has often been attributed to Lewis Strauss, then Chairman of the Atomic Energy Commission, in a 1954 speech, but there is controversy about exactly what he said and even more about what he meant. For more on the issues, visit http://media.cns-snc.ca/media/toocheap/toocheap/.

[8] *The China Syndrome*, starring Jane Fonda, Jack Lemmon, and Michael Douglas.

as old-fashioned, but much of the antinuclear movement embraced wood and other decentralized energy sources. In popular music and in counter-cultural journals, wood was portrayed as a renewable resource that is tradi-tional and natural, and that can be harvested and transformed into energy using simple means, in ways that can be easily understood. Wood was a decentralized resource that encouraged self-sufficiency and supported local economies. Heating with wood was part of taking a home "off-grid." In many circles wood became symbolic of American rugged individualism and curmudgeonly independence. Unfortunately, however, wood is an extremely polluting energy source unless it is burned in stoves that are outfitted with expensive, high-tech scrubbers. Even then, burning wood contributes more to greenhouse gas emissions than burning natural gas, oil, and perhaps even some kinds of coal.[9] In addition, it is difficult to use wood on a large scale without significantly transforming ecosystems.

In an age in which our main concern is increasingly with greenhouse gas emissions, and we look to advanced technology to "deliver us from evil," the value valences of nuclear and wood are beginning to flip again. Nuclear power is being reframed as a clean energy source for a greenhouse world, while wood is increasingly being seen again as a marginal energy source that can be tolerated if used responsibly by a few people living in remote areas. The move towards seeing energy policy through the lens of climate change was pioneered by scientists rather than environmentalists, contrary to what is widely supposed. Environmental organizations were slow to embrace climate change, in part because it disrupted many of the usual associations that went with environmentalism, of which this idea that nuclear is good and wood is bad is an example. This begins to suggest the extent to which environmentalism (certainly in the USA) has operated as a set of tacit commitments, tendencies, stereotypes, and prejudices rather than as a reflective worldview or a coherent philosophy. Indeed, many environmental issues, rather than being subjected to systematic analyses, have tended to migrate along an axis of what is speakable or unspeakable, depending on reigning political and psychological associations, stereotypes, and so on. Until recently, nuclear power was simply not discussed in polite green company, though it is now entering the domain of the discussable. Population, one of the central green issues of the 1960s and 1970s, became

[9] It is often said that burning wood is carbon-neutral because the wood that is burned will eventually release its carbon into the atmosphere anyway. Even if this were true, the rate at which carbon is released and sequestered matters enormously in determining atmospheric concentrations of green-house gases.

unspeakable during the 1980s and 1990s, and the impact of immigration on the environment is in the same position today.[10] The environmental impacts of dietary choices are only now beginning to reach public consciousness, largely through the statements of Rajendra Pachauri, chair of the IPCC and lifelong vegetarian.[11] In these respects environmentalism in the USA has functioned more like an American political party (a large, diverse set of interest groups) than as a set of philosophical commitments or worldviews.

This comes out dramatically when environmentalists are faced with value conflicts. For example, species reintroduction programs are supposed to restore nature to some preferred state, but such programs often make life difficult for individual animals whose interests many environmentalists are concerned to promote. In some cases the interests of protected species can clash; for example, the Sierra big-horned sheep and the mountain lion in parts of California.[12] Perhaps most profoundly, many environmentalists chafe at the plain fact that a poor urban dweller who can't afford to eat much meat lives a much greener lifestyle than the affluent Sierra Club member who "lives close to nature."

To return specifically to energy, the problem that we face is that energy use is at the heart of everything we do or consume. The environmental impact of even a wilderness expedition is to a great extent a function of energy use, because of both the energy involved in producing the gear and the energy used in powering the gearheads. The significance of this second consideration is often overlooked. However, it has been argued that someone who eats a diet heavy in factory-farmed meat could actually decrease their greenhouse gas emissions by giving up meat-fueled walking in favor of gasoline-fueled driving, because the fossil fuel intensity of conventional meat production is so much greater than that of driving.[13]

Energy production necessarily involves transforming nature. Producing energy, whether from fossil fuels or by eating fruits and nuts, leaves nature in a different state than it otherwise would have been in. This matters because environmentalists are in general hostile to the human transformation of nature.

---

[10] This issue has come up in the Sierra Club in elections for officers, but even there the question has been largely whether the issue is discussable rather than being a discussion of the issue.

[11] "This is something that the IPCC was afraid to say earlier, but now we have said it . . . Please eat less meat" (http://lists.mutualaid.org/pipermail/sustainabletompkins/2008-January/003208.html).

[12] For more on these conflicts, see my *Ethics and the Environment: An Introduction* (Cambridge University Press, 2008), ch. 6, and "The Rights of Animals and the Demands of Nature," *Environmental Values*, 17(2) (May 2008): 181–199.

[13] For this argument, see Chris Goodall, *How to Live a Low-Carbon Life* (London: Earthscan, 2007); for a response, see www.pacinst.org/topics/integrity_of_science/case_studies/driving_vs_walking.html.

They typically endorse Paul McCartney's "wistful words of wisdom": "let it be." Moreover, environmentalists value nature and the natural, and these concepts imply little or no human transformation.[14]

My point is not that environmentalism is incoherent, or that characteristic green values and dispositions cannot be sharpened and defended. My point is rather that the work needs to be done. What is needed is a philosophy of environmentalism that makes explicit, clarifies, and defends to the greatest extent possible core green commitments. I suspect that in the end some green dispositions and beliefs will have to be reined in, reformulated, or even jettisoned. However, we won't know until they are clarified and defended in the best way possible.

This much is clear: it is not only humans that transform nature, but also other animals and plants. What is generally different about humans is this conjunction: Humans now are transforming nature on a much larger scale than other animals, and they are capable of modulating these transformations through individual and collective action based on reflection. Some would say that phytoplankton transform the planet in an even more dramatic way than humans. I don't think that it is productive to argue this point one way or another. What phytoplankton lack is the second conjunct: the ability to modulate their impact based on reflection. That is why it is true to say that in an important respect phytoplankton are part of nature while humans are not.[15]

The fundamental question that is too often avoided is what transformations of nature, under what conditions, with what motivations, for what purposes, in what contexts, are morally acceptable? This is the question that is at the heart of debates over energy policy.

2 GRASPING THE NETTLE

When we return to the choices we must make regarding energy policy, we see that each choice involves a nettle: it will sting! In this section I will briefly

---

[14] Robert Goodin, *Green Political Theory* (Cambridge: Polity Press, 1992); Robert Elliot, *Faking Nature* (New York: Routledge, 1997).

[15] Of course in other respects humans are part of nature (e.g., they are constituted by natural materials, they are subject to natural laws, etc.); this brings out the multiplicity of meanings and associations of 'nature' and its cognates. For more on these points, see Bernard Williams, "Must a Concern for the Environment Be Centred on Human Beings?," reprinted in L. Gruen and D. Jamieson (eds), *Reflecting on Nature: Readings in Environmental Philosophy* (New York: Oxford University Press, 1994); and Paul Veatch Moriarty, "Nature Naturalized: A Darwinian Defense of the Nature/Culture Distinction," *Environmental Ethics*, 29 (2007): 227–246.

review the nettles and the stings associated with what are generally regarded as the most environmentally friendly energy sources.

2.1 Energy efficiency

Literally, energy efficiency is not an energy source at all, but rather a strategy for reducing the amount of energy that we must produce and thus the extent to which we must transform nature. Virtually everyone agrees that it is the best energy policy, to the extent to which it can be implemented.

There is little doubt that a great deal of energy is wasted: according to one estimate, 66 percent in the electricity sector, 71 percent in transportation, 20 percent in industry, and 20 percent in residential and commercial buildings.[16] One reason that we have not reached higher levels of energy efficiency is that prices do not adequately reflect the costs of producing and consuming energy. However, even when policies are adopted that promote energy efficiency, there are limits to what they can achieve. Even to speak of energy efficiency implies using energy, and using energy implies transforming nature. In some cases improvements in energy efficiency are free (e.g., making greater use of natural light); in other cases they require new products and technologies that in some cases require large amounts of energy to produce. Imagine, for example, how much energy would be required in order to replace the world's automobile fleet with energy-efficient vehicles, fabricated from the most efficient materials and kitted out with the most fuel-efficient engines.[17]

2.2 Solar

Solar energy can be directly used in various ways. Passive solar systems use the architectural design, natural materials, or absorptive structures of buildings to heat water or homes. Active solar energy systems require solar collectors (such as photovoltaic cells) and can be used to generate electricity. This electricity can be used to directly power a building or can be fed into an existing electrical grid.

---

[16] The *New York Times* attributes these figures to Lawrence Livermore National Laboratory; see www.nytimes.com/imagepages/2008/04/06/weekinreview/06revkin.html.

[17] I owe this example to Ronald Mitchell. Other examples include more fuel-efficient refrigerators and furnaces. As Robert Darst pointed out in conversation, even compact fluorescent bulbs have more embedded energy than incandescents and impose further environmental costs because they contain small amounts of mercury.

One nettle associated with the use of solar energy to produce electricity is cost, but the costs of solar energy have been steadily declining while the costs associated with fossil fuels have been increasing.[18] Still, electricity generated by photovoltaics is 2–5 times more expensive than electricity currently delivered to residential customers. In addition, it has been argued that much of the luster of photovoltaics dissipates when their embodied energy is taken into account (i.e., the energy used in manufacturing, installing, and maintaining the cells).[19] A third concern is that cadmium may be released in producing photovoltaic cells, which are made of cadmium telluride. Cadmium is a toxic heavy metal that concentrates in the food chain and is implicated in lung and kidney disease. In any case there are limits on the amount of energy that solar can deliver given the variable nature of the solar radiation that strikes the Earth's surface both across space and through time. Furthermore, if solar energy were to supply the American energy grid with a significant fraction of demand, large areas would have to be covered with photovoltaic cells, and some people find this possibility aesthetically objectionable.

2.3 Wind

Wind energy is currently more attractive than solar but also confronts us with some of the same nettles. Wind-generated electricity is much more economically competitive than solar, and is even approaching the cost of electricity generation from a new, coal-fired power plant. There is also much less energy embodied in wind turbines than in photovoltaic cells. However, even more than solar, wind energy is both inconsistent and inconstant. While early concerns about the apparently devastating ecological effects of wind turbines have largely been assuaged, this remains an active area of research.[20] A further problem concerns the aesthetic acceptability of wind farms, indicated by controversies about their siting both in North America and Great Britain.[21]

---

[18] A caveat: The price of coal has not been increasing but the "clean coal" that would have to be part of any sustainable energy future is substantially more expensive than "dirty coal."

[19] However, such claims are controversial; see, e.g., Colin Bankier and Steve Gale, "Energy Payback of Roof Mounted Photovoltaic Cells," *Energy Bulletin*, June 16, 2006, www.energybulletin.net/node/17219.

[20] While turbines are no longer seen as "bird blenders," under some circumstances they can seriously threaten bird and bat populations. For a review, see A. Drewitt and R. Langston, "Collision Effects of Wind-Power Generators and Other Obstacles on Birds," *Annals of the New York Academy of Sciences*, 1134 (2008): 233–266.

[21] Perhaps the best known controversy in the USA is over Cape Wind's proposal to site a $900 million wind farm on Horseshoe Shoal in Nantucket Sound off Cape Cod in Massachusetts. For an opinionated account of the controversy, see Wendy Williams and Robert Whitcomb, *Cape Wind: Money, Celebrity, Class, Politics, and the Battle for Our Energy Future on Nantucket Sound* (New York: Public Affairs, 2007).

Issues
• *bad*
ecological
effects

2.4 Hydropower

Hydropower is an even cheaper source of energy than burning fossil fuels and is responsible for virtually no greenhouse gas emissions. In many parts of the world, including the USA, there is still a great deal of potential for developing hydropower resources. Still, by damming wild rivers, hydropower development can have enormously damaging ecological effects. In the USA there is a growing movement to remove dams in the Pacific Northwest that have adversely affected salmon populations. The James Bay Project in Canada and the Three Gorges Dam in China have remade nature on a much greater scale, compromising natural values such as wildness and naturalness and disrupting human communities. Proposals to build dams for producing hydropower have given rise to massive protest movements in both South Asia and South America.

2.5 CCS coal

A favored choice of policymakers is continuing to burn coal, but in plants that capture carbon that can then be sequestered. A recent MIT study concluded that

> CO$_2$ capture and sequestration (CCS) is the critical enabling technology that would reduce CO$_2$ emissions significantly while also allowing coal to meet the world's pressing energy needs.[22]

Issues
• *environmental*
effects
• *hard to*
capture CO$_2$

Coal is cheap and abundant, and is currently the source of more than one-quarter of the world's energy supply. However, not only does burning coal produce acid rain, poor visibility, and various deleterious health effects, but it also produces 50–100 percent more carbon dioxide per BTU than other fossil fuels. We are now able to control the pollutants produced from burning coal, and if we were able to do the same trick with carbon dioxide, coal would be a very attractive fuel indeed. Unfortunately, capturing carbon dioxide is itself extremely energy intensive, requiring a CCS coal-fired generating plant to consume as much as 25 percent more energy than a conventional plant that emits carbon. This drives up costs, and is what led to the cancellation of the American government's first attempt to build a CCS plant ("FutureGen"). Moreover, at this point no one knows whether these technologies could reliably be deployed on a scale that would make a significant difference to atmospheric concentrations of carbon dioxide.

---

[22] Massachusetts Institute of Technology, "The Future of Coal: An Interdisciplinary MIT Study" (Cambridge, MA: MIT, 2007). http://web.mit.edu/coal/.

Even if they could be so deployed, some of the worst effects of the coal cycle would not be eliminated but perhaps would even be exacerbated (e.g., mountain-top removal, occupational safety concerns, etc.).[23]

2.6 Nuclear power

Like hydropower, nuclear power is attractive because it does not directly emit greenhouse gases. Indeed, heavy reliance on nuclear power is part of why Europe has lower per capita emissions of greenhouse gases than the USA.[24] While the figure is not quite so rosy when embodied energy is included, nuclear still does well on this dimension relative to fossil fuels and perhaps even photovoltaics.

The stings of nuclear energy are elsewhere. First and foremost is the problem of storing nuclear waste. Such waste will have to be managed for at least 10,000 years, and no one really knows how to do this. Of course, we may someday come up with methods of storing, decontaminating, or even using these wastes as a resource, but building nuclear plants today entails gambling on other people's futures. In addition, there is a great deal of concern about nuclear plants being "terrorist-magnets." Drawings of American nuclear plants were found in Al-Qaeda documents captured in Afghanistan, and shortly after the 9/11 attacks the nuclear plant at Three Mile Island was temporarily shut down due to a "credible threat." A report from the Union of Concerned Scientists predicted that a terrorist attack on the Indian River Nuclear Plant near Manhattan could kill 44,000 people immediately and as many as 518,000 over the long term.[25] Finally, uranium mining raises many of the same issues as coal-mining, but some additional health risks as well since uranium ore emits cancer-causing radon gas.

2.7 Biofuels

For the last several years biofuels have been a favored alternative energy source in North America and Europe as well as in Brazil. While calculating subsidies is difficult and controversial, we can estimate that the US

[23] For an overview of some of the issues, consult the Appalachian Center for the Economy and Environment. www.appalachian-center.org/.

[24] More than three-quarters of France's electricity, almost half of Sweden's, and more than a third of Europe's is generated by nuclear power. See www.euronuclear.org/info/encyclopedia/n/nuclear-power-plant-europe.htm.

[25] Edwin S. Lyman, "Chernobyl on the Hudson?" (Cambridge, MA: Union of Concerned Scientists, 2004). www.ucsusa.org/nuclear_power/nuclear_power_risk/sabotage_and_attacks_on_reactors/impacts-of-a-terrorist-attack.html.

government provides about $11 billion per year in corn ethanol subsidies while the countries of Europe spend at least half this amount in subsidizing corn ethanol and biodiesel. These policies are remarkably perverse given that recent studies show that corn ethanol could actually double greenhouse gas emissions over the next twenty years and result in further increases for more than a century.[26] This could occur because of the effects of the biofuels markets in changing land use patterns.

Changing land use patterns in response to increasing demand for biofuels are already having deleterious ecological effects in countries such as Indonesia, where tropical rainforests and peat lands are being converted to palm oil plantations, resulting both in increases in greenhouse gas emissions because of the loss of carbon sinks and in the destruction of habitat for endangered species such as the orangutan, the Sumatran tiger, and the Asian rhinoceros.[27] Indigenous people are also being harmed by these land use changes. In response to these unanticipated consequences, in 2007, the European Union (EU) reduced its 2005 goal for biofuels to constitute 20 percent of all vehicular fuels by 2020 to 10 percent, and even that is controversial.

Perhaps the most dramatic consequence attributed to increased demand for biofuels is increases in food prices, resulting in food shortages and political turmoil around the world. In a speech on April 4, 2008, the World Bank president, Robert Zoellick, reported that staple food costs had risen by as much as 80 percent since 2005, that rice had hit a 19-year high, and that the real price of wheat was at a 28-year high. At the time he was speaking, food riots were occurring in several countries including Egypt, Côte d'Ivoire, Burkino Faso, Haiti, and the Philippines. In his speech, Zoellick identified biofuels as one of the causes of the spike in food prices. Since then, food prices stabilized, declined, and have begun rising again.

It is clear that there are multiple causes for increases in the price of food including growing demand for meat and dairy products in developing

---

[26] Timothy Searchinger, Ralph Heimlich, R. A. Houghton, Fengxia Dong, Amani Elobeid, Jacinto Fabiosa, Simla Tokgoz, Dermot Hayes, and Tun-Hsiang Yu, "Use of U.S. Croplands for Biofuels Increases Greenhouse Gases Through Emissions from Land-Use Change," *Science*, 319 (February 29, 2008): 1238–1240. See also Joseph Fargione, Jason Hill, David Tilman, Stephen Polasky, and Peter Hawthorne, "Land Clearing and the Biofuel Carbon Debt," *Science*, 319 (February 29, 2008): 1235–1238. For further discussion, visit www.bioenergywiki. net/Searchinger_Wang_debate.

[27] Center for Science in the Public Interest, "Cruel Oil: How Palm Oil Harms Health, Rainforest and Wildlife" (Washington, DC: May 2005). www.cspinet.org/palm/PalmOilReport.pdf.

countries such as China and India, and an increasing preference for wheat rather than rice or maize among the emerging urban middle classes in many developing countries. For example, per capita meat consumption in China has increased 250 percent in the last twenty-five years and is now more than 100 lbs (45.4 kg) per year, but still less than half the per capita meat consumption of Americans.[28] Local and regional droughts and floods (some probably triggered by climate change) have periodically reduced harvests in parts of the world. For example, a multi-year drought in Australia resulted in wheat exports dropping by 46 percent from 2005 to 2006, and then another 24 percent in 2007.[29] The collapse of the dollar has also contributed to the 2007–8 food crisis since much of the global commodities market is denominated in dollars.

What is the impact of the increased consumption of biofuels? While this is controversial, some things are clear. US ethanol production is increasing at rates from about 15–40 percent per year. According to the the US General Accounting Office, by 2012 almost one-third of the US corn crop will be devoted to producing ethanol.[30] While a spokesman for the US government claimed that increased demand for ethanol was responsible for only about 2–3 percent of the increase in the price of food during 2007–8,[31] the British newspaper, *The Guardian,* reported that an unpublished World Bank Study attributed 75 percent of the increase in food prices to increased demand for ethanol.[32] The United Nations Food and Agriculture Organization identified rising demand for biofuels as one of the causes of the 2007–8 price spike, and cites increasing demand for biofuels as "a leading driver" of the most recent spike in prices.[33] Wherever the exact truth lies, it seems clear that ethanol subsidies and mandates are playing a significant role in pushing up food prices.[34]

[28] BBC, "The Cost of Food: Facts and Figures," October 16, 2008, http://news.bbc.co.uk/2/hi/in_depth/7284196.stm; and Humane Society of the United States, "Farm Animal Statistics: Meat Consumption," November 30, 2006, www.humanesociety.org/news/resources/research/stats_meat_consumption.html.

[29] David Challenger, "Blue Skies, Blue Days in Rural Australia," CNN, October 28, 2008. http://articles.cnn.com/2008-10-28/world/australia.drought_1-blue-skies-worst-drought-drought-condition.

[30] Reuters, "Ethanol to Take 30 pct of U.S. Corn Crop in 2012: GAO," July 11, 2007. www.reuters.com/article/idUSN1149215820070611.

[31] Andrew Martin, "Food Report Criticizes Biofuel Policies," *New York Times,* May 30, 2008. www.nytimes.com/2008/05/30/business/worldbusiness/30food.html.

[32] Aditya Chakrabortty, "Secret Report: Biofuel Caused Food Crisis," *The Guardian,* July 3, 2008. www.guardian.co.uk/environment/2008/jul/03/biofuels.renewableenergy.

[33] Food and Agriculture Organization of the United Nations, "Food Outlook," December 2009. www.fao.org/docrep/012/ak341e/ak341e00.htm.

[34] See the *New York Times* editorial, "Man-Made Hunger," Sunday, July 6, 2008, available at www.nytimes.com/2008/07/06/opinion/06sun1.html.

Many promoters of biofuels put their faith in "second-generation" biofuels that exploit "waste" materials rather than food crops. Champions of other energy sources also look past the problems of the present into a gauzy future of endless possibility. In the next section we will examine these hopes. First, however, I want to restate the main point of this section: using energy transforms nature, and developing energy sources involves grasping nettles.

3 TECHNOLOGY'S GRACE

The idea of American exceptionalism runs very deep in American culture.[35] Part of what makes America exceptional today by industrial world standards is the fact that so many Americans believe that they will be delivered from their sins by divine intervention. Perhaps this belief in God's grace in the next world is mirrored by a belief in technology's grace in this world.

When it comes to energy policy, like Adam after the Fall, we now have knowledge of good and evil. Thanks to climate scientists from Arrhenius to the IPCC, we know at least the broad outlines of what our use of fossil fuels is doing to the climate system. Although we may choose to avert our eyes, we also know that existing alternative energy systems also entail unwanted consequences. This is what the problem of grasping the nettle is all about. The promise of technology is that it will save us from having to grasp a nettle. It will deliver us from our fallen state and return us to the Eden of cheap, unlimited energy that leaves nature pristine, untouched by human hand or intention.

Those who have this faith often talk about the need for something along the lines of a Manhattan or Apollo project to create new energy technologies.[36] They talk about crash programs to develop fusion, "second-generation" biofuels, photovoltaics "beyond conventional silicon," and so on. While I do not want to dismiss the importance of technology development especially over the long term, what is needed most urgently is individual and collective action. This is obscured by the "technofix" mentality and "crash-project" analogies. While all analogies limp to some extent, these analogies are particularly lame in some important respects. The Apollo project (for example) was directed toward producing a single, well-defined

---

[35] The recognition of this is usually attributed to Alexis de Tocqueville's 1835 book, *Democracy in America*. For discussion, see Seymour Martin Lipset, *American Exceptionalism: A Double-Edged Sword* (New York: W. W. Norton & Company, 1996).

[36] See, for example, Martin Hoffert, "An Energy Revolution for the Greenhouse Century," *Social Research*, 73(3) (Fall 2006): 981–1000.

result: American boots on the moon. Our current challenge is the much more thoroughgoing one of transforming the global energy systems which support human life. This is a challenge of diffusion and adoption at least as much as a challenge of technological innovation.

For more than a generation, analysts have been talking about "no regrets" policies and "harvesting the low-hanging fruit."[37] What they mean by this is that significant actions can be taken to address climate change that would either be cost-free or cost-effective, wholly independent of the climate change threat. Yet, to a great extent, these actions have not been taken. Consider, for example, the fact that the hybrid engine, which is only now beginning to seriously penetrate the automobile market, was first developed in 1916.[38] Or consider the fact that the 1908 Model T got 25 miles (40 km) per gallon (3.79 liters), while the 2004 fleet average for all American cars was 21 miles (33.6 km) per gallon.[39] We can develop fancy new technologies but they will do little to solve our energy problems if they are not deployed. As Lovins and others have shown, an enormous amount could be done that is not being done with existing technologies.[40] What explains this failure to act?

Many factors are involved but much of the explanation lies in individual attitudes and values, and collective (including political) responses. It also matters how these responses are layered and how they relate to each other. We need to provide incentives for adopting a coherent and consistent energy policy, and to develop and implement the technologies that support it. Most of all, we need citizens who are willing to change their behavior and to commit to binding themselves in various ways, including not punishing politicians who impose costs on them to spur challenges in their behavior. For example, each of us may prefer to drive while others take public transport, but since acting on this desire leads to worse consequences for each of us than taking public transport, we need to be willing to take public transport and not punish politicians who adopt policies that discourage us from driving.

Technology matters but its grace will not save us from ourselves. We are back to the nettles.

[37] See, e.g., Amory B. Lovins *et al.*, *Least-Cost Energy: Solving the CO₂ Problem* (Andover, MA: Brickhouse Publication, 1981).

[38] www.hybridcars.com/history/history-of-hybrid-vehicles.html.

[39] www.wanttoknow.info/050711carmileageaveragempg.

[40] This is one of the implications of S. Pacala and R. Socolow, "Stabilization Wedges: Solving the Climate Problem for the Next 50 Years with Current Technologies," *Science*, 305 (August 13, 2004): 968–972.

4 CLIMATE CHANGE: THE BIG ENCHILADA

Choices in energy policy implicate a range of issues including cost, national security, ecological destruction, and pollution. Looming over all of these issues is climate change. It is the big enchilada because of the magnitude and scale of the threat. Climate change is a global phenomenon that has the potential to extinguish half the species on the planet, threaten food supplies in much of the world, set off massive refugee flows, and disrupt relations among countries. It is especially difficult for us to act on climate change because it challenges our conceptions of rational self-interest, ethics, and justice among states.

↑ why its hard for us

4.1 Self-interest

Climate change will in aggregate be bad for the people of the world but it is difficult to say exactly how bad and to make this assessment independent of ethical considerations. Nordhaus claims that it will be moderately bad: the optimal policy is a carbon tax of about $17 in 2005, ramping up to $270 in 2100.[41] Stern, on the other hand, holds that future economic damages could be 20 percent of global GDP, and that the optimal carbon tax now is $311.[42] To a great extent the difference between them rests on the choice of a discount rate, what Stern believes is an ethical decision and Nordhaus believes is an empirical one. Questions of ethics also enter because climate change, at least to some level of warming, will produce both winners and losers.

Stern and nordhaus!

Stern - ethical

Nordhaus empirical

Even from an individual point of view it is difficult to know how to think about self-interest and climate change. Some, such as Steven Schneider, have analogized spending on climate change to buying insurance. He has said:

We buy fire insurance for a house and health insurance for our bodies. We need planetary sustainability insurance.[43]

While in some respects this analogy may be illuminating, viewing climate change spending as buying insurance is peculiar in that we have no relevant actuarial tables. Morever, those who can afford planetary insurance – rich people who are now alive – are those who are least likely to be severely

---

[41] *A Question of Balance: Weighing the Options on Global Warming Policies* (New Haven, CT: Yale University Press, 2008).

[42] Nicholas Stern, *The Economics of Climate Change: The Stern Review* (Cambridge University Press, 2007).

[43] Regina Nuzzo, "Profile of Stephen H. Schneider," *Proceedings of the National Academy of Sciences of the USA*, 102(44) (2005): 15725–15727. www.ncbi.nlm.nih.gov/pmc/articles/PMC1276082/.

affected by climate change. Those who are most likely to be severely affected – poor people who will come after us – are not in a position to purchase such insurance.

4.2 Morality

Climate change poses questions of morality since it involves some people harming other people.[44] However, how these people are related and how these harms come about depart significantly from our normal conception of a moral problem.

A paradigm moral problem is one in which an individual acting intentionally harms another individual; both the individuals and the harm are identifiable; and the individuals and the harm are closely related in time and space. A paradigm case of a moral problem is Jack intentionally stealing Jill's bicycle. Jack acts intentionally in harming Jill; Jack and Jill and the harm are clearly identifiable; and Jack and Jill and the harm are closely related in time and space. If we vary the case on any of these dimensions, we may still see the case as posing a moral problem, but its claim to be a paradigm moral problem is weaker. For example, if Jack is part of an unacquainted group of strangers, each of whom acting independently takes one part of Jill's bike, resulting in the bike's disappearance, we may still see Jack as acting wrongly, but this is less clear than in the first example. If we vary the case on several dimensions simultaneously, the view that morality is involved is weaker still, perhaps disappearing altogether. Imagine a case in which, acting independently, Jack and a large number of unacquainted people set in motion a chain of events that prevents a large number of future people who will live in another part of the world from ever having bikes. The core of what constitutes a moral problem remains: some people have acted in such a way that harms other people.[45] However, since most of what typically accompanies this core has disappeared, recognizing climate change as a moral problem may require us to revise or expand our concept.[46]

---

[44] In his paper in this volume, John Nolt calculates that each American is responsible for about two deaths as a consequence of their climate-altering activities.

[45] There is, of course, room to argue that, while some people have made other people worse off, no harm has occurred; I cannot take up that challenge here.

[46] The ideas in this subsection were first developed (at greater length) in my "The Moral and Political Challenges of Climate Change," in S. Moser and L. Dilling (eds), *Creating a Climate for Change: Communicating Climate Change and Facilitating Social Change* (New York: Cambridge University Press, 2007), 475–482. For more on the themes of this and the subsequent subsection, see also my "Climate Change, Responsibility, and Justice," *Science and Engineering Ethics*, 16 (2010): 431–445.

4.3 Global justice

When we view nation states as climate change actors, it is obvious that the rich countries of the North disproportionately emit greenhouse gases while the poor countries of the South disproportionately suffer the damages.

Consider the example of Bangladesh. A sea-level rise of 1 meter will flood one-third of its coastline, creating 20 million environmental refugees. Saltwater will intrude inland, fouling water supplies and crops, and harming livestock. Cyclones and other natural disasters will become more frequent and perhaps more intense, causing even greater damage. Four billion dollars is needed for Bangladesh to begin to adapt to climate change by building embankments, cyclone shelters, roads and other infrastructure, yet Bangladesh's 2007 total national budget was less than $10 billion.[47] Bangladesh will suffer in all these ways from climate change, yet its carbon dioxide emissions per capita are 1/20th of the global average, and about 1/100th of US emissions.

It is facts such as these that lead us to see climate change as posing problems of global justice. Yet climate change strays from the paradigm. The problems posed by climate change are not like those posed by one country unjustly invading another country. The nation state is one level of social organization that is relevant because it is causally efficacious, but it is not the primary bearer or beneficiary of moral responsibilities.[48]

Climate change is largely caused by rich people, wherever they live, and is suffered by poor people, wherever they live. Greenhouse gas emissions vary greatly within countries as well as across countries. For example, the emissions profiles of people in California and New York, on a per capita basis, are more similar to those of Europeans than to other North Americans. Urban people in India emit more than rural people, regardless of income. Viewed globally, half the world's carbon is emitted by the world's richest 500 million people.[49] These 500 million people live disproportionately in North America, Europe, Australasia, or Japan, but they exist in every country of the world. Indeed, there are more of these high emitters in China than there are in New Zealand, and probably more than there are in Australia.

---

[47] Xinhua News Agency, "Bangladesh Gov't to Set Up Fund for Long-Term Disaster Management," July 6, 2008. www.reliefweb.int/rw/rwb.nsf/db900sid/KKAA-7GA9MT?OpenDocument.

[48] For further discussion, see my "Duties to the Distant: Humanitarian Aid, Development Assistance, and Humanitarian Intervention," *Journal of Ethics*, 9(1–2) (2005): 151–170.

[49] I owe this point to a presentation by Steve Pacala.

Moreover, there will be a great deal of variability within nations on who will suffer the damages of climate change. It is likely that more poor people will suffer from climate change in the USA than in many G77 countries due to the high US population, the large number of poor people, and the relatively undeveloped systems of land use management and emergency response in many parts of the country. *Poor people suffer*

4.4 Respect for nature

In addition, to challenging our ideas of self-interest, morality, and global justice, climate change also invokes a concern in some people about dominating nature. This concern does not need to be based on biocentrism or ecocentrism, but rather can be based on a richer conception of what it takes for humans and other sentient beings to flourish. The concern may be seen as prudential, moral, or grounded in some other way. However it figures, it is an important concern for many people although it can be difficult to articulate and defend.

Humans, like other animals, modify the environments in which they live. The extent and degree of anthropogenic changes, measured in any reasonable way, is currently overwhelming and increasing exponentially.

There are various ways of measuring the human impact on nature. In 1986 Vitousek and his colleagues approached this problem by calculating the fraction of the Earth's net primary production (NPP) that is appropriated by humanity, and thus not directly available for other forms of life.[50] What they found is that humanity probably appropriates about 40 percent of Earth's terrestrial NPP.[51] Another approach to assessing the human impact on nature is ecological footprint analysis, pioneered by William Rees and Mathis Wackernagel.[52] The ecological footprint of a nation, community, or individual is the amount of land area required to produce the resources it consumes and to absorb the wastes it generates, given its standard of living and prevailing technology.

---

[50] Peter Vitousek, Paul R. Ehrlich, Anne H. Ehrlich, and Pamela Matson, "Human Appropriation of the Products of Photosynthesis," *BioScience*, 36(6) (June 1986). NPP is the amount of biomass that remains after primary producers (autotrophic organisms such as higher plants or algae) have accounted for their respiratory needs.

[51] Subsequent studies using different methodologies have produced a range of figures, but Vitousek *et al.*'s original claim seems roughly correct. For a review, see Christopher B. Field, "Sharing the Garden," *Science*, 294(555121) (December 2001): 2490–2491.

[52] Mathis Wackernagel and William Rees, *Our Ecological Footprint: Reducing Human Impact on the Earth* (Gabriola Island, CA: New Society Publishers, 1996).

In a 1997 article, a group of distinguished scientists led by Vitousek reviewed the broad range of human impacts on nature.[53] What they found was that between one-third and one-half of the Earth's land surface has been transformed by human action; carbon dioxide in the atmosphere has increased by more than 30 percent since the beginning of the Industrial Revolution; more nitrogen is fixed by humanity than all other terrestrial organisms combined; more than half of all accessible surface fresh water is appropriated by humanity; and about one-quarter of the Earth's bird species have been driven to extinction. Their conclusion was that "it is clear that we live on a human dominated planet."[54] According to the World Wildlife Fund's *Living Planet Report*, some time in the late 1980s humanity began to consume resources faster than the Earth can regenerate them, and this gap is increasing every year.[55] The bottom line from both studies is that we are treating the Earth and its fundamental systems as if it were a toy that we could treat carelessly. It is as if we have scaled up slash and burn agriculture to a planetary scale, as if we could move to another planet once we have exhausted this one.[56]

In dominating the Earth in these ways, we are failing to show respect for nature. But why should we think that we should respect nature? Several reasons can be given but here is one: nature provides a background condition against which our lives have meaning. Respect is a fitting response to the role that nature plays, and also contributes to nature continuing to play this role in the future. Relating oneself to nature in this way is not a necessary or sufficient condition for all lives having meaning at all times and all places, but it is a very important condition for many of us here and now in the societies in which we live in which meaning is often so difficult to find.

Consider the following analogy. Representational painting is not the only kind of valuable painting, but it is one very important kind of valuable painting. Indeed, it seems plausible to regard it as the mother from which other forms of valuable painting emerged. This kind of valuable painting exploits the contrast between foreground and background. What is in the foreground gains its meaning from its contrast with the background. What I want to suggest is that nature provides the background against which we live

[53] Peter M. Vitousek, Harold A. Mooney, Jane Lubchenco, and Jerry M. Melillo, "Human Domination of Earth's Ecosystems," *Science*, 277(5325) (July 25, 1997): 494–499.

[54] Ibid., p. 494.

[55] World Wildlife Fund, "2010 Living Planet Report," www.panda.org/about_our_earth/all_publications/living_planet_report/.

[56] I owe this analogy to Jeremy Waldron.

our lives, thus providing an importance source of meaning. This, I submit, is sufficient reason for us to respect nature. For when we fail to respect nature, we lose this important source of meaning. [← *important*]

Another reason to be concerned about respecting nature is from a concern with psychological wholeness. Respecting the Other as independent and autonomous is central to knowing who we are and respecting ourselves.[57] Whitman, the sage poetic observer of American democracy, had something like this in mind when he wrote: "I swear the Earth shall surely be complete to him or her who shall be complete."[58] [*Respecting nature = respecting ourselves*]

5 SLOUCHING TOWARD THE FUTURE

Now back to the nettles. The current conventional view is that if we are to avoid "dangerous anthropogenic interference with the climate system," then our global greenhouse gas emissions are going to have to peak within the next 10–15 years, decline by at least 50 percent by mid-century, and then move toward virtual elimination by the end of the century. The decisions we must make now have even greater urgency since some poor countries need to increase emissions (e.g., most African countries) and some large emitters have relatively low per capita emissions compared to industrialized countries (e.g., China, India, Brazil).[59] Moreover, energy crises are breaking out all over the world (e.g., China, Chile), and decisions are being made now that commit countries to particular energy policies and emissions patterns for the next generation. This is happening while many of the options that are most discussed (e.g., hydrogen cars, new generations of nuclear plants, "clean" coal, etc.) are not available in the real time in which decisions must be made. According to Jim Hansen, we must actually stabilize emissions in the next decade – not agree to do so, decide to do so, or develop a plan to do so. [*We need to act*]

Given this way of understanding our problem, how do we decide which nettles to grasp? It is reasonable to say that these are political decisions, but

[*need to take action*]

[57] Part of the anxiety we feel about Timothy Treadwell (portrayed in Werner Herzog's Film *Grizzly Man*) is that the special empathy he feels for the bears often slips into psychological appropriation and a failure of respect. I believe that some of these themes relating respect to psychological integrity can be found in the writings of Kant and Freud, but I cannot develop this point here. On this topic I have benefited from conversations with Beatrice Longuenesse.

[58] From his poem, "A Song of the Rolling Earth," available at http://classiclit.about.com/library/bl-etexts/wwhitman/bl-ww-rollingearth.htm.

[59] It is sobering to realize that a single Southern company power plant in Juliette, Georgia, emits more greenhouse gasses annually than Brazil's entire power sector. Juliet Eilperin, "World's Power Plant Emissions Detailed," *Washington Post*, Thursday, November 15, 2007.

I'm skeptical that our political institutions are up to it. The problem we face is unprecedented in its nature and difficulty.[60] Jurisdictional boundaries and competing scales cause multiple, overlapping and hierarchically embedded collective action problems. On a daily basis we witness policy failures and dysfunctions with respect to problems that are much less complex. There is a strong status quo bias of people and institutions, and an even stronger status quo bias is built into our particular form of representative government as opposed to that of Britain or Germany, for example. Finally, the interest-group nature of our system is especially prone to create gridlock.

Consider just two examples in which decision-makers have attempted to grasp nettles: There have been attempts to locate large wind farms off the coasts of Massachusetts and Delaware that have run into enormous opposition from a variety of sources. An ambitious plan to introduce congestion pricing in parts of Manhattan, advocated by the mayor and approved by the city council, was killed by the New York State legislature, which simply refused to vote on it. These experiences do not augur well for the USA implementing an effective "cap-and-trade" system for reducing carbon emissions. Even if we get a cap-and-trade system that is considered a policy success, it may be a substantive failure. In order to be effective, such a system must be sensitive to concerns about the ceiling, whether the permits are auctioned or given away, how they are distributed across sectors, the level at which controls are implemented, and so on. In addition, there are ancillary policies (discussed under the rubrics of "safely valves," "price ceilings," and "competitive policies") that could cripple such a system just at the point at which it might become most effective.[61]

The only way to break through on this problem, which is the world's largest and most complex collective action problem, is through the actions of a morally motivated global citizens' movement that acts as a highly committed political interest group. Such a movement would stigmatize coal, meat eating, trophy houses, overheating and overcooling, large living spaces, and private automobiles. It would celebrate living lightly with dignity and elegance, relying on nature's own energy, rediscovering food and the pleasures of eating, the joys of living with nature and other people, and the satisfaction of effective political activism. I think (and hope) that we may be witnessing the birth of such a movement.

[60] I first began to argue this in "The Epistemology of Climate Change: Some Morals for Managers," *Society and Natural Resources*, 4 (1991): 319–329.

[61] For more on these themes, see my "The Post-Kyoto Climate: A Gloomy Forecast," *Georgetown Journal of International Environmental Law*, 20 (2008): 537–551.

6 CONCLUSION

Climate change is a new issue that presents us with some old problems: How can representative democracy respond to long-term problems that have global reach? How can we integrate our moral and political lives in a way that is consistent with liberal democracy? Our best shot at solving this problem is a highly motivated, global citizens' movement that can create the conditions for political action. If we can solve the problem of climate change, we will have succeeded in solving larger problems that haunt American democracy and global governance.

huge problem, if we can solve this we can solve anything

CHAPTER 2

Is no one responsible for global environmental tragedy? Climate change as a challenge to our ethical concepts

Stephen Gardiner

Today we face the possibility that the global environment may be destroyed, yet no one will be responsible. This is a new problem.

Dale Jamieson

Over the last twenty years, the idea that climate change – and global environmental change more generally – is fundamentally a moral challenge has become mainstream. But most have supposed that the challenge is one of acting morally, rather than to our morality itself. Dale Jamieson is a notable exception to this trend. From the earliest days of climate ethics, he has argued that successfully addressing the problem will involve a funda- mental paradigm shift in ethics.[1]

In general, Jamieson believes that our current values evolved relatively recently in "low-population-density and low-technology societies, with seemingly unlimited access to land and other resources," and so are ill- suited to a globalized world.[2] More specifically, he asserts that these values include as a central component an account of responsibility which "pre- supposes that harms and their causes are individual, that they can be readily identified, and that they are local in time and space."[3] But, he claims, global environmental problems such as climate change fit none of these criteria, so that a new value system is needed, one which addresses "fundamental questions" about "how we ought to live, what kinds of societies we want,

---

[1] The most pronounced statements appear in Dale Jamieson, "Climate Change, Responsibility and Justice," *Science and Engineering Ethics*, 16 (2010): 431–445; and "Ethics, Public Policy and Global Warming," *Science, Technology, and Human Values*, 17 (1992): 139–153, from which the epigraph is taken (p. 149). I focus on the most recent. Also relevant are "The Moral and Political Challenges of Climate Change," in Susan Moser and Lisa Dilling, eds., *Creating a Climate for Change: Communicating Climate Change and Facilitating Social Change* (New York: Cambridge University Press, 2007), 475–482; and "Adaptation, Mitigation, and Justice," in Walter Sinnott-Armstrong and Richard B. Howarth (eds), *Perspectives on Climate Change* (Amsterdam: Elsevier, 2005), vol. 5, 221–253.
[2] Jamieson, 'Ethics, Public Policy,' p. 148.   [3] Ibid.

38

and how we should relate to nature and other forms of life."[4] In particular, he declares that there is little hope of success without direct appeal to a hitherto underappreciated "duty of respect for nature."[5]

In my view, Jamieson is right to question the adequacy of conventional ethical thinking for addressing global environmental problems, but the situation with respect to responsibility is less clear-cut than he suggests.[6] In this chapter, I offer several objections to Jamieson's main arguments, and propose an alternative, less revisionary, account of our predicament. Nevertheless, I also suggest that much of the spirit of Jamieson's position remains intact, so that my proposal is more of a "friendly amendment" to his view than an outright rejection of it. According to this amendment, climate change involves a failure of our attempts to delegate responsibility to political institutions charged with acting on our behalf. Such failures are not a new problem, but they are jarring, since they throw many of our normal practices into doubt, and threaten to restore to us demanding social burdens to which we have become largely unaccustomed. This situation need not cast doubt on our basic account of ethical responsibility. Indeed, that account helps to illuminate what is going on. Still, it does imply that we face a large (although somewhat familiar) moral and political challenge.

Section 1 presents an initial overview of Jamieson's position, and makes two preliminary points about it. Sections 2 and 3 consider Jamieson's main arguments against the applicability of individual and political responsibility, respectively. Section 4 offers a sketch of my alternative account. Section 5 considers how this account compares with Jamieson's invocation of a "duty of respect for nature." Section 6 offers a conclusion, and considers some objections.

I PRELIMINARIES

Jamieson observes that climate change is often seen through the lens of responsibility, in prudential or ethical terms, where the ethical includes both individual and political responsibility. But he challenges this practice, claiming that climate differs substantially from paradigm cases in these realms, and that this explains what he calls the lack of "urgency" people feel about the problem. In essence, Jamieson's thought is that "we tend not to conceptualize [climate change] as an urgent moral problem because it is

---

[4] Ibid., p. 147.   [5] Jamieson, "Climate Change, Responsibility."
[6] Stephen M. Gardiner, *A Perfect Moral Storm* (Oxford University Press, forthcoming).

climate change is not a moral problem

not accompanied by the characteristics of a paradigm moral problem," and in particular does not present a clear case of ethical responsibility. In addition, he maintains that, though a plausible argument might be made based on responsibility, this would require a significant extension or revision of central concepts that we would have to show there is good reason to engage in, and whose efficacy is dubious.[7]

people need to notice this to actually take change though

Despite this, Jamieson believes that matters are not hopeless, because climate change also engages a third, less discussed "value" – "the duty of respect for nature" – which "cannot easily be taken up by concerns of global justice or moral responsibility," but which "should motivate people to acknowledge a responsibility to respond to climate change." Indeed, he claims that "unless [this value] is widely recognized and acknowledged, there will be little hope of successfully addressing the problem."[8]

Before taking up Jamieson's arguments, I want to make two preliminary points. The first concerns his discussion of prudential responsibility. Jamieson agrees with the mainstream thought that present climate policy "is contrary to most reasonable notions of enlightened self-interest."[9] But he says that the dominant ways of expressing this idea, through appeals to insurance and economic estimates of aggregate costs and benefits, face important obstacles. Although Jamieson makes several points, the idea that ties together the most important seems to be that an appeal to enlightened self-interest requires a self and set of (enlightened) interests that are readily identifiable. In the case of global environmental problems, this "self" is presumably humanity considered as such. But this raises a challenge. Not only is "humanity" a complex abstract entity, but it is not obvious from a theoretical point of view what counts as its interests. Presumably they are either reducible to, or at least dominated by, the interests of individual human beings. But this implies that talk of "humanity's interests" implicitly invokes some kind of integration over space and time of a very large population of individuals whose values and projects are quite diverse. Clearly, such integration is a substantial and nontrivial task involving important value judgments. Given this, it is doubtful that "the case for responding aggressively to climate change can be made simply in terms of prudential responsibility,"[10] in large part because in this case the "prudential" quickly turns into the ethical. We might add that the appeal to

humanity can be pushed down to individual's interests

Then becomes an ethical issue

[7] Jamieson, "Climate Change, Responsibility," pp. 439–440.
[8] Jamieson suggests this value system should focus on fostering and developing a set of "twenty-first century" and "green" virtues, including "humility, courage . . . moderation," "simplicity and conservatism" (Jamieson, "Ethics, Public Policy," p. 294).
[9] Jamieson, "Climate Change, Responsibility," p. 434.   [10] Ibid., p. 435.

the interests of humanity as such brings on significant motivational worries. In addition to the question of whether and to what extent people will in fact be motivated to promote the interests of humanity in general, especially if they perceive its interests to conflict with their own, we should note that part of the appeal of couching the problem of climate change in terms of enlightened self-interest is the assumption that self-interest brings with it no special motivational problems – people are automatically motivated to pursue their own interests – but here the abstract nature of the entity whose self-interest is said to need promoting threatens that assumption.

The second preliminary point concerns Jamieson's use of "urgency" to describe what is lacking in the case of climate change by contrast to a "paradigm moral problem."[11] What is meant by "urgency" is not totally clear. But two possibilities naturally spring to mind. The first is that it refers to the moral severity of the problem; the second that it describes the strength of the motivation of agents to address it. These interpretations have different implications for how we understand Jamieson's charge against conventional moral values. Read the first way, the complaint is that climate change is not as morally severe by normal moral lights as the usual paradigm cases, in the sense that the underlying reason for moral complaint is less serious. Read the second way, the thought is that agents are not motivated to act in the same way towards climate change as they are in those more standard cases, in the sense that their underlying motive for action is not as strong. On one way of putting it, the difference is that the first reading concerns *justifying reasons* for action, and the second *motivating reasons*.

Justifying and motivating reasons are different things, and raise distinct questions.[12] Jamieson likely does not distinguish between them because he assumes that there is a tight connection between appreciating moral severity (justifying reasons) and being motivated to act in accordance with them (motivating reasons), so that to some extent they stand or fall together. In particular, according to a popular and mainstream view in contemporary metaethics ("internalism about moral motivation," or simply "internalism"), if one really appreciates a justifying reason, then one will automatically have a corresponding motivating reason to act accordingly.[13]

---

[11] Ibid., p. 436.

[12] See, e.g., Michael Smith, *The Moral Problem* (Oxford: Blackwell, 1985); David Brink, *Moral Realism and the Foundations of Ethics* (Cambridge University Press, 1989). For a recent overview, see Connie Rosati, "Moral Motivation," *Stanford Encyclopedia of Philosophy*, 2006.

[13] In "Adaptation," Jamieson says: "Surely there is some connection between seeing an act as morally right and performing it. That something is the morally right thing to do is a powerful consideration in its favor" (p. 240). This is most naturally read as an internalist claim, and in discussion Jamieson has indicated that he accepts this characterization.

externalists vs. internalists

Nevertheless, it is worth pointing out that internalism is controversial. Specifically, according to another standard position (unsurprisingly called "externalism"), there need be no tight connection between justification and motivation. They are simply different issues. Specifically, externalists maintain that agents might grasp the moral severity of a particular action perfectly well – and so possess a justifying reason not to do it – and yet not be motivated accordingly. In the most obvious cases, they see what the right thing to do is, but just don't want to do it.[14]

Externalism is worth mentioning because it reveals one way to avoid Jamieson's conclusion about the need for a conceptual paradigm shift. If we think that the problem is that people are not *motivated* to act by the fact that climate change is a moral problem, this creates trouble for our moral concepts under internalism that it need not under externalism. Under internalism, a lack of motivating reasons suggests a lack of appropriate justification. But this is not so under externalism. On that view, we might genuinely appreciate the moral severity of the problem, and so the justifying reasons, and yet still not be motivated to act. Perhaps we are simply not interested in responding to such justifications. This might show that there is something wrong with us (our motivations), but not with morality (our moral concepts). Perhaps we are just bad or imperfect moral agents. This need not imply that we need a conceptual paradigm shift, only that we ought to be morally better than we (currently) are. *no need for a change in values.*

This is a good idea – concept

some people focus on motivation

This escape route from Jamieson's argument is important because some people's concerns about the relevance of moral discourse to climate change do seem to revolve around questions of motivation rather than justification. Still, it will not be sufficient for everyone. For one thing, it relies on a rejection of internalism, and many will think that this is too heavy a price to pay. For another, it does nothing to address the worry about whether justifying reasons also fail to gain traction, and this is clearly at least one part of Jamieson's concern. Given these points, I shall focus most of my discussion below on justifying reasons – or questions of moral severity – rather than on motivating reasons. But, of course, if one accepts internalism, most of the discussion will generalize to motivation as well.

taking an alternative look on his own ability to discern – really cool

[14] Most contemporary externalists want to allow for some kind of connection between justifying and motivating reasons, including the possibility of a strong contingent connection based in facts about human nature. But the basic theoretical point remains, and the extent to which this concession should be made might be disputed. See Brink, *Moral Realism*.

2 INDIVIDUAL RESPONSIBILITY

Jamieson's main target is the idea that the concept of ethical responsibility clearly applies to climate change.[15] He divides ethical responsibility into individual moral responsibility and political responsibility. In this section, I discuss his views about individual responsibility.

Jamieson claims that paradigm cases of individual responsibility are ones where "an individual acting intentionally harms another individual; both the individual and the harm are identifiable; and the individual and the harm are closely related in time and space."[16] For the sake of argument, let us grant this assumption. Jamieson offers as a paradigm example the case of "Jack intentionally stealing Jill's bicycle" (*Jack 1*). But he argues that climate change is more like this:

Jack 6: "Acting independently, Jack and a large number of unacquainted people set in motion a chain of events that prevents a large number of future people who will live in another part of the world from ever having bicycles."

Moreover, he argues that *Jack 6* helps to "explain why many people do not see climate change as a moral problem; or if they do see it as a moral problem, it fails to have the urgency of a paradigm moral problem." While acknowledging the possibility that part of the core remains because some people have acted in a way that harms other people, Jamieson asserts that, at a minimum, "most of what accompanies this core has disappeared . . . it is difficult to identify the agents and the victims or the causal nexus that obtains between them; thus, it is difficult for the network of moral concepts (for example, responsibility, blame, etc.) to gain traction."

I have some concerns about the Jack analogy. First, more could be done to make the case closer to climate change cast as an ethical issue. Jamieson himself says that the climate case involves three main elements: the rich (i) appropriate more than their fair share of a global public good, thereby (ii) harming the poor through imposing risks of heavy damages on them, and (iii) do so for unnecessary reasons. I agree that these are central features, but doubt that the Jack analogy really reflects this. Recall that Jamieson begins with the paradigm, "Jack intentionally steals Jill's bicycle." One issue is that, for most people, a stolen bicycle is a minor loss, whereas climate change involves risks of severe harm.[17] Another is that no reason is offered for why

---

[15] Jamieson, "Climate Change, Responsibility," p. 439.
[16] The quotations in this paragraph are from Jamieson, "Climate Change, Responsibility," pp. 435–436.
[17] There is also no sense in which Jack has a share in the bicycle, so that the two of them are already involved in a common endeavor.

the bicycle is stolen. But presumably it would make a difference if Jack were a doctor urgently in need of transport in order to save a life, or if he just wanted a joy ride.

For such reasons, a better core case of individual responsibility (parallel to *Jack 1*) to invoke might be something like:

George 1: "George steals Sanjay's smoke alarm and then sets fire to Sanjay's house while Sanjay is asleep inside. He does this because he is bored and would like a little excitement."

This seems at least as clear a paradigm as the Jack case, but preserves the loss of resources, the risk of severe harm, and the sense that the agent's reasons are relatively trivial.[18] Moreover, if anything, the moral seriousness of this case seems stronger. This is interesting in itself, as it may imply that *George 1* is a more central paradigm of moral responsibility than *Jack 1*. But it may also turn out to be important as we progress. For even if subsequent departures from this new paradigm do detract from the initial severity of the case, if one sets out from a more serious starting-point, there is a better chance that more urgency will survive,

Of course, like *Jack 1*, *George 1* is not a direct analogue to climate change. So, how might a more appropriate analogy go? One possibility is to transform *George 1* so as to more closely resemble Jamieson's *Jack 6*. Hence, we might say:

George 6: "Acting independently, George and a large number of unacquainted people set in motion a chain of events that causes a large number of future people who will live in another part of the world never to have smoke alarms, and to have their houses set on fire."

Two things strike me as salient about this example. First, *George 6* appears to be more morally serious than *Jack 6*. Whether this makes enough difference to overcome Jamieson's worries, I am not sure. But the question is worth asking. Second, and more importantly, the analogy can be improved. *George 6* takes features from *Jack 6* that do not seem quite apt for climate change, and may sway our perceptions of urgency. Hence, perhaps we can do better still by correcting this. (Though we cannot expect a perfect fit, getting a little closer may be instructive.)

Two initial disanalogies between climate change and *George 6* actually seem to help Jamieson's thesis that moral severity is reduced. The first is that there is some scientific uncertainty surrounding the magnitude, timing and

---

[18] One deficiency of *George 1* is that stealing a smoke alarm is not quite parallel to taking more than one's "fair share of a public good." *George 7* (below) does a little better.

precise nature of future climate impacts. For such reasons, it seems better to say that our actions impose a serious risk of significant negative and perhaps catastrophic impacts, rather than simply that they cause them (as *George 6* suggests).

The second disanalogy is one Jamieson mentions in passing but does not include as a feature of *Jack 6*. Agents in the climate case do not directly intend to inflict harm on others, in the sense that this is neither the sole nor prominent aim of their actions, nor indeed any direct aim at all.[19] From their point of view, climate change is only an unwelcome by-product of their activities, not any part of their purpose. One sign of this is that (presumably) if emitters could instantly and costlessly eliminate the indirect effects of their emissions, they would see no reason not to do so.

[handwritten margin note: not accepting blame – they didn't mean to]

These two points help Jamieson's case. Climate change is unlike *George 6* in that the agents do not directly intend the results of their actions, and these actions impose a serious risk rather than simply causing the impacts mentioned. Presumably, other things being equal, these features of the situation reduce the moral urgency of climate change with respect to the paradigm. However, I doubt that they alter the case dramatically, and in such a way as to seriously undermine urgency. More importantly, there are further disanalogies that pull in the opposite direction, as we shall now see.

The third disanalogy concerns the claim that climate change primarily affects "future people in another part of the world" (so that action "will mainly benefit poor people who will live in the future in some other country").[20] First, concern about the vulnerability of the poor is not restricted to other countries. As Jamieson himself mentions, there are plenty of poor people in rich countries who share many of the same vulnerabilities, and may even be in worse shape because of a lack of appropriate social infrastructure. National wealth does not always coincide with less vulnerability. Second, concern about vulnerability is not restricted only to the poor. In the medium to long term, humanity in general looks vulnerable. If we are considering an ice-age magnitude shift in only a century or so (as the IPCC projects), then it is not at all obvious that richer people will be able to protect themselves adequately with extra resources. Third, vulnerability does not arise only in the future. Prominent studies tell us that some severe impacts are likely happening now, and that others are coming in the near future due to climate change to which we are already committed.[21] Hence,

[handwritten margin note: There are still poor people in the world]

[19] Jamieson, "Ethics, Public Policy," p. 149.   [20] Jamieson, "Climate Change, Responsibility," p. 434.
[21] Global Humanitarian Forum, *Human Impact Report: Anatomy of a Silent Crisis* (Geneva: Global Humanitarian Forum, 2009).

how to deal with unavoided impacts is a serious issue, and one that arises now and in the next few decades, affecting many people who are currently alive and about to be born. In short, it does not seem that our behavior affects only "future (poor) people in another part of the world." Instead, if we consider the long-term effects of climate change and the need for adaptation, we affect people in both the present and the future, poor and rich. Since more are affected, this increases the moral severity of our actions.

The fourth and fifth disanalogies concern Jamieson's claim that the agents in *Jack 6* are "unacquainted" and "act independently." Surely matters are much more complicated. First, we contribute to climate change as individuals through our involvement in complex social systems where the choices we make are not strictly "independent" of the choices of others but rather framed by them and by the choices of the past. These yield to us certain kinds of infrastructure and cultural expectations surrounding energy use. One way to illustrate this is to think about individual carbon footprints. Clearly, there are wide international differences here. In 2005, the global average was 1.23 metric tons of carbon per capita. In the USA, the average was 5.32, in the UK 2.47, in China 1.16, in India 0.35, and in Bangladesh 0.08.

Of course, individual emissions vary considerably from these averages within countries.[22] Still, in a country like the USA it is quite difficult for an individual to reduce her emissions beyond a certain level, and a level that is higher even than the average in some other developed countries. Even more importantly, there is more at stake for the individual in doing so. An American who tries to move to the global average (1.23) is in a different situation than someone in China, for example, or Bangladesh. This is not to say that she should be excused (though surely it makes some difference). Instead, the point is that it is difficult to see individual Americans as "acting independently" when they bring about greenhouse gas emissions. Their actions are tied together in deep and important ways by structural facts about their economies and lifestyles.

Second, a related claim should surely be made about the idea that emitters are "unacquainted." Though it is true that most Americans, British, and Chinese have met very few of their compatriots, this does not mean that they are morally equivalent to mere strangers. Instead, their agency is to some extent unified by their roles as citizens. Indeed, at least as a matter of theory, the state is widely understood to be *their* agent (however much it may fail to live up to this ideal in practice). This connection should not be ignored (see Sections 5–6).

[22] Jyoti Parikh, Manoj Panda, A. Ganesh-Kumar, and Vinay Singh, "CO$_2$ Emissions Structure of Indian Economy," *Energy* (2009), doi:10.1016/j.energy.2009.02.014.

Third, though they do not require it, the bicycle analogies tend to suggest a number of single and isolated incidents of independent action that result in the nonexistence of bicycles in the future. By contrast, climate change involves long-term *patterns* of action. On the one hand, some groups of agents have emitted far more than others over time and continue to do so. Moreover, they have benefited substantially from doing this. On the other hand, these groups are part of an international community that has officially recognized the problem for nearly twenty years, and repeatedly attempted to deal with it (but failed). Surely this suggests a deeper connection to the problem than that implied by Jamieson's example.

Given these points, it seems plausible to revise *George 6* along the following lines:

George 7 (or *George and his buddies*): "George and his buddies like to have big firework displays over the river. These shoot burning debris into the air, predominantly over the poorer neighborhoods on the other side. This has already imposed, and continues increasingly to impose, a serious risk on many people in the area that their houses will catch on fire. George and his buddies are aware of this risk, keep saying that they will cut back, buy safer fireworks, contribute funds to the fire department in the poorer neighborhoods, and so on. But they don't. Instead, they keep making the displays bigger. They like fireworks. (They could like other things too. But they are used to fireworks.)"

In my view, this example conveys a sense of moral severity substantially beyond *Jack 6* or *George 6*. George and his buddies are clearly seriously morally irresponsible and blameworthy. Since the case is designed to be closer to climate change than the alternatives, this is an important result. The question then arises as to whether the increase in urgency is sufficient for, as Jamieson puts it, "the network of moral concepts (for example, responsibility, blame, etc.) to gain traction."[23] My own opinion is that it is. If this is right, then Jamieson's claim that our concepts of individual moral responsibility must be extended or revised requires further defense, as does his claim that a paradigm shift in ethics is required.

There is one last feature of the case that deserves special mention. Climate change is a severely lagged phenomenon, such that many of the cumulative effects of current behavior will be imposed in the future, on people who are very young now, or not yet born. As stated, *George and his buddies* seems to downplay this characteristic of the case (saying only that the fireworks "increasingly" imposes a serious risk); so, what happens if we emphasize this feature (as *Jack 6* does), by saying that the risks are

---

[23] Jamieson, "Climate Change, Responsibility," pp. 435–6.

Kind of confused

predominantly imposed on future people? Is this enough to dissolve the sense of urgency?

I agree that the intergenerational dimension of the climate problem is central, and also that it offers the best explanation for continued political inertia.[24] However, I doubt that the root of the problem lies with our concept

Should he really be so concerned about the future?

of responsibility. On the contrary, we seem to have no problem grasping the moral severity of imposing severe risks on future generations, and in regarding this as an especially serious ethical matter. Suppose, for example, that someone plants a time bomb with a 100-year fuse underneath an elementary school.[25] Does it matter that most of those at serious risk from this bomb do not yet exist? It seems not. The motivational efficacy of such reasons can, of course, be questioned. But even here we should note that appeal to the interests of the future is a perennially popular political strategy, and one often used to justify major sacrifices. So, it is not so clear that this is the problem either.

3 POLITICAL RESPONSIBILITY

An obvious initial objection to *George and his buddies* is that in rejecting the claims about "independence" and "unacquainted agents" we have moved to an area of ethical responsibility that Jamieson considers separately, that of political responsibility. Climate change is often seen as an issue of global justice, and Jamieson mentions several issues that make this framing plausible. These include that "most of the emitting is done by the rich countries of the North, but most of the climate-change-related dying is done in the poor countries of the South," and that some poor countries (such as Bangladesh) are especially vulnerable. Jamieson notes that there are complications, including that there are more high emitters in developing nations

Political views of Jamieson 2 phase

such as China and India than in some developed countries (such as Australia and the Netherlands). But his main objections to the political responsibility model are (first) that "in several important respects, causing climate change is not like one country unjustly invading another country," and (second) that "the nation state is one level of social organization that is relevant to addressing climate change because it is causally efficacious, but the nation state is not the primary bearer or beneficiary of ethical responsibilities in this regard."[26]

I think that there is something right about these claims, but that they do not license a move away from the political responsibility model. In the rest

---

[24] See Gardiner, *Perfect Storm*.   [25] Derek Parfit, *Reasons and Persons* (Oxford University Press, 1985).
[26] Jamieson, "Climate Change, Responsibility," p. 439.

of this section, I confront Jamieson's arguments directly; in Section 5 I explain how we can accommodate what is persuasive about them.

Let us begin with the comparison with military invasion. Although greenhouse emissions are not exactly like tanks crossing a border, they are not completely different. Think of acid rain as an intermediate case. This is an issue of transboundary pollution (and hence might be thought of as a kind of "invasion"), and is legally recognized as such. The relevant differences between sulphur dioxide emissions which cause acid rain and carbon emissions which cause climate change seem to be as follows: first, carbon emissions are quickly and globally diffused throughout the atmosphere; second, they have negative impacts on the home nation as well as on others; third, these impacts are much less readily isolated and attributed than those of acid rain; and fourth, many nations (and especially all powerful nations) are also contributing.

These differences undoubtedly complicate the situation; but it is not clear that they make it radically different conceptually. On the one hand, some nations produce very little carbon pollution of their own, and yet are profoundly at risk. Hence, they seem relatively similar to victims of acid rain. For example, Bangladeshis produce only 0.08 metric tons of carbon per year per capita. This is orders of magnitude less than the global average and the averages for countries such as the USA and Canada. It is also much lower even than would be required by the 60–80 percent cuts in global emissions suggested by many scientists.

On the other hand, we can easily imagine a world in which many powerful nations were producing sulphur emissions, causing acid rain in many other nations and even (to a much lesser extent) their own, but where it was difficult to tell precisely whose emissions were causing which impacts. This problem could be very similar to climate change, and yet would not (in my view) imply the need for a conceptual shift.

Given such concerns, we need to know more about the ways in which military invasions and climate change are disanalogous. Of course, Jamieson's intent here may be to appeal to the features of the Jack and Jill cases. But we have already argued that these are not sufficient to undermine the responsibility model in the individual case. So, further argument would be needed to show why things are different in the political case. Until such argument is forthcoming, the first objection is not compelling.

Let us turn now to Jamieson's second claim against the political responsibility model, that states are not the primary bearers or beneficiaries of ethical responsibility for climate change. This is a complex issue. Let me make one quick point, and then move to a more general discussion.

The quick point is just that the idea that the nation state should not be the locus of concern does not seem to disqualify the climate issue as one of global justice as conventionally understood. Many paradigm examples in contemporary theories of global justice also resist a focus on the nation state. Moreover, they often specifically do so through claims about moral and political responsibility. Consider, for example, two of the most influential approaches in the literature. With regard to (what Jamieson calls) moral responsibility, Peter Singer has argued over many years that our poor record of assistance to the global poor violates a morally minimal duty of aid that ultimately applies to individuals.[27] With regard to political responsibility, Henry Shue and Thomas Pogge have both argued extensively that powerful nations have violated negative duties not to harm the more vulnerable through the manipulation of international agreements and institutions.[28] Given the prevalence of such moves, it does not seem true that the claim "climate change involves issues of political responsibility" is out of step with mainstream work in global justice. Indeed, quite the contrary.

There is, of course, a natural response to this point. It could (plausibly) be argued that much contemporary work on global justice has itself failed to gain traction, and that this is because it, also, fails to communicate "urgency", in terms of either justifying or motivating reasons. This is a different argument than the one Jamieson explicitly offers. However, it would fit well with his claim that even though it may (in the end) be plausible to think of climate change as posing issues of ethical responsibility, "this argument would have to be revisionary," and would "have to show that there are good reasons for extending or revising our concepts of ethical responsibility in such a way that problems posed by climate change would fall under them."[29] Perhaps the same could be said about global justice more generally.

I will not try to address this new argument directly here, but instead try to say something about the more general issue that it raises by offering a sketch of my own position. I then consider the implications of my position for the kind of approach that Jamieson thinks stands the only chance of working, that of appealing to duties to nature. Again, my claim will be that there may be something right about this; but my gloss on what that is will be different from Jamieson's.

---

[27] E.g., Peter Singer, "Famine, Affluence, and Morality," *Philosophy and Public Affairs*, 1 (1972): 229–243.

[28] Henry Shue, *Basic Rights* (Princeton University Press, 1980); Thomas Pogge, *World Hunger and Human Rights* (Cambridge: Polity, 2002).

[29] Jamieson, "Climate Change, Responsibility", p. 439.

4 MY APPROACH

Some kinds of ethical claims seem to stand in conflict with existing social structures, and in potentially radical ways. Many claims of global justice and climate ethics are of this sort. In such cases, it seems plausible to think that there might be deep inconsistencies between real-world practices and what (morally speaking) needs to be done. Often, the situation is of the following sort. First, the problem at hand is fundamentally one of collective behavior, and it seems unlikely that it can be resolved simply by appealing for changes in the actions of individuals, given the world as it is. This is for familiar reasons associated with collective action problems: perhaps, by itself, my contribution is (or appears to be) infinitesimally small, or perhaps others won't change no matter what I do, or perhaps ceasing my own problematic behavior will only make others do more damage, and so on. Second, the negative impacts of any individual's behavior are not adequately captured by low-level moral practices and appraisal. Perhaps they are simply invisible to them, or perhaps they are seen but are very difficult to take seriously in context, especially relative to the more immediate and local benefits of the behavior. Third, nevertheless, the activities in question remain seriously inconsistent with fundamental moral convictions (either those of the individuals whose behavior causes them, or those of their society, or both). So, there is a serious ethical problem.

Jamieson claims that the situation with respect to climate change is a new kind of problem. But the general issue as I have described it is not new. It has arisen in the past with many social issues, such as the abolition of slavery, the civil rights movement, and the emancipation of women.[30] Still, Jamieson is right to emphasize a particular feature of this kind of situation. Part of the issue is that conventional practices of appraisal do not capture the immediate manifestations of, and causal contributions to, a serious moral wrong. In particular, at the individual level, the relevant behaviors often seem (to conventional eyes) to be "innocent,"[31] or at least not "unusual";[32] and at the collective level, it is either not clear who has the relevant responsibility to deal with the problem, or, if it is clear but they have failed, whose responsibility it is to address this failure.[33] We might say that the crux of the issue is that conventional practices – at the individual and social

---

[30] See also, John Nolt, Chapter 3, this volume. [31] Jamieson, "Ethics, Public Policy," p. 149.
[32] Walter Sinnott-Armstrong, "It's Not *My* Fault," in Walter Sinnott-Armstrong and Richard Howarth (eds), *Perspectives on Climate Change* (New York: Elsevier, 2005), pp. 285–307.
[33] See Sinnott-Armstrong's example of the bridge, p. 287.

level – "fail to grasp, or get a grip" on the issue at hand, so that it "slips through the cracks." Let us call this "the grasping problem." One possible cause of this problem, as Jamieson suggests, is that ethical concepts fail to "gain traction." But this is not the only candidate explanation. Hence, it is worth distinguishing the more general phenomenon.

When confronted by an instance of the grasping problem, two initial reactions are tempting. One is to assert that the individual behavior that produces bad results should be moralized. Another is to refuse to accept responsibility at the individual level, complaining that there would be something bizarre or unfair about doing so, given conventional practices, and arguing that solutions ought to be pursued elsewhere. As it happens, in the case of climate change, Jamieson emphasizes the first path – advocating that we should adopt a set of "green virtues" at the individual level that are nonresponsive to consequences,[34] while Walter Sinnott-Armstrong advocates the second, arguing that the problem is primarily political, and should be dealt with there.[35]

Still, we should be careful here about jumping to conclusions too quickly. First, it is far from clear that we should be in the business of making *a priori* pronouncements. For one thing, particular manifestations of the grasping problem are likely to have different features, so that it would be rash to conclude that a "one size fits all" response is desirable (i.e., that we should always emphasize the political or the individual level). Instead, we would need to know more about the specific case before choosing any particular path. For another, the actual pronouncements are likely to be driven as much, if not more, by empirical assumptions about how agents and institutions actually act and interact as by philosophical claims. (Note that Jamieson and Sinnott-Armstrong are both officially consequentialists, but offer very different pronouncements about what is needed.)

Second, we might expect that the more interesting questions are not at the extremes. Moreover, there are signs of this even in Jamieson and Sinnott-Armstrong. For example, as it turns out, Jamieson is not unconcerned about political issues. And though he claims that ultimately political change will be primarily driven by individual value change, he also thinks that one of the individual virtues that matters here is "cooperativeness."[36] Similarly, Sinnott-Armstrong emphasizes political commitments, but nonetheless believes that it would be good if individuals did not engage in frivolous high-consumption activities either; he just thinks that it is better

---

[34] Jamieson, "When Utilitarians Should Be Virtue Theorists", *Utilitas*, 19(2) (2007): 160–183.
[35] Ibid.   [36] Jamieson, "When Utilitarians," p. 182, note 65.

not to get involved in blaming them.[37] Given this, we should expect that the real action in responding to the grasping issue lies in the details of working out the appropriate relationship between various different levels of individual and social action.[38]

From these observations, it seems that the most pertinent philosophical question is really how we are to understand the relationship between individual and collective responsibilities, and whether this can help us to address the grasping problem when it arises. This is the conceptual issue that seems most important.

As it turns out, I am inclined to think that the conceptual issue is not very difficult, and that it does not cast doubt on the appropriateness of looking at climate change as an issue of ethical responsibility, but rather illuminates that approach. Here is my proposal.

According to a long tradition in political theory, political institutions and their leaders are said to be legitimate because, and to the extent that, citizens delegate their own responsibilities and powers to them. The basic idea is that political authorities act in the name of the citizens in order to solve problems that either cannot be addressed, or else would be poorly handled, at the individual level, and that this is what, most fundamentally, justifies both their existence and their specific form.

This simple model suggests an equally simple account of failures of ethical responsibility. Take climate change as an example. First, it seems to follow straightforwardly from the model that the most direct responsibility for the current failure of climate policy falls on recent leaders and current institutions. If authority is delegated to them to deal with global environmental problems, then they are failing to discharge the relevant responsibilities and are subject to moral criticism for this failure.

Of course, against this, it might be argued that such institutions were not designed to deal with large global and intergenerational problems; hence, the assignment of responsibility is unfair. There is some truth to this claim. Nevertheless, we should not concede too much too quickly. After all, existing leaders and institutions have not been slow to take up the issues and assume the mantle of responsibility, making many fine speeches, organizing frequent meetings, promising progress, making the topic a campaign issue, and so on. Hence, even if this role was not originally envisioned, many political actors have acted as if it did belong to them, and that they were capable of discharging it. They did not, for example, simply declare to their constituencies that the topic was outside of their

[37] Ibid. [38] Jamieson, "Adaptation."

purview or competence, nor did they advocate for fundamentally new or different institutions (e.g., by declaring the need for a new global council on the topic, or even a global constitutional convention). Given this, it is far from clear that they cannot be held at least partly responsible for assuming the role, and for their subsequent failure to deliver. They can hardly claim to be ignorant of, or to have refused, the responsibility.

Nevertheless, second, the more important issue is the following: Suppose that it is true that humanity simply lacks the appropriate institutions to deal with global environmental change. What follows? If political institutions normally operate under delegated authority from the citizens, the answer seems clear. This is a case where the delegation has either not happened or has failed to be successful. How do we think about this? Again, there is a natural answer. If the attempt to delegate effectively has failed, then the responsibility falls back on the citizens again, either to solve the problems themselves or, if this is not possible, to create new institutions to do the job. If they fail to do so, then they are subject to moral criticism, for having failed to discharge their original responsibilities.

At first glance, this move may seem startling. If the world's leaders and institutions are failing to deal with climate change, the average person might ask, how does that suddenly become *my* problem? Moreover, isn't that deeply unfair?

In response, let me make two comments. First, although the move is startling, it is a traditional one in political theory, and often made in mainstream arguments about rights of civil disobedience, revolution, and the like.[39] In short, this is not a foreign, or even unusual, model of political responsibility. Indeed, arguably, it is built into the foundations of democratic thinking and institutions more generally, as a natural consequence of their basic rationale. Hence (again), if there is a problem, it is not a new one, and not specific to climate change. The whole idea that citizens might be politically responsible for the behavior of their institutions is in some respects a radical and demanding one.

Second, the fact that the move seems startling to many contemporary readers may itself be the consequence of a certain vision of modern political justification. Some democratic thinkers believe that the role of social and political institutions is to discharge as many ethical responsibilities as possible for the citizenry, so that under an ideal system individuals would

---

[39] I am not advocating these measures. How to respond to political failure is a complex and difficult question. Moreover, one must be sure not to overlook either the successes of conventional institutions or the potential for certain kinds of intervention to make matters (much) worse.

not have to worry at all about such responsibilities, but would instead be *2 ways* maximally free to engage in their own pursuits (subject to the external *to view* constraints set out by the system). But here it is noticeable that success *institutions* breeds the elimination of responsibility at the individual level. The better the rest of the system is at discharging responsibilities on behalf of individuals, the fewer direct demands such responsibilities make on the individual. Hence, it is likely that the demands themselves become unfamiliar, and indeed perhaps invisible to the individual herself. If this is right, it seems plausible to think that the more effective a social system is (or is perceived to be) in discharging responsibilities in general, the more demanding any significant unmet responsibilities will seem. Or, to put the point in another way, for those used to very wide freedom to pursue their own ends without worrying about broader responsibilities, the emergence of a serious failure to discharge is likely to be deeply jarring. The issues will seem very unfamiliar and the nature of the responsibilities extreme. But this may say more about the past successes of the delegated responsibility paradigm than its likelihood of current or future failure.

In summary, in this section, I have argued that we should not be too quick to conclude that conventional accounts of political responsibility are inadequate to deal with global problems such as climate change. If the problem is ultimately one of a failure to delegate responsibility, there need be no conceptual mistake. The grasping problem is predictable, and not specific to this case, or even to global environmental problems more generally. Moreover, it does not threaten the basic political theory surrounding responsibility, and indeed that approach can help to explain the jarring nature of some appeals to that notion. We are simply not used to thinking about what our responsibilities are, how demanding they might be, and the role they already play in justifying contemporary social life. Climate change brings these matters into sharp relief. But it does not create them. *we create them... ominous*

5 A RIVAL APPROACH TO NATURE *Jamieson's suggestion*

How then should we deal with the climate problem? Jamieson suggests that we must appeal to a duty of respect for nature. I am more optimistic about mainstream invocations of ethical responsibility (individual and political), especially in so far as they take seriously issues of intergenerational ethics, political legitimacy and global justice. However, I also think that there is something to Jamieson's core concern. Hence, in this section, I will try to offer a different account of it.

Jamieson believes that climate change "puts at risk" a value of respect for nature, and that this explains "why some people are so passionate about the issue." He does not offer a direct account of the value, but claims that "human domination violates it," and that anthropogenic climate change is a "central expression" of this. He suggests the following account of domination. First, "rather than being governed by its own laws and internal relations, nature is increasingly affected by human action ... so thoroughgoing that it can be said to constitute domination." Second, this domination is expressed substantively, by facts such as "one third and one half of the Earth's land surface has been transformed by human action," and, attitudinally, by behaviors that "treat Earth and its fundamental systems as if they were toys that we can treat carelessly, as if their functions could easily be replaced by a minor exercise of human ingenuity."[40]

By contrast, my perspective does not begin by emphasizing the value of nature, in the way that talk of a duty of *respect for* nature might suggest. Instead, it considers the matter as a question of agency, from which thoughts about responsibility naturally follow. The basic idea is that we should consider our *relationship* to the wider world, and in particular the natural world, as part of an understanding of our own agency. This idea is compatible with (say) assigning nature some independent value that ought to be respected, or claiming that it has large instrumental or constitutive value for flourishing human lives, but it does not require such claims as a first step. At this level of explanation, such questions may be deferred.

One way to illustrate this perspective is through a heuristic. Suppose we begin with a claim about the situation facing us as human beings:

Humanity's challenge: "Humanity is, in geological and evolutionary terms, a recent arrival on the planet, and is currently undergoing an amazingly rapid expansion, in terms of sheer population numbers, technological capabilities, and environmental impact. A basic question that faces us as humans, then, is whether, amidst all this, we can meet the challenge of adapting to the planet on which we live."[41]

The views represented in the challenge are, I suspect, fairly common. Like most ethical views, they rest on a number of claims that might be contested. However, this need not trouble us here. For current purposes, the relevant points are that these views might provide some insight into the values of passionate environmentalists, and that these values connect rather straightforwardly with the account of delegated responsibility discussed above.

---

[40] The quotations in this paragraph are from Jamieson, "Climate Change, Responsibility," p. 441.
[41] Stephen M. Gardiner, 'Rawls and Climate Change: Does Rawlsian Political Philosophy Pass the Global Test?,' *Critical Review of International Social and Political Philosophy*, forthcoming.

If we accept the perspective of the challenge, it is easy to see why global [*important*] environmental issues such as climate change pose an ethical problem, and why this raises issues of responsibility. Climate change is one of a number of issues that suggest that currently we are failing the challenge. Moreover, in this case the threat is serious enough to imply that if we don't do something substantial soon, the failure may become severe, or even catastrophic. But, interestingly, we do not (yet) need controversial claims about the value of nature in order to make this point.

6 WHERE DO WE STAND?

Jamieson argues that global environmental change poses a new moral problem. Our normal concepts of ethical responsibility fail to "gain traction" when confronted with issues such as climate change because these do [*Jamieson*] not have the features of a paradigm moral problem, and this undermines how we understand the urgency of the case. He concedes that our concepts might be extended or revised to meet the challenge, but suggests that we would do better to invoke a duty of respect for nature.

I have offered five main responses. First, externalists about moral motivation might locate the problem in our motives rather than our ethical concepts. Perhaps the concepts apply; but we just don't want to act on them. Second, climate change is more compatible with an alternative paradigm case (*George 7*), and this rescues the sense of urgency (if it needs rescuing). Third, the problem of attributing responsibility is not new, but is shared with a range of other major social issues. Fourth, a traditional model of delegated political responsibility can account for such problems, and also for the jarring effect when they come to light. Fifth, this kind of approach can ground the concern that Jamieson calls "a duty of respect for nature," but in a different, and possibly less controversial, way.

These differences seem significant. In the end, however, they may result only in a reframing of Jamieson's concerns. One sign of this is that it might be argued that our two accounts are closer than they initially appear, for two reasons. [*Jamieson's view*]

First, Jamieson sometimes makes comments that suggest the delegated responsibility model. For example, in one paper he shows sympathy for a "deliberative ideal" of politics resting on the idea that "the best society is one that is a democratic expression of the reflective views of its citizens, based on their most fundamental values," and explains this in terms of a belief (attributed to Benjamin Franklin and the other Founding Fathers) that

delegated
responsibility

"a political system is a set of institutions designed by people to serve their deepest purposes."[42]

Second, it might be thought that Jamieson's talk of conceptual change is actually equivalent to my talk of changes in practices. He defines a system of values as something that "specifies permissions, norms, duties, and obligations ... assigns blame, praise, and responsibility; and ... provides an account of what is valuable and what is not ... [and] a standard for assessing our behavior and that of others."[43] But this suggests that any change in these things counts as "conceptual" change for him, so that the difference between us dissolves.

Despite these potential convergences, I suspect that there is more between our views than this. First, recall that Jamieson also attributes the conceptual problem to the fact that our current values evolved relatively recently in "low-population-density and low-technology societies." This is an interesting claim, and one that the delegated responsibility model might accommodate. But notice that it goes beyond my basic statement of the grasping problem.

Second, Jamieson also suggests that the conceptual problem has deeper, more psychological roots. Recent work in social psychology, he claims, implies that, although our value system takes harm to be central to ethical responsibility, other considerations are equally important. Moreover, the various factors can come apart, and may do so in the case of climate change. To illustrate this, he quotes with approval the following passage:

> Global warming doesn't ... violate our moral sensibilities. It doesn't cause our blood to boil (at least not figuratively) because it doesn't force us to entertain thoughts that we find indecent, impious or repulsive. When people feel insulted or disgusted, they generally do something about it, such as whacking each other over the head, or voting. Moral emotions are the brain's call to action. Although all human societies have moral rules about food and sex, none has a moral rule about atmospheric chemistry. And so we are outraged about every breach of protocol except Kyoto. Yes, global warming is bad, but it doesn't make us feel nauseated or angry or disgraced, and thus we don't feel compelled to rail against it as we do against other momentous threats to our species, such as flag burning. The fact is that if climate change were caused by gay sex, or by the practice of eating kittens, millions of protesters would be massing in the streets.[44]

This passage implies that the way in which we categorize matters as moral or immoral is profoundly influenced by emotional reactions that are not engaged by global environmental problems and the like. Moreover, it

---

[42] Jamieson, "Moral and Political Challenges."   [43] Jamieson, "Ethics, Public Policy," p. 147.
[44] Daniel Gilbert, "If Only Gay Sex Caused Global Warming," *Los Angeles Times*, July 2, 2006.

suggests a deep obstacle to dealing with these problems: our psychologies prevent us from processing climate change as an ethical issue in the right way. This claim goes far beyond my general diagnosis that there is a grasping problem, supposing that in this case it has a very deep root, and may be completely intractable. It also suggests a fundamental way in which the problem is conceptual, rather than merely one of our practices.

This second difference suggests the severe worry that we are simply psychologically incapable of responding to global environmental problems like climate change. This is not Jamieson's own view, and I will not try to address it here. Instead, I will just quickly say why I think that (paradoxically) even this severe concern tends to support the persistence of our normal concepts of ethical responsibility. Imagine for a moment that it turns out to be true that humanity just cannot manage to process climate change as an ethical problem. Suppose also that this leads to global environmental tragedy, inflicting catastrophic harms on future generations and the natural world. How do we process that thought ethically? Many philosophers would say that no blame can be assigned. Since the assumption is that humanity cannot see the problem, then they cannot be held responsible. ("Ought implies can," the slogan has it; hence, it cannot be true that humanity failed to do what it should have done, and so it cannot be blamed.) But this response seems too quick. Even if there is a sense in which it would be inappropriate to *blame* humanity for its failure in the usual way – after all, there would not be much point in punishment, and so forth – an alternative (perhaps deeper) sense of responsibility remains. The thought that humanity might really not be able, even in principle, to rise to the challenge, is a profoundly disturbing one. It suggests a sharply limited vision of who we are and are capable of being. Moreover, most people I talk to say that this is a vision that makes them feel "sick to their stomachs." If so, even this severe case seems to speak to one, and perhaps distinctive, sense of urgency. Perhaps many of us would hate it to be true that humanity is simply incapable of responding adequately to the challenge, and this might help to motivate us to try to show that we are (as the expression goes) "better than that." If, as I suspect, the problem of ethical action is not as deep as this pessimism about humanity suggests, this distinct kind of ethical assessment and its associated sense of urgency may add extra weight to more standard approaches.[45] Indeed, it may be just the consideration to which Jamieson is helpfully drawing our attention, albeit understood in a rather different way.

[45] I am grateful to audiences at the University of Edinburgh, the University of Oregon, and Delft Technical University, and especially to Elizabeth Cripps, Lauren Hartzell, Dale Jamieson, and Ted Toadvine.

CHAPTER 3

Greenhouse gas emission and the domination of posterity

John Nolt

I INTRODUCTION

Domination takes many forms: tribe over tribe, faith over faith, nation over nation, class over class, race over race, gender over gender. This chapter concerns a novel form: the present generation's domination of posterity via the medium of greenhouse gas emissions.[1] I contend that our emissions of greenhouse gases constitute unjust domination, analogous in many morally significant respects to certain historic instances of domination that are now almost universally condemned, and, further, that no benefits that we may bequeath to the future can nullify that injustice.

2 DEFINITION OF DOMINATION

There is a substantial literature on domination, much of it devoted to sociological or anthropological analyses of the forms of domination and resistance to it.[2] When domination occurs among contemporaries, it often exhibits features that will be irrelevant to our discussion: the dominated are compelled to obey the dominators or are humiliated by them; the dominated resist with various subterfuges; the dominated may exhibit false consciousness; and so on. Such features, however prominent in certain contexts, are nevertheless inessential to domination itself.

Iris Marion Young (whose account of domination is concerned exclusively with its political and institutional forms) defines domination as "institutional conditions which inhibit or prevent people from participating

---

[1] By "posterity" I here mean human posterity, though a broader view would include all biotic entities. Elsewhere I take such a broader, non-anthropocentric view; see my "Sustainability and Hope," in *Sustainability Ethics: 5 Questions*, Evan Selinger, Ryan Raffelle, and Wade Robison, eds. (Copenhagen, Automatic/VIP Press, 2010), pp. 139–165; and "Hope, Self-Transcendence and the Justification of Environmental Ethics," *Inquiry*, 53, 2 (April 2010), pp. 162–182.

[2] See, for example, D. Miller, M. Rowlands, and C. Tilley, eds., *Domination and Resistance* (London, Unwin Hyman, 1989).

in determining their actions or the conditions of their actions." "Persons," she adds, "live within structures of domination if other persons or groups can determine without reciprocation the conditions of their action."[3] Philip Pettit offers a similar definition of domination as the "power of interference on an arbitrary basis."[4] Neither of these definitions requires of domination any of the features mentioned in the previous paragraph.

Francis N. Lovett has combined, broadened, and augmented features of Young's and Pettit's definitions to arrive at more widely applicable and more substantive definition. According to Lovett, a subject is dominated by an agent (may be either an individual or group) if, and only if, the following three conditions are met:

1 *Imbalance of power*: the agent wields "some degree of superior power over the subject."

2 *Dependency*: the subject is "not free to exit" the relationship without incurring costs.

3 *Absence of rules*: there are no "established and commonly known rules, laws, or conventions effectively governing the use of power" by the agent. The agent wields power over the subject arbitrarily.[5]

[margin: excludes gov't]

The absence of rules condition prevents, for example, the legitimate authority of just government from being regarded as domination. All three conditions, of course, can be satisfied to a greater or lesser degree, and this accurately reflects the fact that domination itself is a matter of degree.

Lovett's definition intentionally leaves open the possibility of benevolent domination, such as might occur between a parent and child if the parent had arbitrary power over the child but exercised it only for the child's benefit.[6] Benevolent domination is, however, not at issue here; and, in my idiolect at least, it has the ring of an oxymoron. I therefore find it natural and helpful to add a fourth condition:

4 *Harm*: the agent wields power in ways that harm the subject.

We could, of course, call relationships meeting conditions 1–3 "domination" and those meeting conditions 1–4, "harmful domination." But because only harmful domination is of interest here and because frequent repetition of the qualifier "harmful" would be tedious, I will here use "domination" to mean "harmful domination" – that is, the sort of relationship between an agent and a subject that satisfies conditions 1–4.

[margin: concept of harmful domination]

---

[3] Iris Marion Young, *Justice and the Politics of Difference* (Princeton University Press, 1990), p. 38.
[4] Philip Pettit, *Republicanism: A Theory of Freedom and Government* (Oxford, Clarendon Press, 1997), p. 52.
[5] Francis N. Lovett, "Domination: A Preliminary Analysis," *Monist*, 84, 1 (January 2001), sec. I.
[6] Ibid., p. 99.

3 OUR DOMINATION OF POSTERITY

By this definition our emissions of greenhouse gases constitute domination of posterity. If we take the agent to be contemporary emitters of greenhouse gases (i.e., us, both individually and collectively), the subject to be posterity over the time during which our greenhouse emissions cause harm, and the social relationship to be succession of generations, all four conditions are met:

1 *Imbalance of power*: through our greenhouse gas emissions, we wield considerable power over posterity. Future generations have no comparable power over us.

2 *Dependency*: by the very structure of time, posterity is not free to exit its relationship with us. Future generations cannot exist without having us as their predecessors.

3 *Absence of rules*: there are no established and commonly known rules, laws, or conventions effectively governing our use of power with respect to distant future generations. We are able to wield power over them arbitrarily.

4 *Harm*: the effects of current greenhouse gas emissions will be predominantly harmful: rising global average temperatures, increased drought and flooding, biodiversity loss, shortages of fresh water, disruption of agriculture (especially at low latitudes) and consequent famine, increasing intensity of hurricanes and typhoons, increasing frequency and intensity of other forms of extreme weather, and so on.[7] Such disruptive phenomena are likely to spark secondary violence as stressed populations are forced to migrate and/or struggle with one another over scarce resources.

Thus, we (both individually and collectively) are by our emission of greenhouse gases dominators of posterity.

It might be thought that our domination of posterity is nothing new – that past generations have always so dominated their successors. But many past generations have not significantly harmed their successors. Some, it is true, have harmed their successors locally by depleting resources in their area. But many have not significantly depleted resources and (since some resources are renewable or fungible) not all those that have depleted resources have harmed their posterity. Moreover, although there has always been some degree of power imbalance between the current generation and posterity, it has never been as great as it is today. Past generations that

---

[7] Intergovernmental Panel on Climate Change (IPCC), "Climate Change 2007: Synthesis Report," sec. 3.2.

harmed their successors did so haphazardly and almost entirely unwittingly. This hardly counts as "wielding power" (condition 1). The consequences of our emission of greenhouse gases are, by contrast, global and long-enduring, and contemporary science gives us a reasonably clear idea of what they are. We *do* wield power over posterity – knowing power whose enormity is unprecedented.

4 THE OBJECTION OF LACK OF INTENTION

It might be thought, however, that we cannot be dominating posterity because we have no intention of doing so. Can domination occur without intent? The contemporary philosophical literature is, so far as I can see, unanimous in its answer: Of course it can.[8] To see how, it is useful to distinguish between domination motivated by the dominators' hatred, fear, or resentment of the subjugated and domination motivated by, for example, caprice, greed, or power lust. The former aims to suppress the dominated, the latter to elevate the dominators. We may call these two forms, respectively, the *domination of antipathy* and the *domination of self-aggrandizement*. In the domination of antipathy the dominators derive satisfaction directly from acts of suppression. Often, maybe even always, they aim to dominate. In the domination of self-aggrandizement, however, domination is not the aim but merely a by-product. Self-aggrandizing dominators need not intend to dominate, yet they do. Often the two forms are combined, but our domination of posterity is overwhelmingly of the self-aggrandizing and unintentional kind.

5 THE OBJECTION THAT INDIVIDUALS
DO NOT CAUSE HARM

Some may deny the role of *individuals* in this domination. Walter Sinnott-Armstrong, for example, while conceding that our *collective* greenhouse emissions are harming posterity, claims nevertheless that *individual* emissions are not. If he is right, individuals fail to meet condition 4; hence, their emissions do not contribute to the domination. Using the example of joy riding, Sinnott-Armstrong argues as follows:

Driving does not cause harm in normal cases ... Global warming will still occur even if I do not drive just for fun. Moreover, even if I do drive a gas-guzzler just for

---

[8] In addition to the sources already cited, see Jean Harvey, "Social Privilege and Moral Subordination," *Journal of Social Philosophy*, 31, 2 (Summer 2000), pp. 177–188, and Jean Harvey, *Civilized Oppression* (New York, Rowman and Littlefield, 1999).

fun for a long time, global warming will not occur unless lots of other people also expel greenhouse gases. So my individual act is neither necessary nor sufficient for global warming.[9]

It is true, of course, that an individual's emissions are neither necessary nor sufficient for global warming. But rarely are the causal contributors to complex phenomena necessary or sufficient for those phenomena. (Think, for example, of the causes of wars or market fluctuations.) An individual's emissions clearly are, however, causal contributors to global climate change. The main causal pathway is well known: molecules of the greenhouse gases that we cause to be released disperse into the atmosphere where they trap heat and increase (albeit very slightly) the global average temperature. Since the Earth is too hot already, further temperature increases are harmful. Thus, each individual's greenhouse gas emissions contribute causally to harm.

Surprisingly, Sinnott-Armstrong denies this:

It is not unusual to go for joyrides. Such drivers do not intend any harm. Hence, we should not see my act of driving on a sunny Sunday afternoon as a cause of global warming or its harms.[10]

The *non sequitur* is stunning. Consider this analogous inference: it is not unusual for people to smoke; smokers intend harm no more than drivers do; hence the act of smoking is harmless – or, at least, should not be seen as a cause of harms.[11] Moreover, Sinnott-Armstrong's first premise is false: joy riding in fossil-fuel-powered vehicles *is* unusual *in an intergenerational context*. Such an act has been possible only to a tiny fraction of humanity, and only for a little more than a century. Posterity is likely to view it as an aberration.

Later, however, Sinnott-Armstrong takes a different tack:

There is no individual person or animal who will be worse off if I drive than if I do not drive my gas guzzler just for fun. Global warming and climate change occur on such a massive scale that my individual driving makes no difference to the welfare of anyone.[12]

---

[9] Walter Sinnott-Armstrong, "It's Not *My* Fault: Global Warming and Individual Moral Obligations," in *Perspectives on Climate Change: Science, Economics, Politics and Ethics, Advances in the Economics of Environmental Resources*, vol. 4, Walter Sinnott-Armstrong and Richard B. Howarth, eds. (Amsterdam, Elsevier, 2005), p. 289.

[10] Ibid., p. 290.

[11] Sinnott-Armstrong could reply here that smoking *is* unusual in the relevant sense, since it is a departure from natural or background conditions. However, anthropogenic burning of fossil fuels is the same sort of departure from natural or background conditions. If smoking is unusual, so is driving.

[12] Ibid., p. 293.

This is an empirical claim, but Sinnott-Armstrong gives no evidence for it. Its truth is hardly obvious. Even one Sunday afternoon drive (like one cigarette) slightly increases the probability of harm, and the number of people and animals in the world during the centuries in which the emissions from the drive will persist in the atmosphere is (like the number of potential mutation sites in the individual strands of DNA in our lungs) vast.

Still, we need not split hairs. The issue is neither the single cigarette nor the single Sunday afternoon drive. It is, for both cigarettes and driving, the harmful tendency of the habit. Sinnott-Armstrong gives no good reason to doubt the harmfulness of the *habit* of driving, whether for fun or not. And driving, of course, is not the only source of greenhouse gas emissions.

How much suffering and death will, for example, an average US citizen cause by her or his participation in the greenhouse-gas-intensive American economy?[13] Any guess will necessarily be very crude, but we ought to make the attempt. Current US emissions are about one-fifth of the world's. There are about 300 million Americans. This means that the average American's share is roughly 1/1,500,000,000 of current emissions. But since excessive greenhouse emissions have been occurring since the beginning of the Industrial Revolution, not all the future harm is a result of emissions of people who are alive today. The contemporary average American's share will vary depending on precisely when this person lives. To give some definition to the problem, let's suppose that this person lives from 1965 to 2040. From these dates we can estimate her proportion of the overall harm due to excess atmospheric carbon dioxide through 2040. (I omit other greenhouse gases from this estimate, but they would not much affect the result.) Roughly three-quarters of the anthropogenic atmospheric increase in carbon dioxide concentrations will have occurred during this person's lifetime.[14] Thus, her share is about 0.75 × 1/1,500,000,000, or about one part per 2 billion of the total increase through 2040.

But billions of people may be harmed by those emissions.[15] If, say, 4 billion people suffer and/or die as a result of current global emissions, then

[handwritten margin note: It's not all our fault — interesting pt.]

[13] Harm for Canadians and Australians will be similar, though slightly lower.

[14] This figure is obtained as follows: the concentration at the beginning of her life in 1965 was 320 ppm. Projecting the current growth rate of about 2 ppm/year into the near future, we obtain an estimate of 450 ppm for 2040. Thus total increase during her lifetime comes to 450 – 320 = 130 ppm. The preindustrial concentration of CO_2 was about 280 ppm. So before this person's birth, the atmospheric CO_2 concentration had increased by 40 ppm. During her lifetime it will increase by an additional 130 ppm. The total anthropogenic increase until the time of her death therefore is 40 + 130 = 170 ppm, of which about 76 percent (about three-quarters) will have occurred during her lifetime. (Data on atmospheric concentrations were obtained from: ftp://ftp.cmdl.noaa.gov/ccg/co2/trends/co2_mm_mlo.txt.)

[15] The IPCC predicts, for example, that "climate change over the next century is *likely* to adversely affect hundreds of millions of people through increased coastal flooding, reductions in water supplies,

the average American causes through her participation in her nation's carbon emissions the suffering and/or deaths of two future people. There are great uncertainties here, of course, but the harm we are likely to be causing, even as individuals, is not trivial.

6 THE INJUSTICE OF THE DOMINATION

Even though all domination is (by our definition) harmful, not all domination need be unjust. If, for example, one nation unjustly attacks another and the second emerges victorious to dominate the first, then, conceivably, that domination may for a time be just. Thus, we may not infer that our domination of posterity is unjust merely from the fact that it is domination. It is unjust, but to show that requires supplementary considerations. Eight features of the domination contribute to this injustice:

[margin note: additional conditions]

1 *It is domination of a majority by a minority.* It is likely that each future generation that suffers the effects of global climate change will be larger than the present generation and that there will be many such generations. The effects of past, present and near future carbon dioxide emissions will, according to the Intergovernmental Panel on Climate Change (IPCC), "continue to contribute to warming and sea-level rise for *more than* a millennium, due to the time scales required for the removal of this gas from the atmosphere" (italics mine).[16] Assuming a lifespan of 75 years and 25 years per generation, there are on average three generations alive at any given time. If the average world population over the next millennium is 7.5 billion (a figure we will probably reach well before mid-century), then there will be on average $7.5 \div 3 = 2.5$ billion people per generation. A millennium is 40 generations. Thus at least 2.5 billion $\times 40 = 100$ billion people will live during times affected by our greenhouse gas emissions. This estimate is rough and uncertain, of course; but, given the Earth's limited carrying capacity, the true figure is very unlikely to be more than a few times that, and, barring some enormous catastrophe, unlikely to be many times less.

We do not know how many of these roughly 100 billion people will be harmed by our greenhouse emissions. Probably a large percentage will suffer at least minor harms (e.g., increased difficulty or expense of

increased malnutrition and increased health impacts"; source is IPCC (Intergovernmental Panel on Climate Change), "Climate Change 2007: Synthesis Report," p. 65; italics in original. Many of the potential harms of global climate change are not accounted for in this statement. Moreover, the time span is only the next century; but greenhouse gas emissions will continue to increase global temperatures and sea levels for more than a millennium (IPCC report, p. 47).
[16] Ibid., p. 47.

obtaining life's necessities), and some smaller number (but perhaps still in the billions) will suffer severely and/or die. All of those harmed will suffer our domination. There are about 6.8 billion people alive at the time of this writing, but the poorest billion or two contribute very little to climate change.[17] Therefore, it is almost certain that our emission of greenhouse gases constitutes domination of a majority by a minority.

2 *The subordinated are innocent of any harm to their dominators.* In many historical instances of domination (as, for example, in the violent struggles between nations or religious groups), the subordinated were not entirely innocent of harm to the dominators. But our posterity is utterly innocent of harm to us.

3 *The subordinated are voiceless, powerless and without recourse against the domination.* In historical cases of domination (slavery, racism, sexism, etc.) the subordinated nearly always had some power, however small, to resist or retaliate against the domination. Future people have no power to resist or retaliate against our emission of greenhouse gases.[18]

4 *The motives for the domination are often frivolous.* In this respect, the USA is one of the world's worst offenders. Our per capita greenhouse gas emissions are, for example, over twice those of the UK or Japan.[19] It is sometimes claimed that, as a result of our sparse settlement over a vast continent, we must use more energy than other nations. At most this special pleading justifies a time of transition to denser cities and lower carbon emissions. It does not justify exceptionally high emissions in perpetuity. Moreover, it is not merely patterns of settlement that account for the higher emissions. US emissions are also elevated by popular practices that are plainly frivolous: building and inhabiting energy-inefficient trophy houses, over-illuminating cities, driving excessively in excessively large vehicles, maintaining in all weathers an unnecessarily narrow range of indoor temperatures, consuming fossil fuels for mere entertainment, and so on.

5 *The subordinated have reason to expect beneficence, not harm, from their dominators.* Because we have received great benefits (e.g., knowledge,

---

[17] See Figure 10, "Projected distribution of global emissions across the world's individuals in 2030," in Chapter 8 in this volume.

[18] John O'Neill argues, to the contrary, that future generations can benefit or harm us. (See "Future Generations; Present Harms," *Philosophy*, 68, 263 (1993), pp. 35–51.) But they can harm us, he argues, chiefly by causing our projects to fail. Yet it is doubtful that our continuing projects are constitutive of ourselves and thus doubtful that we are harmed by the posthumous failure of those projects.

[19] Based on 2006 figures from the United Nations Statistics Division, unstats.un.org/unsd/environment/air_greenhouse_emissions.htm. (These were the most recent UN figures available at the time of this writing.)

democratic institutions, fine art and literature, useful technologies) and few, if any, substantial and knowing harms from our predecessors, we have, in a metaphorical but mundane sense, acquired a debt from the past that we can pay only to the future. Indeed, some of our predecessors have intentionally given us gifts of great value. In the USA, the National Park Service Act of 1916, for example, declared that the purpose of the parks is to:

> conserve the scenery and the natural and historic objects and the wild life therein and to provide for the enjoyment of the same in such manner as will leave them unimpaired for the enjoyment *of future generations.*[20]

We are among those future generations. The Preamble to the US Constitution lists among its purposes to "secure the Blessings of Liberty to ourselves and our *Posterity*" (emphasis mine). We who are citizens of the USA belong to that posterity. Analogous forethought has occurred elsewhere. Since all of us have received much benefit and little harm from our predecessors, our posterity has reason to expect similar beneficence from us.

6 *Many of the dominators have systematically denied the domination.* Some coal and oil corporations (e.g., ExxonMobil), certain members of the Bush/Cheney administration, and several prominent "conservative" talk show hosts and their followers have denied or belittled the problem.[21]

7 *The domination is worst for those who are in other ways already the most disadvantaged.* The IPCC predicts that many of the worst effects of near-term global climate change will occur among the poor in southern Asia and Africa.[22]

8 *The harmful consequences of the domination will worsen rather than improve over time, diminishing long-term hope.* While nearly all other forms of harm fade relatively rapidly, the harms of greenhouse gas emissions will worsen for centuries. We have already noted the IPCC's prediction that carbon emissions will continue to contribute to increasing sea levels and global temperatures for more than a millennium. Moreover, says the IPCC, "the net costs of impacts of increased warming are projected to

[20] Quoted in Daniel S. Peirce, *The Great Smokies: From Habitat to National Park* (Knoxville, TN, University of Tennessee Press, 2000), pp. 47–48; italics mine.
[21] "Smoke, Mirrors and Hot Air: How ExxonMobil Uses Big Tobacco's Tactics to Manufacture Uncertainty on Climate Science" (Cambridge, MA, Union of Concerned Scientists: January 2007); Tarek Maassarani, "Redacting the Science of Climate Change: An Investigative and Synthesis Report" (*sic*), (Washington, DC, Government Accountability Project, March 2007).
[22] IPCC (note 15 above), pp. 48–50, 52, 65.

increase over time."[23] Some negative impacts, such as extinctions, will be irreversible. Loss of the Greenland ice sheet (which "could occur on century time scales," causing enormous sea-level rises that would inundate coastal regions and cities worldwide) might likewise be irreversible.[24] Moreover, according to the IPCC, "there is *high confidence* that neither adaptation nor mitigation alone can avoid all climate change impacts" (italics in original).[25]

The result may be long-term diminishment of hope. We who are middle-aged or older now will see little of the damage of global climate change. Our children will see more. Our grandchildren will see still more, and so on, for many, many generations. Thus, as time passes, there will (unless countervailing tendencies prevail) be reason for ever-diminishing hope. Though subtler than the more overt effects, such a progressive loss of hope may not be less profound. It may even exacerbate the overt effects, since hopelessness can breed violence, which in turn further diminishes hope, in a downward spiral.

In these eight respects, then, our domination of posterity by emission of greenhouse gases is unjust.

How will posterity experience this injustice? One can only speculate, groping for analogies. Here is a guess: since electronic and print media will preserve a cultural memory of the world as it was and could have been, those who suffer the damages of global climate change will experience their condition as a fall from a world of hope and plenty to a world of misery. They may feel estranged, disinherited, or exiled. Perhaps they will experience something akin to the bitterness recorded by the writer of Psalm 137: "By the rivers of Babylon we sat down and wept when we remembered Zion."

7 THE COMPENSATORY BENEFITS OBJECTION

But, it might be thought, our domination of posterity is *not* unjust, since we are also doing much that will benefit posterity. Science, technology, medicine and the arts, for example, continue to advance in ways that will benefit the future. This objection consists of two claims: (1) that the good we are doing for posterity is comparable in magnitude to, or greater than, the harm we are doing to them, and (2) that because of this we are doing posterity no injustice. Claim (1) is becoming increasingly difficult to defend. The worsening problems not only of global climate change but also of resource depletion (and consequent resource wars), biodiversity loss, agricultural

[23] Ibid., p. 65.   [24] Ibid.   [25] Ibid.

crises, population growth, burgeoning debt, and widening inequities between rich and poor suggest that the overall balance of harm and benefit is not in posterity's favor. But to assess the validity of claim (1) would require an accounting of costs, benefits and probabilities far more comprehensive than can be attempted here.

Suppose, then, for the sake of argument that (1) is true – that we *are* doing posterity more good than harm. Still it does not follow that we are not doing posterity an injustice. Slaveholders provided food, clothing and shelter to slaves, and men who oppressed women often provided them with material or cultural benefits. Probably there have even been some slaves or oppressed women who received from their dominators more good than harm. It does not follow, however, that the domination was not unjust. Our present case is analogous. We are knowingly and unjustly harming posterity, and the injustice is not nullified by compensatory benefits.

Suppose that I break your arm for my benefit (perhaps I am doing a medical experiment that will bring me fame) but at the same time give you a million dollars. There are, no doubt, people who would agree to such an exchange if their consent were sought. But suppose I have not obtained your consent – just as we have not obtained posterity's consent to disrupt the planetary climate. Surely, then, I have committed an injustice against you – an injustice not erased by the compensatory benefit, not even if you yourself account the benefit greater than the harm. The injustice consists not merely in the infliction of harm on you, but in harming you without your consent. That injustice would be deepened if I dominated you in such a way that you had no recourse against me – and it would be deepened still further if any of the other seven conditions mentioned in the previous section held.[26] Moreover, many of the harms of global climate change will be far worse than broken arms.

Statements to the effect that the benefits bequeathed to posterity by our fossil-fuel-powered economy outweigh the costs of global climate change – or, worse, that global climate change itself will be good for posterity – are disturbingly reminiscent of nineteenth-century claims to the effect that slaves benefited from their slavery or that women benefited from their subjugation. It is easy now to see those claims for what they were: flimsy rationalizations of injustice. Future people may take a similar view of current claims that they will benefit from our legacy to them.

---

[26] For some of the conditions to hold would require changing the case slightly. Thus, for example, regarding condition 1 (it is domination of a majority by a minority), "you" would have to be many people, all of whose arms I break.

8 THE NON-IDENTITY OBJECTION

I have argued that our greenhouse gas emissions constitute unjust domination of posterity – and that the injustice cannot be nullified by compensatory benefits. But, it might be objected, there is one compensatory benefit that *does* nullify the injustice: the very existence of the people whom we are dominating.[27] This objection is, of course, a version of the non-identity problem, made prominent by the work of Derek Parfit.[28] The idea is that any major contrary-to-fact reduction of greenhouse gas emissions would displace many people who are now involved in industries that emit these gases. These people, or their children, would consequently find different mates than they actually will find. Hence, if they were subsequently to have children, these would not be the children they will in fact have. Their children would in turn pair with others and so on, until after many generations an entirely different population would exist than will in fact exist. Hence all those who will actually exist after that time would under the counterfactual greenhouse gas reduction program never be born. Ironically, therefore, the only way to save them from the harm of our greenhouse gas emissions is to prevent them from ever existing. The objection, then, is that our domination cannot be unjust to them if the only possible alternative to it entails their never having existed at all.

Parfit thinks that for such large-scale choices as our counterfactual greenhouse gas reduction program, the actual and counterfactual populations would become totally disjoint (i.e., nobody would be a member of both) "after one or two centuries."[29] The figure is merely an educated guess, but presumably there is some future time beyond which under a counterfactual greenhouse gas reduction policy none of those who will in fact live would remain and other people would have been born instead.

Still, the non-identity objection succeeds, if at all, only for some of the people we are dominating; for global climate change is already underway, and some of its victims already exist, or will exist regardless of what we do. The objection does, however, grow in force the farther we look into the future; and the millennium or more during which anthropogenic climate

---

[27] It is debatable, however, whether to be caused to exist is a benefit. See Derek Parfit, *Reasons and Persons* (Oxford, Clarendon Press, 1984; reprinted with corrections 1987), Appendix G. Regardless of how we resolve this question, the burden of the remainder of this section is to show that a person can be done an injustice by a practice that is necessary for her existence.

[28] Ibid., ch. 16. Parfit himself does not advocate the objection, though he discusses the problem extensively. See also Gregory Kavka, "The Paradox of Future Individuals," *Philosophy and Public Affairs*, 11 (Spring 1982), pp. 93–112.

[29] Ibid., p. 361.

change will persist is, no doubt, time enough for many who are dominated to owe their lives to the domination.

But from the fact that they owe their lives to the domination, it does not follow that we are not unjustly harming them. Many of them will suffer and die as a result of our greenhouse gas emissions. Suffering and death are harms. And the dependence of their lives on the policies that produce these harms does not eliminate the injustice of these harms. This is obvious for people who are made so miserable by climate change that it would be better for them never to have lived. But it is also true in less extreme cases. Consider, by way of analogy, a child born into slavery whose parents could have met only as a result of their enslavement. Such a child would not have existed were it not for the institution of slavery. Her life, moreover, need not, on balance, be miserable; we may imagine that she is freed as a young adult and lives happily thereafter. Still, it does not follow that slavery was not an injustice to her. Likewise, it does not follow from the premise that future people owe their lives to our domination that the domination is not unjust.

I see no reason, moreover, why the injustice of the domination must be understood merely as a gap between what will be and could have been for the (rigidly designated) people who comprise our actual posterity. It need not, in other words, be conceived solely in what Parfit calls *person-affecting* terms. Injustice might also reside in the gap between what will be for them and what could have been for a posterity that would have emerged had we not been so careless. (The injustice in that case might be to posterity conceived non-rigidly, as consisting of whoever on various alternative futures our successors over the next millennium or so would be.) There might also be injustice to transgenerational communities or institutions.[30]

9 DOMINATION, EXPLOITATION, AND CONTRACT THEORY

It is sometimes suggested that greenhouse gas emissions are unjust because they *exploit* future generations. But the case for the charge that we are unjustly *dominating* future generations is stronger than the case for the charge that we are *exploiting* them. The latter has been made by Christopher Bertram. Exploitation, as Bertram conceives it, "occurs where people who are engaged in co-operation together, often, but not necessarily, in a common institutional scheme, act so as to derive a benefit that is greatly

---

[30] There are various ways in which such an account might be developed. See, for example, Edward Page, *Climate Change, Justice and Future Generations* (Cheltenham, UK, Edward Elgar, 2006), pp. 150–158.

out of proportion to their contribution."[31] But global climate change lasts so long (on the order of a millennium) that future generations affected by it may have cooperative projects with us, if at all, only in a very attenuated sense. (Are we engaged in cooperation with the Chinese or Mayans of the year 1000?) Hence, exploitation, though it may occur intergenerationally, does not fully account for the injustice of our greenhouse gas emissions. That injustice may be deepened by, but does not presuppose, the existence of common projects or institutions from which we benefit unfairly. At its center, rather, is this crude fact: we are inflicting undeserved harm and premature death on future people. Whether or not this is exploitation, it clearly is domination.

More generally, the charge that we are unjustly dominating posterity does not rely on any notion of reciprocity – or on contract theory, all extant versions of which, as Stephen Gardiner has shown, face serious difficulties in the intergenerational context.[32]

IO OUR RESPONSIBILITY

Who is responsible for the injustice? We are, of course – we, both collectively and individually. That we are responsible collectively is indisputable. That we are responsible individually has been disputed by Sinnott-Armstrong. Part of Sinnott-Armstrong's case is built on the claim, addressed above, that individuals do not cause climate change. But Sinnott-Armstong also challenges proponents of individual responsibility to state a plausible moral principle from which that responsibility can be derived. He considers and rejects many such principles, but he does not consider this one: members of a group that is unjustly dominating others have, individually, an obligation not to exacerbate that domination. That, it seems to me, is a minimal requirement. A case could be made for an obligation of passive, or even active, resistance to the domination.

An individual's contribution to global climate change is, of course, only a minuscule fraction of the total effect. But the same is true in the cases of racism and sexism. And because the harm will in many cases be serious (people will die, lose their homes, suffer hunger, thirst, injuries or illnesses) and of long duration, individual responsibility, though less direct and more

---

[31] Christopher Bertram, "Exploitation and Intergenerational Justice," in Axel Gosseries and Lukas H. Meyer, eds., *Intergenerational Justice* (Oxford University Press, 2009), p. 164.
[32] See Stephen Gardiner, "A Contract on Future Generations?," in Gosseries and Meyer, note 31 above, pp. 77–118.

diffuse than in the cases of racism and sexism, is, as I have argued above, nevertheless substantial. Moreover, we are responsible *knowingly*. The evidence that anthropogenic climate change is occurring is unequivocal and available to us all. So is the knowledge that our electricity is produced primarily by the burning of coal and that our transportation system is powered almost exclusively by oil, and that the burning of these fossil fuels is a major cause of the problem.

II THE OBJECTIONS FROM CONCEPT STRETCHING AND DEMANDINGNESS

Yet it might be objected that the connection between cause and effect (emission and harm) is too diffuse, distant, or indirect to support claims of moral responsibility – and especially of *individual* moral responsibility. The very concept of responsibility, it may be said, is here stretched too thin – removed too far from its ordinary context.[33] Similarly, but in a more pragmatic vein, it might be objected that the requisite notion of responsibility is too demanding to be widely accepted or to motivate changes in individual behavior.

The concept of responsibility may be stretched by the recognition of our responsibility for the domination of posterity, but such stretching is a normal feature of moral progress and conceptual development. There is nothing especially novel in responsibilities via causal connections that are diffuse, distant, and indirect. It is by now obvious, for example, that sexist or racist talk is morally culpable, not only because of its potential to demean people within hearing but also because it perpetuates sexist or racist attitudes among the speakers and listeners – attitudes that may cause subsequent harm. But here the concept of moral responsibility has been stretched well beyond what once had been its domain. The potential victims of attitudes fostered by such speech may be distant and completely unknown to the speaker. The causal connections may be impossible to trace. Yet these contingencies do not abrogate our responsibility not to use racist or sexist language, nor do they make that responsibility pragmatically too demanding. In the case of racist and sexist language, the necessary concept stretching has in fact been accomplished – and to good effect for many people – by protest, argument, song, story, sermon, and other forms of moral education.

---

[33] Dale Jamieson raised this objection when I presented the original version of this paper at the University of Tennessee's Energy and Responsibility Conference, Knoxville, April 10–12, 2008.

One can, no doubt, cast the net of responsibility too far. To find oneself responsible for too much of the world's misery is not moral sensitivity but neurosis. Conceivably, the ethical enterprise could lose credibility and undermine itself by demanding too much. But it is also a mistake to demand too little. Science and technology have given us the power knowingly to harm future people over long spans of time. An ethic that does not hold each of us, and all of us collectively, responsible to those people for our use of that power is an ethic that has failed the test of the times.

To summarize: I have argued that we are, by our carbon emissions, unjustly dominating posterity. Any person or group that contributes to greenhouse gas emissions by consuming electricity; using cars, trucks, trains, or aircraft; or buying products embodying fossil-fuel-based energy knowingly contributes to this injustice and hence is morally responsible for some portion of it. No desire for pleasure or convenience, no appeal to popularity or tradition, and no mere consideration of wealth or profit can excuse such a person or group. The only morally tenable excuse is necessity.

12 THE EXCUSE OF NECESSITY

But what counts as necessity? Probably the world's best-known study of necessity is Henry David Thoreau's *Walden*. Thoreau's definition is worth quoting:

> By the words, *necessary of life*, I mean whatever, of all that man obtains by his own exertions, has been from the first, or from long use has become, so important to human life that few, if any, whether from savageness, or poverty, or philosophy, ever attempt to do without it . . . The necessaries of life for man in this climate may, accurately enough, be distributed under the several heads of Food, Shelter, Clothing, and Fuel.[34]

Thoreau's list expresses a minimalist ideal. But, for better or for worse, to live a decent human life (especially if one is raising a family) in the twenty-first century requires a good bit more than this minimum.

Still, it requires a good bit less than the average American or Canadian or Australian (see note 13) has. Consider a more congenial measure: the average standard of living in the UK or Japan. Surely what the average citizen of these nations has cannot by any reasonable standard count as "less than necessary." Yet their per capita carbon emissions are less than half those of the USA.

[34] Henry David Thoreau, *Walden*, in Joseph Wood Krutch, ed., *Walden and Other Writings* (New York, Bantam, 1981), p. 114.

13 OUR DUTY

So what is our duty – those of us who participate in greenhouse-gas-intensive economies – with regard to greenhouse gas emissions? The total injustice is in many ways analogous to the great social injustices of racism and sexism. All of us, individuals and institutions, who have been in this world for half a century or more have had to change our thinking and behavior significantly to address our complicity in these injustices. We should expect no less with respect to the great injustice of our domination of posterity.

We may not contribute to this injustice without necessity. Preferring certainty to exactness, I have not attempted to specify what counts as "necessity," but have instead proposed as a reasonable upper limit the average per capita emissions of Western Europe, Great Britain, and Japan. Probably the limit should be lower. *At minimum*, then, it is our duty, both individually and collectively, to maintain carbon emissions at or below that level. For the average American (or Canadian or Australian), this implies a reduction of at least half.

solutions

CHAPTER 4

Climate change, energy rights, and equality[1]

Simon Caney

It is now widely recognized that the Earth's atmosphere is undergoing profound changes. The most recent report of the Intergovernmental Panel on Climate Change (IPCC) states that temperatures have increased in the last hundred years. It writes, for example, that "[t]he total temperature increase from 1850–1899 to 2001–2005 is 0.76°C ± 0.19°C," adding ominously that "[t]he rate of warming averaged over the last 50 years (0.13°C ± 0.03°C per decade) is nearly twice that for the last 100 years."[2] In addition to this, temperatures are projected to increase in the future. All of the six scenarios considered by the IPCC found that temperatures will rise by 2090–2099 as compared to the temperatures between 1980 and 1999. According to the best estimate of the B1 scenario, temperatures will increase by 1.8°C. If on the other hand we turn to the A1F1 scenario, its best estimate is that temperatures will increase by 4.0°C.[3] And if we examine the "likely range," then the lower limit is 1.1°C and the higher limit is 6.4°C.[4]

Sea levels, too, are projected to increase. According to one scenario (the B1 scenario), sea levels are projected to rise by 0.18–0.38 meters and according to another (the A1FI scenario), the increase is projected to be 0.26–0.59 meters.[5] These projections, it is important to add, do not include "future rapid dynamical changes in ice flow."[6] They omit, that is, the massive sea-level

[1] This paper was presented at a conference on "Global Egalitarianism: Theory and Practice" at San Diego State University (April, 10, 2008) and a conference on "Energy and Responsibility" at the University of Tennessee (April 12, 2008). I am grateful to audiences at both for probing questions. I am especially indebted to Ted Stolze (my respondent at San Diego). I am also grateful to Richard Arneson, Carol Gould, Aaron James, Bruce Landesman, Nicole Hassoun, and Henry Shue. I wrote a first draft of this paper while I was a Leverhulme Research Fellow and completed it while holding an Economic and Social Research Council (ESRC) Climate Change Leadership Fellowship. I am grateful to the ESRC and the Leverhulme Trust for their support.
[2] S. Solomon et al., "Technical Summary," in Climate Change 2007: The Physical Science Basis. Contribution of Working Group I to the Fourth Assessment Report of the Intergovernmental Panel on Climate Change (Cambridge University Press, 2007), edited by S. Solomon, D. Qin, M. Manning, Z. Chen, M. Marquis, K. B. Averyt, M. Tignor, and H. L. Miller, p. 36.
[3] Ibid., p. 70.   [4] Ibid.   [5] Ibid.   [6] Ibid.

rises that might occur because of the melting of ice sheets. In addition to these temperature increases and sea-level rises, climate change is projected to result in a rise in the severity and the frequency of severe weather events.

All of these changes, of course, pose grave threats to humanity, to non-human animals, and to the natural world. If we focus solely on human impacts, we can say that anthropogenic climate change will jeopardize the following fundamental interests:

(a) health (because of water-borne diseases, vector-borne diseases, and heat stress)
(b) life (because of freak weather events – such as storm surges, hurricanes, tornadoes, and violent flooding)
(c) subsistence (because of crop failure, flooding of agricultural land, and the loss of land because of salinization and sea-level rise)
(d) the capacity to attain a reasonable standard of living (because of damage to infrastructure, roads, and buildings).

Given this, humanity has, other things being equal, good reason to cut greenhouse gas emissions so that the total concentration of greenhouse gases is not so great as to result in "dangerous anthropogenic interference" (Article 2, United Nations Framework Convention on Climate Change) with the Earth's climate. There are a number of different greenhouse gases – carbon dioxide, methane, nitrous oxide, hydrofluorocarbons (HFCs), perfluoro-carbons (PFCs), and sulfur hexafluoride (SF6). The question we thus face is what level of greenhouse gases in the atmosphere is acceptable. One common way of thinking about this has been to hold that we should aim to prevent the global average temperature from rising by more than 2°C (from the Industrial Revolution). This is, for example, the climate goal that has been adopted by the European Union.[7] To avoid this rise, the concen-tration of greenhouse gases in the atmosphere may not exceed a certain level of concentration (a figure in the region of 450 ppm is often proposed). Specifying the exact threshold is, however, fraught with problems.[8] Determining what constitutes "dangerous anthropogenic interference"

---

[7] The European Council of Ministers stated at the 1,939th Council Meeting in Luxembourg on June 25, 1996 that "global average temperatures should not exceed 2°C above pre-industrial level" (cited on p. 3 of the European Commission's Communication "Winning the Battle Against Global Climate Change" (issued on February 9, 2005 and available at http://eur-lex.europa.eu/LexUriServ/site/en/com/2005/com2005_0035en01.pdf).

[8] James Hansen argues that a 1°C rise in temperature (from 2000 levels) would result in dangerous climate change. See Hansen et al., "Dangerous Human-Made Interference with Climate: A GISS ModelE Study," Atmospheric Chemistry and Physics, 7, 9 (2007), pp. 2287–2312; Hansen et al., "Target Atmospheric CO_2: Where Should Humanity Aim?," Open Atmospheric Science Journal, 2 (2008), pp. 217–231.

requires both normative criteria as to what constitutes a "dangerous" level as well as highly complex scientific analysis.[9] Given this, it bears noting that some hold that an increase of less than 2°C is also ethically troubling. In a comprehensive survey, Rachel Warren has documented many adverse effects that have been projected to occur if temperatures increase by less than 2°C. These include the loss of some species, crop failure in Africa, increased exposure to disease, the melting of the Greenland ice sheet, and an increased probability of the collapse of Atlantic thermohaline circulation.[10]

It is thus difficult to specify the upper limit by which temperatures may increase. Nonetheless, however one frames the appropriate target, it is clear that there is a target of greenhouse gas concentration above which we should not go. Now suppose that one has a target level of the concentration of greenhouse gases in mind. We can then try to determine from this roughly what volume of greenhouse gases can safely be released into the atmosphere. This figure tells us how much people across the world may jointly emit. Given this, we then need to know how the right to emit this total volume should be distributed among humanity. This is the question that this chapter shall examine.

Prior to answering it, we might consider some challenges to the preceding line of reasoning. First, some, of course, might contest the argument presented above, arguing that it presupposes that the appropriate response to anthropogenic climate change is to seek to "mitigate" it and, they argue, this is wrong. We should instead concentrate on adaptation and need not mitigate. If we, as a species, continue to grow economically, then we can afford to adapt at a later stage. This, it might be said, is more rational than cutting down the economy now.[11]

There are, however, at least two distinct reasons why this challenge fails. First, this answer assumes that increased economic growth will enable future generations to adapt but we have good reason to doubt that this is the case. (i) Our understanding of exactly when and where climatic harms will occur is too poor for us to know exactly where preventive measures should be taken in advance of their occurring. In addition to this, (ii) adaptation requires account-able and responsive governance. Increased wealth, on its own, is unable to help

[9] For an interesting discussion see Michael Oppenheimer and Annie Petsonk, "Article 2 of the UNFCCC: Historical Origins, Recent Interpretations," *Climatic Change*, 73, 3 (2005), pp. 195–226.

[10] "Impacts of Global Climate Change at Different Annual Mean Global Temperature Increases," in *Avoiding Dangerous Climate Change* (Cambridge University Press, 2006), edited by Hans Joachim Schellnhuber, Wolfgang Cramer, Nebojsa Nakicenovic, Tom Wigley, and Gary Yohe, pp. 93–131, especially pp. 95–96.

[11] See Bjørn Lomborg, *The Skeptical Environmentalist: Measuring the Real State of the World* (Cambridge University Press, 2001), ch. 24.

people living in countries with unaccountable political structures to cope with dangerous climate change. (iii) We also have no assurance that the future wealth will be distributed in such a way as to enable the vulnerable to adapt. Increased wealth is quite compatible with having radical inequalities. It would require an act of faith to think that the rich will be motivated to spend these resources enabling the global poor to adapt. For these reasons, economic growth may not enable people to adapt with dangerous climate change.

In response to this it might be argued that even though adaptation may not be possible in all or many cases this does not mean that one should mitigate now. It would be better, so the argument runs, for humanity to experience economic growth even if this results in the creation of certain harms (such as the loss of small island states) because with the increased wealth one can remunerate the victims of these dangerous climatic events. Even if money cannot ensure that people adapt, it can be employed to compensate people for their loss.

As a number of critics have pointed out, however, the assumption that the loss of a good can always be compensated by the granting of another good or goods is highly disputable. Individuals have rights to a decent environment and one cannot then violate any of them with a view to compensating later.[12] In addition to this some goods are not "substitutable" and loss of them cannot be replaced by supplying another. They are irreplaceable.[13] It would therefore be wrong to destroy them now with a view to "compensating" those ill affected by this at a later stage.[14]

A second, quite separate, reason to reject the "grow-then-adapt" (or "grow-then-compensate") argument is that it is widely recognized that not mitigating now with a view to adapting later is by far the most expensive option. This is one of the key findings of the Stern Review.[15] For that reason alone it should be rejected.

---

[12] See Henry Shue, "Bequeathing Hazards: Security Rights and Property Rights of Future Humans," in *Global Environmental Economics* (Malden, MA: Blackwell, 1999), edited by Mohammed H. I. Dore and Timothy D. Mount, pp. 40–42; and Clive L. Spash, *Greenhouse Economics: Value and Ethics* (London and New York: Routledge, 2002), pp. 231–236. Shue grounds his argument on an appeal to the inalienability of certain rights. I do not think that inalienability is required to ground this objection. Even if a right is alienable (by the rights-holder deciding to waive it) it is wrong for others to trespass on it without the rights-holder's consent.

[13] Robert E. Goodin, *Green Political Theory* (Oxford: Polity, 1992), pp. 57–61. See further Eric Neumayer, *Weak Versus Strong Sustainability: Exploring the Limits of Two Opposing Paradigms* (Cheltenham: Edward Elgar, 2003), 2nd edn.

[14] For a further discussion of these issues see Caney, "Climate Change and the Future: Time, Wealth and Risk," *Journal of Social Philosophy*, 40, 2 (2009), pp. 163–186.

[15] Sir Nicholas Stern, *The Economics of Climate Change: The Stern Review* (Cambridge University Press, 2007).

A second challenge to the argument sketched above would be to maintain that cutting the use of fossil fuels is not necessary because humanity can adopt technological solutions which prevent dangerous climate change. Geo-engineering, it might be argued, is required – not mitigation. There are a variety of different geo-engineering proposals that have been mooted.[16] Some propose inserting sulfur in the atmosphere to enhance the Earth's reflectivity.[17] Others have explored the possibility of putting mirrors in space to reflect the Sun's rays.[18] A third proposal is to insert vertical pipes in the ocean pumping water from the bottom to the surface in order to stimulate algae. This would both absorb carbon dioxide and generate dimethyl sulfide, which, in turn, helps to produce clouds which can reflect the Sun.[19] Fourth, and finally, some have proposed iron fertilization of the oceans – the intention being to encourage algae to grow and absorb carbon dioxide.

There are a number of well-known objections to many of these geo-engineering projects. First, the science is highly uncertain and there is disagreement about the efficacy and cost of such measures. The Fourth Assessment Report, for example, reports that geo-engineering ventures "remain largely speculative and unproven."[20] To give one specific example: Klaus Lackner of Columbia University argues that iron fertilization of oceans may not succeed. He claims that "[i]f the detritus does not sink deeply enough, it would be available through long-term mixing in a relatively short time – years to decades. This could put the CO_2 back in circulation in the next 50 years – precisely when things are most critical."[21]

---

[16] For excellent discussions see David W. Keith, "Geoengineering the Climate: History and Prospect," *Annual Review of Energy and the Environment*, 25 (2000), pp. 245–284; Keith, "Geoengineering," *Nature*, 409, 6818 (2001), p. 420; David W. Keith and Hadi Dowlatabadi, "A Serious Look at Geoengineering," *Eos, Transactions, American Geophysical Union*, 73, 27 (1992), pp. 289–293; and Alan Robock, "20 Reasons Why Geoengineering May Be a Bad Idea," *Bulletin of the Atomic Scientists*, 64, 2 (2008), pp. 14–18, 59. See also the more recent Royal Society, *Geoengineering the Climate: Science, Governance and Uncertainty* (London: The Royal Society, 2009).

[17] Paul Crutzen, "Albedo Enhancement by Stratospheric Sulfur Injections: A Contribution to Resolve a Policy Dilemma?," *Climatic Change*, 77, 3–4 (2006), pp. 211–219.

[18] Roger Angel, "Feasibility of Cooling the Earth with a Cloud of Small Spacecraft Near the Inner Lagrange Point (L1)," *Proceedings of the National Academy of Sciences*, 103, 46 (2006), pp. 17184–17189.

[19] James E. Lovelock and Chris G. Rapley, "Ocean Pipes Could Help the Earth to Cure Itself," *Nature*, 449, 7161 (2007), p. 403.

[20] Terry Barker et al., "Summary for Policymakers," *Climate Change 2007: Mitigation. Contribution of Working Group III to the Fourth Assessment Report of the Intergovernmental Panel on Climate Change* (Cambridge University Press), edited by B. Metz, O. R. Davidson, P. R. Bosch, R. Dave, and L. A. Meyer, p. 15.

[21] Quoted in Victoria Gill, "Seeds of Doubt," *Chemistry World*, December 18, 2007 (available at www.rsc.org/chemistryworld/News/2007/December/18120701.asp).

Second, there are risks attached to each of these options. For example, iron or phosphate fertilization of the oceans may lead to an increase in methane hydrate, which would make climate change even worse.[22] In addition, releasing sulfur particles into the atmosphere runs a considerable risk of causing ozone depletion.[23] Furthermore, a recent study by Kevin Trenberth and Aiguo Dai concludes that one cost of sulfur injections into the atmosphere is that "cutting down solar radiation is apt to reduce precipitation," thereby "[c]reating a risk of widespread drought and reduced freshwater resources."[24] Furthermore, Alan Robock of Rutgers University argues that all geo-engineering projects will have a cost because by adopting a response that does not involve cutting carbon dioxide emissions they "would allow continued ocean acidification, because some of the carbon dioxide humans put into the atmosphere continues to accumulate in the ocean."[25]

Third, as Dale Jamieson and Stephen Gardiner have argued, there are considerable ethical problems with this kind of meddling with the natural world. It can be seen as a kind of hubris and a lack of an appropriate sense of humility. What right does humanity have to disfigure the natural environment in this way?[26]

In addition to all of the above, there are political problems, too. For example, what if the wrong people got hold of the techniques?[27] Furthermore, geo-engineering would require extensive and long-term global cooperation – a condition that seems unlikely to be met. Moreover, it would be likely to generate disputes. Suppose that some countries experienced climate problems: They would likely blame those engaging in geo-engineering for these.[28]

---

[22] See Keith, "Geoengineering the Climate: History and Prospect," p. 270.

[23] Ibid., p. 271.

[24] Trenberth and Dai, "Effects of Mount Pinatubo Volcanic Eruption on the Hydrological Cycle as an Analog of Geoengineering," *Geophysical Research Letters*, 34, August (2007), p. 4.

[25] Robock, "Whither Geoengineering?," *Science*, 320, 5880 (2008), p. 1166. See also Robock, "20 Reasons Why Geoengineering May Be a Bad Idea," p. 15.

[26] Dale Jamieson, "Ethics and Intentional Climate Change," *Climatic Change*, 33, 3 (1996), pp. 323–336; Stephen Gardiner, "Is Geoengineering the 'Lesser Evil'?," *Environmental Research Web*, 18 April (2007) (available at http://environmentalresearchweb.org/cws/article/opinion/27600); and Gardiner, "Is 'Arming the Future' with Geoengineering Really the Lesser Evil? Some Doubts About the Ethics of Intentionally Manipulating the Climate System," in *Climate Ethics: Essential Readings* (New York: Oxford University Press, 2010), edited by Stephen Gardiner, Simon Caney, Dale Jamieson, and Henry Shue.

[27] Keith, "Geoengineering the Climate: History and Prospect," p. 275.

[28] For the last two political problems see Stephen Schneider, "Geoengineering: Could – or Should – We Do It?," *Climatic Change*, 33, 3 (1996), p. 299.

With all this duly noted, it is advisable to disaggregate different kinds of geo-engineering. It would be wrong simply to assume that all geo-engineering projects are liable to the same problems. However, even if we do this and even if some geo-engineering projects can meet the preceding four challenges, it would be highly implausible to think that a policy of geo-engineering would entirely obviate the need for lowering greenhouse gas emissions.[29] For all these reasons, geo-engineering does not remain a viable *alternative* to mitigation.[30] The case for cutting the emissions of greenhouse gases thus remains.

A third challenge to this proposition comes from those who argue that people have an interest in development and that this takes priority over the need to combat climate change. It would therefore be wrong to insist that people cut their emissions of greenhouse gases. This critic might quite rightly emphasize the dire poverty of those who live in developing countries. The poor living in China and India, they maintain, have a right to develop and thus no reason to cut emissions. The claim is that one cannot jointly realize both (A) the goal of mitigating climate change and (B) the legitimate development of the global poor. Let us call this challenge the development challenge. The development challenge raises many complex problems which cannot be resolved in a short space.[31]

Several points can nonetheless be made to defuse the objection. First, it bears noting that dangerous climate change is harmful to the economic development of countries like China and India. Rising sea levels, increased temperatures and more freak weather events will wreak havoc with the aspirations of those in Asia, China, and Africa to develop economically.

---

[29] For this reason some propose a combination of geo-engineering and mitigation. See T. M. L. Wigley, "A Combined Mitigation/Geoengineering Approach to Climate Stabilization," *Science*, 314, 5798 (2006), pp. 452–454.

[30] An additional point is made by Roger Angel. Having described how the very many sunshades could be launched into space to create a large sunshade, he concludes his analysis by saying, "[i]t would make no sense to plan on building and replenishing ever larger space sunshades to counter continuing and increasing use of fossil fuel. The same massive level of technology innovation and financial investment needed for the sunshade could, if also applied to renewable energy, surely yield better and permanent solutions. A number of technologies hold great promise, given appropriate investment." Angel, "Feasibility of Cooling the Earth with a Cloud of Small Spacecraft Near the Inner Lagrange Point (L1)," p. 17189.

[31] The development challenge has been advanced in Thomas Schelling, "Some Economics of Global Warming," *American Economic Review*, 82, 1 (1992), pp. 7–9; Schelling, "The Cost of Combating Global Warming," *Foreign Affairs*, 76, 6 (1997), pp. 13–14; Bjørn Lomborg, *The Skeptical Environmentalist*, pp. 322–324; and the views of the panel of experts reported in *Global Crises, Global Solutions* (Cambridge University Press, 2004), edited by Bjørn Lomborg, pp. 605–644.

In this sense we do not face a choice between development or protection of the Earth's atmosphere. The latter is a precondition of the former.[32]

Second, to see why (A) and (B) are jointly compatible it is important to focus on who should mitigate. Now my suggestion is that on any plausible account of who should combat climate change, the larger portion of the burden rests with the members of affluent industrialized countries like the USA.[33] If, for example, one affirms a polluter pays principle, this entails that members of Western industrialized economies (who are after all responsible for 75 percent of total emissions) should pay. The per capita emissions of the overwhelming majority of people in India or China are still very low. For instance, in 2006 the average emissions of a Chinese person were just over a quarter of those of the average American. More precisely, the per capita emissions of the Chinese were 4.7 tonnes of CO_2 whereas the per capita emissions of Americans was 19.3 tonnes.[34] If we consider the emissions of the developing countries as a whole, then in 2004 we find that these countries (including China and India) emitted only 41 percent of the world's emissions although they constituted 80 percent of the world's population.[35] The world's wealthiest 20 percent thus cause 60 percent of the world's current annual emissions. The polluter pays principle would thus allow China and India to develop. Consider now a second principle – an ability to pay principle. This, too, would ascribe the largest portion of the burden to members of the industrialized world. So, on the basis of this very cursory and necessarily truncated examination,[36] we can see that on any plausible account of moral responsibility the duty to

---

[32] Proponents of the development challenge might then argue that greater economic development will ensure adequate adaptation. This takes us back to the first challenge raised and the problems I noted with it.

[33] The case is well made by Henry Shue, "Global Environment and International Inequality," *International Affairs*, 75, 3 (1999), pp. 531–545. Shue considers three separate rationales and shows that they all lead to the same conclusion: Industrialized economies should foot the bill for combating climate change.

[34] "Gas Exchange: CO_2 Emissions 1990–2006," *Nature*, 447, 7148 (2007), p. 1038.

[35] Michael R. Raupach, Gregg Marland, Philippe Ciais, Corinne Le Quéré, Josep G. Canadell, Gernot Klepper, and Christopher B. Field, "Global and Regional Drivers of Accelerating CO_2 Emissions," *Proceedings of the National Academy of Sciences*, 104, 24 (2007), p. 10292. This article also finds that developing countries and the least-developed countries are only responsible for 23 percent of total emissions since the Industrial Revolution (p. 10292).

[36] For more on this see Caney, "Cosmopolitan Justice, Responsibility and Global Climate Change," *Leiden Journal of International Law*, 18, 4 (2005), pp. 747–775 (reprinted in *The Global Justice Reader* (Oxford: Blackwell, 2008), edited by Thom Brooks); Caney, "Environmental Degradation, Reparations and the Moral Significance of History," *Journal of Social Philosophy*, 73, 3 (2006), pp. 464–482; Caney, "Climate Change, Justice and the Duties of the Advantaged," *Critical Review of International Social and Political Philosophy*, 13, 1 (2010); and Derek Bell and Simon Caney, *Global Justice and Climate Change* (Oxford University Press, forthcoming).

mitigate does not preclude the needy from developing (either simply because the needy are the least able to pay or because the needy have not caused the problem).

A critic might respond to the last argument that this shows that (A) and (B) are jointly satisfiable in theory but, they add that, in practice, the two will collide. They will clash because some, probably very many, will not comply with their duty to mitigate and so others then face the choice between either, on the one hand, acting so as to prevent dangerous climate change or, on the other hand, furthering their own development. Now it is hard to gainsay the empirical assumption made by this objection. High emitters, like the USA, show no sign of complying with their duties. However, this does not unsettle the arguments of this chapter. For the focus of this chapter is on how energy rights *should* be distributed. One can answer this even if it is the case that we know that some are not doing their bit. That some are not adhering to their duty to cut their own energy use does not entail that this project is somehow misconceived for we still need to know (a) how energy rights should be distributed in principle. It is just that, in the face of noncompliance, we now need also to ask (b) how should some respond if others use more than their fair share of energy rights?

A third point bears noting, namely that the development challenge presupposes that development must take the highly industrialized path adopted by the West but there is no necessity for it to do this. If we conceive of development as the promotion of human interests, then industrialization is one way of furthering these, but it is not the only one.

We are left then with the problem identified earlier. There is a need to limit the concentration of greenhouse gases in the atmosphere and thus a need to fix an upper limit on the amount of greenhouse gases that we, the current generation, emit. A number of different kinds of policy instruments have been proposed to lower the concentration of greenhouse gases in the atmosphere. It is common to distinguish between (i) price restrictions (e.g., carbon taxes), (ii) quantity restrictions (e.g., place a cap on how much people can emit carbon), and (iii) regulations (e.g., insist on certain standards for buildings or vehicles). These all concern people's energy rights. In what follows, I am going to assume that the appropriate tool involves a system of carbon permits with trading. It would be possible to think of the fair distribution of emission rights by using taxes – those with a lesser right to emit greenhouse gases could be subject to higher taxes and there could be tax exemptions on those with greater rights to use greenhouse gas-intensive products. Nonetheless, for simplicity's sake (and because many political systems have adopted or are planning to adopt emissions trading schemes rather than carbon taxes) I shall discuss the distribution of emissions in terms of the distribution of tradeable emissions permits.

Before doing so, it is important to stress that mitigation does not simply concern people's energy rights. A successful mitigation policy might also include various other components including, for example, (i) incentivizing research into the development of clean technologies and the transfer of these technologies to developing countries, (ii) facilitating carbon capture and storage, (iii) using education to encourage environmental virtues, (iv) using "carbon disclosure" and utilizing people's concern for their reputation and social standing to encourage people to lower their own emissions, and (v) developing carbon sinks. So, mitigation involves much more than the distribution of energy rights. Nonetheless the latter is one key component and worthy of exploration in its own right.

I WHO SHOULD OWN ENERGY RIGHTS?

The fair distribution of energy rights raises at least two separate questions. First, who should be allocated rights to use fossil-fuel-intensive energy sources? What kinds of entities are the rights-bearers? Is the right to emit greenhouse gases to be held by individuals, or firms, or states, or some other entity? A second key question is: How should rights to emit greenhouse gases be distributed? Should they be allocated equally or according to some other distributive criterion?

Let us consider the first question. Rights might be distributed to states, or to firms, or to individuals. Many who write about the allocation of emissions assume that states are the rightful owners of emission rights. Michael Grubb, to take one prominent example, once held that governments should each be allocated a set of permits to use greenhouse gases and they can then choose to use this permit or to sell it to other states.[37]

A system of carbon trading in which states possess rights to emit greenhouse gases and can buy and sell them is, however, potentially very unfair. Such a scheme would, in principle, allow an unscrupulous government (or group of politicians responsible for buying and selling carbon permits) to abuse this position in several ways. Consider two possibilities. First, under such an arrangement, a government could sell the emission rights that some of its citizens might desperately need to members of another state and pocket the revenue. Second, the government could distribute these rights

---

[37] Michael Grubb, *The Greenhouse Effect: Negotiating Targets* (London: Chatham House, Royal Institute of International Affairs, 1992), 2nd edn.

to some citizens but not to others of that state (members of a disfavored ethnic group, for example). Without the right to emit greenhouse gases, their ability to support themselves may be severely compromised. Ascribing the right to emit greenhouse gases to states may thus be unfair toward the members of those states.[38]

Someone who is motivated by the argument of the last paragraph might prefer a more individualistic option. Some have argued that the right to emit greenhouse gases should be distributed to individuals. There are, at least, two different ways of doing this. First, some have proposed a system under which individuals own personal carbon accounts (PCAs) and would have to use these when purchasing carbon-intensive products (gasoline, say, or a plane flight). Individuals are, in effect, given ration cards and can either use them or sell some of their quota to others. This kind of approach has been canvassed by many and, in 2007, was endorsed by the then British Minister for the Environment, David Miliband.[39] A second kind of individualistic approach adopts what has been called "cap and share."[40] Under this scheme, firms which emit greenhouse gases are required to buy permits in order to do so. However, the permits to emit greenhouse gases are allocated (equally) to individual citizens. Each individual is at liberty to sell their permit to a bank, which will in turn sell them to those firms that wish to buy the permissions necessary for the amount of greenhouse gases that they wish to emit.

Such individualistic approaches have a precedent. As proponents of such schemes note, this is a form of rationing; and rationing has been common during times of war. One issue concerns the scope of such schemes. It is possible to envisage such a scheme working within a nation state. However, it is highly utopian to think that this could be applied on a transnational, let alone a global level. So, although such an individualistic scheme might be applicable, it cannot constitute a comprehensive approach to energy rights.

---

[38] See "Justice, Morality and Carbon Trading," *Ragion Pratica*, 32, June (2009), p. 215. See also William Nordhaus, *A Question of Balance: Weighing the Options on Global Warming Policies* (New Haven, CT and London: Yale University Press, 2008), pp. 159–161.

[39] See David Fleming, "Tradable Quotas: Using Information Technology to Cap National Carbon Emissions," *European Environment*, 7, 5 (1997), pp. 139–148; Fleming, *Energy and the Common Purpose: Descending the Energy Staircase with Tradable Energy Quotas (TEQs)*, 3rd edn (September 2007) (available at www.teqs.net/book/teqs.pdf); Mayer Hillman with Tina Fawcett, *How We Can Save the Planet* (London: Penguin, 2004); Richard Starkey and Kevin Anderson, "Domestic Tradable Quotas: A Policy Instrument for Reducing Greenhouse Gas Emissions from Energy Use" (Tyndall Centre for Climate Change Research: Technical Report No. 39, December 2005) (available at www.tyndall.ac.uk/research/theme2/final_reports/t3_22.pdf). See also Brian Barry, *Why Social Justice Matters* (Cambridge: Polity, 2005), pp. 268–269.

[40] For details see www.capandshare.org/. See, in particular, the document on "How Cap and Share Works," www.capandshare.org/howitworks.html.

Thus far, we have considered "states" and "individuals" as the possible rights-bearers. As we shall see later, there are other possibilities (Section 5). At this point, however, it is appropriate to turn to the question of which distributive principle should be applied to energy rights.

2 DISTRIBUTIVE CRITERIA — GRANDFATHERING

How should energy rights be distributed? In practice, many schemes adopt "grandfathering." This holds that the distribution of emission rights should reflect the distribution of emissions now or at some fixed point in the recent past. Heavy emitters would be awarded larger emission rights and those with small emissions would be awarded less. So the USA and European countries would be heavily favored by this scheme. How would members of countries like China or India fare? It is common to think that these two, especially China, are high emitters and, therefore, that they, too, would benefit. This perception is, however, misplaced once we examine the emissions per person in such countries. To adopt grandfathering is to adopt an approach which will give greater emissions rights per person to Americans and Europeans than to those from China or India.

I shall not discuss "grandfathering" at any length. As a matter of justice, it has very little, if anything, to recommend it. It rewards people in accordance with their contribution to the creation of a problem and that seems perverse. Many people have a deep and strong commitment to the polluter pays principle. If someone creates an environmental hazard, then, other things being equal, we expect them to remedy that situation. If, however, we accept this, then we have good reason to reject grandfathering for the two are diametrically opposed. Instead of saying the polluter should pay, grandfathering dictates that the polluter should be rewarded! Far from being required to pay for the problem created, it says that they should get preferential rights. This is clearly perverse.

In addition to this it bears observing that grandfathering is at variance with other morally relevant criteria such as how needy people are. It is, for instance, indifferent as to whether people require emissions to function at an adequate level. So it may distribute enormous rights to those who have no need of them and very few, perhaps none, to some who are in desperate need.

The only case for accepting grandfathering is a pragmatic one – namely that it should be adopted at the start of trading scheme to get major emitters to take part and sign up to the scheme, with a view to then altering it over time until emission rights are distributed more fairly.[41] The assumption here is that once agents are participants in a scheme they have an incentive to stay in it for leaving it has reputational costs. So once they are part of a scheme they may accept a distributive rule that if it had been part of the scheme when they thought of joining would have deterred them from becoming a member. With this in mind, grandfathering might be permissible as a means to bringing high emitters on board. This is an important goal, but one might reasonably be skeptical of the prospects of this kind of strategy.[42] Furthermore this pragmatic approach is necessarily incomplete: since its claim is that grandfathering is justified insofar as it is the necessary first step in a transition toward a fair distribution, it needs to be supplemented with an account of what constitutes a fair distribution. The question then remains: What constitutes a fair distribution of energy rights?

3 DISTRIBUTIVE CRITERIA – EQUALITY

Let us consider now a second proposal. One suggestion made by a variety of different people is that each person has a right to emit an equal per capita amount of greenhouse gases. This view then takes an egalitarian approach to the distribution of one kind of energy right. This view is remarkably popular. It was, for example, expressed by Anil Agarwal and Sunita Narain in their seminal *Global Warming in an Unequal World.*[43] It also underpins the proposal known as "contraction and convergence," which has been developed and defended by the Global Commons Institute.[44] Other defenders included Paul Baer and Tom Athanasiou in *Dead Heat* (though

---

[41] For discussion, see Caney, "Justice and the Distribution of Greenhouse Gas Emissions," *Journal of Global Ethics*, 5, 2 (2009), pp. 128–130; and Axel Gosseries, "Cosmopolitan Luck Egalitarianism and the Greenhouse Effect," *Canadian Journal of Philosophy*, Supplementary Volume 31 (2005) on "Global Justice, Global Institutions," edited by Daniel Weinstock, pp. 300–301.

[42] An important test case that may support a more optimistic stance is the EU ETS. When it was first instituted (on January 1, 2005) it operated according to a grandfathering principle. However, on January 23, 2008, the Commission proposed a directive to amend Directive 2003/87/EC. In particular, it proposed that after 2012, the majority of permits should be auctioned. It kept a grandfathering component in that states were accorded permits roughly in proportion to their earlier emissions (Articles 9, 9a, and 10 of the proposed directive). However, a key innovation is that the states then auction off these permits (Article 10). Not all emission rights are to be auctioned, however, and a certain proportion will still be allocated freely (Article 10a). However, the accompanying Explanatory Memorandum to the directive claims that "it is estimated that, at least two thirds of the total quantity of allowances will be auctioned in 2013" (p. 8). For the proposed directive see http://ec.europa.eu/environment/climat/emission/pdf/com_2008_16_fn.pdf. The precise nature of the EU ETS in its third phase has yet to be fully determined.

[43] Anil Agarwal and Sunita Narain, *Global Warming in an Unequal World: A Case of Environmental Colonialism* (New Delhi: Centre for Science and Environment, 1991).

[44] See www.gci.org.uk/.

they have now adopted a different approach, which will be discussed in Section 5) and Eric Neumayer.[45] In his recent book, *One World*, Peter Singer also defends it and it is also affirmed by Brian Barry.[46] Though this approach is widely supported, it is subject to a number of grave difficulties. Since the egalitarian approach has appealed to very many I shall spend some time evaluating it. I believe that it suffers from three deep problems.

3.1 Objection 1

The first problem is a methodological concern. The "equal per capita" view treats the permission to emit greenhouse gases on its own and separated off from someone's access to other goods. But why should we do so? Why not consider someone's entitlements as a whole, taking into account their access to all the relevant "primary goods" (Rawls) or "resources" (Dworkin)? In other words, why should this one particular good (emitting greenhouse gases) have its own principle regulating it and why should this principle be equality? Why single this out for special (equal) treatment? Why not put it into the general pot of all goods to be distributed and then have a distributive rule applying to the whole package of goods contained therein?[47]

The force of this point can be seen by considering a defense of equal per capita emissions offered by Axel Gosseries. Gosseries defends a modified kind of per capita egalitarianism by invoking luck egalitarianism.[48] Luck egalitarianism entails that no one should be worse off than others through no fault of their own. Different luck egalitarians will give different accounts

---

[45] See Tom Athanasiou and Paul Baer, *Dead Heat: Global Justice and Global Warming* (New York: Seven Stories Press, 2002), especially pp. 76–97; Eric Neumayer, "In Defence of Historical Accountability for Greenhouse Gas Emissions," *Ecological Economics*, 33, 2 (2000), pp. 185–192. See also Paul Baer, John Harte, Barbara Haya, Antonia V. Herzog, John Holdren, Nathan E. Hultman, Daniel M. Kammen, Richard B. Norgaard, and Leigh Raymond, "Equity and Greenhouse Gas Responsibility," *Science*, 289, 5488 (2000), p. 2287. For a qualified commitment to the equal per capita view, see Gosseries, "Cosmopolitan Luck Egalitarianism and the Greenhouse Effect," pp. 279–309. Baer and Athanasiou, note, have subsequently rejected the equal per capita view. They now defend what they term a "Greenhouse development rights" approach (www.ecoequity.org/).

[46] See Peter Singer, *One World: The Ethics of Globalization* (New Haven, CT and London: Yale University Press, 2002), p. 43; and Barry, *Why Social Justice Matters*, pp. 268–269.

[47] This point has also been forcefully made by Derek Bell and David Miller. See Bell, "Carbon Justice? The Case Against a Universal Right to Equal Carbon Emissions" in *Seeking Environmental Justice* (Amsterdam: Rodolphi, 2008), edited by Sarah Wilks, pp. 239–257, especially p. 250; and Miller, "Global Justice and Climate Change: How Should Responsibilities Be Distributed?," *The Tanner Lectures on Human Values*, delivered at Tsinghua University, Beijing, March 24–25 (2008), pp. 142–143. This is available at: www.tannerlectures.utah.edu/lectures/documents/Miller_08.pdf.

[48] "Cosmopolitan Luck Egalitarianism and the Greenhouse Effect," pp. 279–309. Gosseries' view allows two temporary deviations from equality in the distribution of per capita emissions (pp. 297–303) and one non-temporary deviation (pp. 303–304).

of what it is that people should have equal shares of. Ronald Dworkin, for example, famously thinks that there should be equality of "resources"; Amartya Sen canvasses equality of "capability to function"; Richard Arneson once canvassed equal "opportunity for welfare"; and G. A. Cohen defends equal "access to advantage."[49] They share the view, however, that no person should be disadvantaged because of factors beyond their control. Now Gosseries both affirms his commitment to luck egalitarianism[50] and then maintains that luck egalitarianism entails equal per capita emissions.[51] It is on this basis that he maintains that "there is no reason for a Chinese or a Peruvian not to have a right to emit CO_2 equal to that of a Canadian or a Hungarian."[52]

This argument, however, fails as a defense of equal per capita emissions because on each account of "equality of what," carbon emissions would form only part of what it is that people should have fair shares of, *and there is no reason to affirm equality of each individual component part of the total package.* For example, a resource egalitarian will favor equality of an individual's set of all resources – where this would include the use of their talents and the use of external resources (including, but not restricted to, the use of fossil fuels). Similarly, someone who adheres to Cohen's equality of access to advantage will wish to equalize each person's combined bundle of advantages, not every single ingredient that goes up to make up their access to advantage. Gosseries' defense of equal per capita emissions is thus vulnerable to the objection leveled above – namely that luck egalitarians are not necessarily committed to equality of each individual component of a collection of goods. An equal per capita approach would not follow from, or be endorsed by, *any* luck egalitarian metric.[53]

Gosseries dissents from this line of reasoning and explicitly affirms what he, following Jon Elster, terms a "local justice" approach, where this holds

---

49 See Richard Arneson, "Liberalism, Distributive Subjectivism, and Equal Opportunity for Welfare," *Philosophy and Public Affairs*, 19, 2 (1990), pp. 158–194; G. A. Cohen, "On the Currency of Egalitarian Justice," *Ethics*, 99, 4 (1989), pp. 906–944; Ronald Dworkin, *Sovereign Virtue: The Theory and Practice of Equality* (Cambridge, MA: Harvard University Press, 2000), pp. 65–119; and Amartya Sen, "Equality of What?," *Choice, Welfare and Measurement* (Cambridge, MA: Harvard University Press, 1982), pp. 353–369. Alternatively a luck egalitarian might endorse Rawls's concept of "primary goods." See John Rawls, *A Theory of Justice* (Oxford University Press, 1999), rev. edn, pp. 54–55, 78–81, 348; Rawls, *Justice as Fairness: A Restatement* (Cambridge, MA: Harvard University Press, 2001), edited by Erin Kelly, pp. 57–61, 168–176.
50 "Cosmopolitan Luck Egalitarianism and the Greenhouse Effect," pp. 280–281.
51 "Cosmopolitan Luck Egalitarianism and the Greenhouse Effect," especially p. 296.
52 "Cosmopolitan Luck Egalitarianism and the Greenhouse Effect," p. 296.
53 For a fuller discussion, see my "Equality in the Greenhouse?" (unpublished) and Bell's instructive analysis in "Carbon Justice? The Case Against a Universal Right to Equal Carbon Emissions."

that particular distributive principles should apply to different goods.[54] Why should we adopt a local justice approach for greenhouse gas emissions? Gosseries offers two replies. His main point is that, from a "methodological" point of view, it is better "to clarify things at this level first, before getting involved in a more complex 'all-things-considered' enterprise."[55]

To say, however, that we should "clarify things *at this level* first" (emphasis added) requires an argument showing why *this* level is an appropriate or natural one. Why is "the emission of greenhouse gases" the right level to focus on? Consider someone who said that each person should have rights to equal per capita emissions of methane, equal per capita emissions of nitrous oxide, equal per capita emissions of carbon dioxide, and so on. That would seem highly implausible. What is the point of disaggregating the greenhouse gases at this level? None. This is clearly not the right level at which to focus. Suppose then that someone says that each person should have equal per capita emissions of greenhouse gases (as a whole). Is this the right level? Again, it is not clear why.

To see why, it is useful to take a step back from this issue and ask why we should care about people's ability to engage in activities which emit greenhouse gases, such as carbon dioxide. The answer is surely that we care about this because we care about people being able to do things such as (i) heat themselves and keep warm, (ii) cook, (iii) construct buildings, (iv) travel from one place to another, (v) use technology for recreation and to communicate with one another, and, no doubt, many other activities and goals. Now underlying these "intermediate" interests are some more fundamental or basic interests. Of course, people may differ about what these fundamental interests are. Some might say, following John Rawls, that there is a higher-order interest in being able to *form, revise and rationally pursue a conception of the good.*[56] On this basis they might say that goods like (i)–(v) matter because they serve this more basic interest in choice and revisability. Others might say that the basic interest is *well-being* and then

---

[54] "Cosmopolitan Luck Egalitarianism and the Greenhouse Effect," pp. 282–283. See Elster, *Local Justice: How Institutions Allocate Scarce Goods and Necessary Burdens* (New York: Russell Sage Foundation, 1992) (cited by Gosseries at p. 282, footnote 12).

[55] "Cosmopolitan Luck Egalitarianism and the Greenhouse Effect," p. 283. His second reason is a more strategic one. The thought is that it might be politically advantageous to divorce this issue – the causes and consequences of climate change – from other issues of global economic, political, and cultural justice because one is more likely to be able to persuade people to take action on the climate change issue on its own than on a total package of global justice. The assumption here is that climate justice requires less dramatic action than global justice as a whole would (p. 283). Whether it would be politically astute to treat greenhouse gases and climate change separately from other aspects of global injustice is beyond my remit here. See, however, my "Equality in the Greenhouse?"

[56] John Rawls, *Justice as Fairness*, pp. 19, 21–23.

argue that (i)–(v) matter because they serve this basic interest. A third view might hold that persons have both an interest in well-being (and thus in keeping warm, preparing food, and using technology) and an interest in autonomy (and thus in being able to move from one place to another at one's own convenience). We do not need to arbitrate between these three (or any other) accounts of persons' fundamental interests. The key point here is that the interest in being able to engage in activities that emit greenhouse gases has significance because it furthers some more fundamental human interests. One can go further. Indeed one might say that the interest in being able to engage in activities which emit greenhouse gases has significance *only* if it contributes to some more fundamental interests. Suppose, for example, that persons do not need to use energy sources that emit high levels of greenhouse gases then and suppose further that there are other reasonably priced energy sources. In these circumstances, then – especially given the harmful nature of greenhouse gas emissions – it is hard to see why persons would have an interest in using greenhouse gas-emitting energy sources at all. If this is true, then we can say that persons can have an interest in using carbon-intensive practices *if and only if* this is necessary for some higher-order interest.[57]

Now how is this relevant? The answer is that if it is the case that the ability to engage in carbon-intensive activities has value only because it contributes to some general basic interest, then we should not treat that ability in isolation from all the other phenomena that also serve this basic interest. All of these phenomena (whether it is the opportunity to engage in carbon-emitting activities or the opportunity to have health care, or the opportunity to have material support in times of need) have value as contributors to some higher-order interest. The emission of greenhouse gases thus derives its importance simply from being one element in a vast system of elements which jointly serve higher-order interests. There is, therefore, absolutely no reason to single it out for special treatment, and to focus on carbon emissions (or even greenhouse gas emissions) is to focus at the wrong "level." Put more succinctly: the *arguments* for attributing normative significance to greenhouse gases emissions entail that the latter should not be distributed according to their own distributive principle.

---

[57] This point is also persuasively made by Tim Hayward in his illuminating "Human Rights Versus Emissions Rights: Climate Justice and the Equitable Distribution of Ecological Space," *Ethics and International Affairs*, 21, 4 (2007), especially pp. 441–444. See also Miller, "Global Justice and Climate Change: How Should Responsibilities be Distributed?," especially p. 142. I develop the point further in "Equality in the Greenhouse?"

The charge then remains. There is no reason to treat the distribution of greenhouse gas emissions separately from an understanding of their moral general entitlements. Therefore the equal per capita perspective is implausible.

3.2 Objection 2

I want now to press a second objection to carbon egalitarianism. This second objection starts with the observation that the equal per capita view is a kind of *resourcism*. That is, it holds that (in the case of climate change) justice requires the distribution of a fair share of a specific sort of resource, where the resource in question is that of being able to engage in carbon-intensive activities. Everyone should have a fair, understood as equal, share of the Earth's carbon budget: they all have the right to consume the same amount of carbon.

Now, as such, the equal per capita view is vulnerable to the objections standardly leveled against resourcism. In particular, it is vulnerable to Amartya Sen's objection that resourcism is guilty of "fetishism." Resources such as wealth and income are means to an end or set of ends. It is therefore misconceived to focus one's concern on the means as if it they were what ultimately mattered. To do so is to give them a status that they do not deserve. Justice should instead, on this view, be concerned with what people are able to do and whether people are actually capable of achieving certain goals.[58] To see the force of this objection, focus on the permission to emit a certain amount of carbon dioxide. This matters only insofar as it enables one to do certain things – like travel or heat oneself. In itself it has no value and so if it were substituted by another energy source that achieved the same outcomes at the same price there could be no legitimate complaint.

Note that this objection is distinct from the first one, for one can press the first objection without rejecting resourcism. The problem highlighted by the first query is that of singling out one good/resource for special treatment. The problem highlighted by the second query is that of why we should think that justice requires focusing on "resources." Someone committed to global resource egalitarianism would agree with the first objection but not the second.

3.3 Objection 3

Let us turn now to a third, related problem with the egalitarian proposal. This problem follows on from the last but gives it a more specific content.

---

[58] See Sen, "Equality of What?," p. 366. The objection is levelled at Rawls's concept of primary goods.

The worry is simply stated: namely that granting people equal emission rights is unfair when some need more emission rights than others.[59] The equal per capita view, the argument runs, is defective because it is indifferent to people's different needs and vulnerabilities. There are a number of different ways of fleshing out this objection. Consider, for example, the following scenarios:

Example I: The case of the needy

Compare, on the one hand, the plight of the elderly and the extremely young who live in a freezing climate with able-bodied people living in warmer climes. The former have greater health needs than the latter. Here, an equal per capita view is very harsh on the needy, the poor and those suffering more generally from fuel poverty.

This example illustrates and develops Sen's argument introduced above because it draws attention to the fact that we are not concerned with resources per se but with what we can do with them and in this case equal emission rights do not entail equal ability to further key interests.

Consider now a second case:

Example II: The case of unequal energy sources

Compare two societies. The first has easy access to non-carbon-intensive energy sources, including, for example, renewable energy sources like wind power (and, one might add, nuclear energy). The second, however, lacks access to renewable energy. It is not windy there and it does not have the mountain streams that the first can use to generate hydroelectric power. (We might suppose, too, that it is in an earthquake zone and so cannot build nuclear power plants.) Now to ascribe the members of these two societies equal rights to emit carbon dioxide is to be insensitive to a morally relevant difference – namely that some need to use carbon more than others. They need to because they have less opportunity to use other energy sources.

Examples I and II thus both show that an equal per capita view is unfair on those with greater needs – whether they have greater needs because of the

---

[59] The same concern is raised by Miller, "Global Justice and Climate Change: How Should Responsibilities Be Distributed? Lecture 2." For an empirical study of the correlation between emissions and geographical factors, see Eric Neumayer, "National Carbon Dioxide Emissions: Geography Matters," *Area*, 36, 1 (2004), pp. 33–40. Neumayer finds a strong correlation between (i) emission levels and the coldness of the climate, and between (ii) emission levels and the opportunity to use renewable energy sources. He also finds a weaker correlation between emission levels and normal transportation costs. Interestingly, Neumayer recognizes that these correlations undermine the fairness of an equal per capita view (p. 39).

kinds of environment they live in or whether they have greater needs because of the lack of other energy sources.

Some might resist this line of reasoning. Consider two counter-arguments.

First, Gosseries discusses those with more needs than others – including, for example, those who need more fuel to heat their houses because they live in cold climes and those who need more energy to travel because they live in an area where the population is dispersed. He argues, however, that, as time progresses, we are entitled to ignore these greater health needs because they are no longer part of one's "circumstances" but reflect their "choices."[60] People could choose not to live in sparsely populated areas where they "have" to travel to get basic amenities, and they could choose not to live in cold climates. And since they can choose to do so, they are not entitled to the extra support from others.

But this counter-argument encounters some serious problems. Most obviously, some will have needs which they cannot alter through their own choice. In some cases this will arise because of *physical* factors: Those who are weak and infirm and who need more electricity to heat themselves cannot simply choose to be strong and healthy. In other cases, the disadvantage will stem from *financial* factors: The elderly pensioner living in an inhospitably cold environment may be too poor to move to warmer climes. And in some cases, the disadvantage will arise from *political-legal* factors. Consider again those living in freezing climates. They may be unable to migrate to a warmer country because of immigration restrictions. There are thus physical, financial, and political factors which entail that some of those with greater needs cannot be said to have chosen to have those needs.

The objection, thus, remains intact: equalizing the right to emit carbon dioxide is unfair on those with greater needs.

Consider now a second argument that might be made. This second argument draws on a response made by Peter Singer to this problem. In *One World*, Singer considers the observation that Canadians might need more resources than some (Mexicans in his example) because of the cold winters they endure. Singer's response is not to call for a more-than-equal allocation of emission permits to those with more needs but to say that this problem is taken care of by making the permits tradeable.[61] Canadians can

---

[60] Gosseries, "Cosmopolitan Luck Egalitarianism and the Greenhouse Effect," pp. 301–303. The choice/circumstance terminology comes from Ronald Dworkin (cf., for example, *Sovereign Virtue*, pp. 322–324).

[61] *One World*, p. 47.

then buy their permits from elsewhere. But this response is unsatisfactory. In the first place, some people may not be able to buy a sufficient number of permits. Consider, for example, economically disadvantaged Canadian citizens. Or, more generally, it might be useful to change the example from contemporary Canadians (whom one might imagine to be well off) to poor Russians or Poles or Siberians.[62] Second, even if a needy person can actually afford to pay for sufficient carbon permits to heat herself, this ignores the bigger picture: namely that under a scheme which allocates everyone the same quota of carbon emissions she is relatively worse off than an able-bodied person. Carbon egalitarianism leads to inequality in one's ability to pursue one's goals and aspirations. As such, the scheme penalizes the disadvantaged even if it is the case that the disadvantaged can buy permits to cover their most basic needs.

The third objection thus stands. The equal per capita view is unfair on the needy and vulnerable. It is perhaps worth noting here that when rationing schemes have been introduced in the past they have often recognized this point. For example, as Ina Zweiniger-Bargielowska documents in her fascinating book on rationing in Britain in World War II and the postwar period, extra rations were allocated to some who were perceived to be especially needy.[63] Furthermore, she reports that where there were flat rate equal allocations of goods these were often regarded as unfair. For example, the overwhelming majority of male manual workers argued that a system of equal food rations was unfair because it did not reflect their greater need for nutritious food.[64]

4 MEETING VITAL NEEDS AND ENABLING DEVELOPMENT

The analysis so far has been critical. I have rejected two approaches to distributing greenhouse gas emission rights. In this section, I wish to draw attention to a third, more promising, possibility. This third approach builds on the objections leveled against the two preceding approaches. More precisely, it starts from the objection that both the egalitarian approach and grandfathering are insensitive, in different ways, to people's basic needs.

---

[62] Note that Singer concludes by saying that "[t]he claim of undue hardship therefore does not justify allowing rich countries to have a higher per capita emissions quota than poor countries" (*One World*, p. 47). This is true but does not deal with the claim of undue hardship suffered by the disadvantaged.

[63] Ina Zweiniger-Bargielowska, *Austerity in Britain: Rationing, Controls, and Consumption, 1939–1955* (Oxford University Press, 2000), p. 15, footnote 23. Zweiniger-Bargielowska adds that Germany's rationing system was more accommodating to people's different needs than Britain's (p. 38).

[64] Ibid., pp. 76 and 262. The reason that the proposal to distribute greater food to those with physically demanding jobs was resisted was, it seems, a purely practical one (p. 17).

Clearly the egalitarian approach is more sensitive to this concern but it, too, I have argued, does not take sufficient account of people's needs, and in particular the differing levels of needs between different people. Drawing on this one might then propose that, as a *minimum* condition of a fair world, the right to emit greenhouse gases should be distributed in such a way that people can enjoy what Henry Shue terms "a minimally decent standard of living."[65] Domestically this requires distributing emissions to those in fuel poverty who need more energy to satisfy their core needs. At the international level, this means distributing emission rights in such a way that one enables the development of the world's poor and disadvantaged in ways which do not trigger dangerous climate change.

It is important to note here that there are two quite different ways of achieving the core moral ideal affirmed by this third approach. One involves distributing greater emission rights to the most disadvantaged. A second proposes auctioning emissions permits and then distributing the proceeds to the disadvantaged. The first kind of approach is advanced by Paul Baer, Tom Athanasiou, and Sivan Kartha. They advocate what they term the "greenhouse development rights" approach.[66] As its name suggests, this proposes distributing emission rights in such a way as to enable the poor of this world to develop while also limiting the total volume of emissions so as to ensure that it does not exceed a dangerous level. It would thus grant extra emission rights to the most disadvantaged so that they can realize the right to develop.

---

[65] Shue, "Subsistence Emissions and Luxury Emissions," *Law and Policy*, 15, 1 (1993), p. 42. Shue has argued powerfully in defence of this approach in an important series of articles including "Subsistence Emissions and Luxury Emissions," "Avoidable Necessity: Global Warming, International Fairness, and Alternative Energy," in *Theory and Practice: Nomos XXXVII* (New York and London: New York University Press, 1995), edited by Ian Shapiro and Judith Wagner deCew, especially pp. 250 ff., and in his chapter in this volume. This approach is also affirmed by Wilfred Beckerman and Joanna Pasek, *Justice, Posterity, and the Environment* (Oxford University Press, 2001), pp. 183–184. Note, too, that some advocates of an equal per capita approach also agree that subsistence needs must be met first and then propose sharing the remainder equally: see Steve Vanderheiden, *Atmospheric Justice: A Political Theory of Climate Change* (New York: Oxford University Press, 2008), pp. 226–227, 243–247, 249. I address Vanderheiden's arguments elsewhere ("Equality in the Greenhouse?").

For concerns about how one could adequately specify the appropriate standard of living, see Stephen Gardiner, "Survey Article: Ethics and Global Climate Change," *Ethics*, 114, 3 (2004), pp. 585–586. For some thoughts on how to address this kind of concern, see Caney, "Human Rights and Global Climate Change," in *Cosmopolitanism in Context: Perspectives from International Law and Political Theory* (Cambridge University Press, 2010), edited by Roland Pierik and Wouter Werner, pp. 43–44.

[66] See Paul Baer Tom Athanasiou and Sivan Kartha, *The Right to Development in a Climate Constrained World: The Greenhouse Development Rights Framework* (Berlin: Heinrich Böll Foundation, 2007). See also their instructive website www.ecoequity.org/. Baer *et al.*'s promising proposal merits more attention that I can devote to it here. I hope to address it further at a later date.

The second kind of approach has been canvassed (in different ways) by Peter Barnes and Oliver Tickell.[67] They call for a scheme with three distinctive features. First, permits to emit greenhouse gases should be distributed via a regular (e.g., annual) global competitive auction. Firms bid against each other for these permits and the permits are distributed to the highest bidder. Second, the volume of greenhouse gases to be emitted declines over time in order to prevent the concentration of greenhouse gases from reaching unacceptable levels. Third, the proceeds of the auction can then be spent on just causes – either to combat dangerous climate change (the narrow version) or to further global principles of economic and environmental justice (the wider version). The key feature of this approach is that it seeks to achieve equity *not* by focusing *solely* on the distribution of emission rights but by focusing on both (a) the revenues raised by the emissions auctions and (b) the principles according to which rights to emit are allocated.[68]

Note that one might combine these approaches. Where an upstream auction is already in place (as it will be in the European Union) it might make sense to continue to use an auction approach but to reform it so as

---

[67] See Peter Barnes, *Who Owns the Sky?: Our Common Assets and the Future of Capitalism* (Washington, DC: Island Press, 2001); Peter Barnes, *Capitalism 3.0: A Guide to Reclaiming the Commons* (San Francisco: Berrett Koehler Publishers, Inc., 2006), especially Part II; Peter Barnes, *Climate Solutions: A Citizen's Guide* (White River Junction, VT: Chelsea Green Publishing, 2008), especially Part III; Peter Barnes, Robert Costanza, Paul Hawken, David Orr, Elinor Ostrom, Alvaro Umaña, and Oran Young, "Creating an Earth Atmospheric Trust," *Science*, 319, 5864, February 8 (2008), p. 724; and Oliver Tickell, *Kyoto2: How to Manage the Global Greenhouse* (London: Zed Books, 2008). See also www.earthinc.org/ and www.kyoto2.org/page1.html.

Note that an auction scheme is also defended by leading economists such as Ross Garnaut (in *The Garnaut Climate Change Review: Final Report* (Cambridge University Press, 2008), ch. 14, especially pp. 331–332); and Cameron Hepburn, Michael Grubb, Karsten Neuhoff, Felix Matthes, and Maximilien Tse, "Auctioning of EU ETS Phase II Allowances: How and Why?," *Climate Policy*, 6, 1 (2006), pp. 137–160.

Finally, one might note here a comparison with Article 10 of the new directive passed by the European Commission on January 23, 2008. The latter proposes that at least 20 percent of the revenues raised in the reformed EU Emissions Trading Scheme should be spent on encouraging renewable energy sources (clause b), CCS (clause c), paying people, especially those in the least-developed countries, not to engage in deforestation (clause d), and adaptation (clause e). See http://ec.europa.eu/environment/climat/emission/pdf/com_2008_16_fn.pdf.

[68] It might be worth noting that the proposal made in the text has affinities with some existing schemes. Consider, in particular, the Alaskan Permanent Fund (a scheme under which a proportion of the revenues from the sale of oil and gas in Alaska is disbursed to all Alaskan citizens in the form of an annual dividend). For details about the Alaskan Permanent Fund see Jonathan Anderson, "The Alaska Permanent Fund: Politics and Trust," *Public Budgeting and Finance*, 22, 2 (2002), pp. 57–68. Comparisons might also be made to Thomas Pogge's Global Resources Dividend (which levies a small dividend on the use of natural resources and then disburses that globally). See Pogge, *World Poverty and Human Rights: Cosmopolitan Responsibilities and Reforms* (Cambridge: Polity, 2008), 2nd edn., ch. 8.

best to enable the realization of key interests. In other contexts, the first kind of approach may be more suitable.[69]

Now both of these schemes can be designed in ways which enable people to satisfy their most fundamental needs and which allow the poor to develop. This is particularly clear in the case of the greenhouse development rights since it is of the essence of this approach that it distributes emission rights so that the disadvantaged can enjoy a right to develop. It is less clear in the case of the auction-based approach. Some might worry that the auction is unfair on the global poor for they and their firms will not be able to afford to bid for global emission rights. Three points are worth stressing in reply. First, this objection has most force when the auction is held for the first time. However, it is important to stress that the proceeds of the auction can be dedicated to the needs of the most vulnerable. They will benefit in absolute terms from the sale of auctions and are certainly better off than they are under the status quo.[70] Moreover, one might propose, as Peter Barnes and his co-authors (who include the 2009 Nobel Laureate for Economics, Elinor Ostrom) do, that half of the proceeds of the auction should be distributed in the form of an annual lump sum to everyone. They argue that if each tonne of carbon dioxide equivalent is auctioned at a price of $20 to $80, then this will yield a revenue of $0.9 to $3.6 trillion dollars a year, which would in turn result in an annual dividend of $71 to $285 per head.[71] Alternatively one might decide (a) to devote more than 50 percent of the auction revenue to this kind of annual dividend, and one might also (b) distribute it wholly to the world's poor and not on an equal per person basis. A key point is that by the time of the second auction the purchasing power of the world's disadvantaged will be augmented, and it will be improved further in each successive auction as long as resources are diverted to the most vulnerable. Second, one may simply stipulate that certain kinds of activities (needed to cover one's most basic needs) should not be part of the auction. In this way one could ensure that the global poor are not disadvantaged. Finally, if neither of these two responses proves adequate one could, of course, alter the auction to ensure that a certain quota of emission rights is set aside for developing countries. These three points, I believe, should dispel any concerns that the scheme being proposed is harmful to the global poor.

Note that neither the distribution of emissions nor the auctioning of emissions is likely to be sufficient to mitigate climate change. Four other

---

[69] For another way of combining the two methods, see Beckerman and Pasek, *Justice, Posterity, and the Environment*, p. 184.

[70] Tickell, *Kyoto2*, p. 229.   [71] Barnes *et al.*, "Creating an Earth Atmospheric Trust," p. 724.

means merit mention. First, governments can employ *regulations* to lower greenhouse gas emissions. They can, for example, stipulate that new buildings meet certain standards of insulation. They can also prohibit the use of some particularly powerful greenhouse gases. Second, governments can use carbon disclosure programs to motivate people to lower their emissions. People care about what others think of them and often avoid some courses of action for fear of the stigma and public disapproval attached to those courses of action. Consider, for example, the stigma attached to drunk driving. Now governments may employ this strong desire in an attempt to lower emissions. If, for example, they require firms to disclose how much greenhouse gas they emit and how much greenhouse gas was emitted to produce each product, then (a) firms may lower their emissions for fear of the stigma attached to high emissions. In addition to this (b) individuals may wish not to be seen to consume goods with a large carbon footprint. Obviously this kind of measure succeeds only where there is social disapproval of high emissions and this is a considerable limitation. However, where there is such an ethos, carbon disclosure can exert a powerful effect on human behavior.[72] Third, governments may seek to develop and implement carbon sequestration – either capturing carbon dioxide as it is emitted or by trying to capture it from the air.[73] Fourth, governments can, and should, foster the development and transfer of new clean energy sources. This is important to enable the global poor to develop without bringing about dangerous climate change.

Each of these measures can, and should, be designed to mitigate climate change and, crucially, should be designed to do so without compromising people's capability to attain a decent standard of living.

---

[72] See on this the interesting discussion by Richard H. Thaler and Cass R. Sunstein, *Nudge: Improving Decisions about Health, Wealth, and Happiness* (New Haven, CT and London: Yale University Press, 2008), pp. 188–192. For an illuminating discussion of the mechanism under consideration, see Geoffrey Brennan and Philip Pettit's excellent *The Economy of Esteem: An Essay on Civil and Political Society* (Oxford University Press, 2004). One final point bears noting here. Carbon disclosure may have an additional benefit noted by Sunstein and Thaler. Individuals care not simply about how they appear to others but also how they appear to themselves. They may want to be good green citizens. Policies that reveal to people just how much energy they are consuming may thus result in lower emissions. As Sunstein and Thaler rightly observe, how much greenhouse gas an activity produces is to a large extent "invisible" (*Nudge*, p. 104). Given this, as long as people care about having lower emissions, then increasing their visibility will encourage lower emissions. See *Nudge*, pp. 193–194.

[73] On the latter see Klaus S. Lackner, Hans-Joachim Ziock, and Patrick Grimes, "The Case for Carbon Dioxide Extraction from Air," *Sourcebook*, 57, 9 (1999), pp. 6–10. See also the discussion of the work by Klaus Lackner by Robert Kunzig and Wallace Broeckner, *Fixing Climate: The Story of Climate Science – and How to Stop Global Warming* (London: GreenProfile Books, 2008), especially chapters 14 and 15.

5 THE PLACE OF EQUALITY

Earlier in this chapter I criticized one specifically egalitarian analysis of global climate change. I now want, in this section, to draw attention to several ways in which equality can, and should, feature in our understanding of the problem of climate change. Three points in particular are worth making.

First, and obviously, the position developed here is quite compatible with defending an egalitarian approach to the distribution of burdens and benefits at the global level. Indeed this chapter is motivated by a commitment to global equality – the point is simply that global equality does not entail an equal per capita approach. So to apply this point to the global auction outlined in the preceding section, one might argue that the proceeds should be spent so as to minimize existing global inequalities.

Second, it bears noting that fair climate policies require a more egalitarian distribution of power. International negotiations on climate change, as well as those on trade and development, are marked by inequalities of power. There are at least two different aspects to this. First, wealthier economies have far greater bargaining power and can threaten to withhold benefits. Second, wealthier countries are able to send a large and well-informed negotiating team to climate negotiation whereas many less developed countries cannot afford to do so.[74] The latter are thus less able to defend their interests. Global inequalities thus lead to inequitable climate negotiations. This makes the reform of international institutions (to even out countries' influence) and, more generally, the more equal distribution of wealth matters of paramount concern.

Third, it is important to recognize that the earlier critique of an equal per capita critique does not preclude one from condemning the fact that some citizens of the world (notably North America and Europe) have emitted far more greenhouse gases than others (those from China, India, and Africa). At first glance, one might think that one cannot both condemn the existing inequalities in emissions and make the argument I made earlier against an equal per capita view. This, however, is clearly mistaken. My arguments earlier show, if correct, that inequalities in emission rights are not *in principle* objectionable *if those with fewer emission rights have commensurately more goods in other domains to make up for their shortfall in emissions*. But in

---

[74] J. Timmons Roberts and Bradley C. Parks, *A Climate of Injustice: Global Inequality, North-South Politics, and Climate Policy* (Cambridge, MA: MIT Press, 2007), pp. 14–19. Roberts and Parks' emphasis is, however, on another aspect of global inequality – namely how global inequalities lead rich and poor countries to lack a common perspective on global climate change and to distrust each other.

our world they clearly do not. The citizens of Africa or India or Bangladesh or China are all low emitters in per capita terms and among the poorest people in the world in terms of their possession of other goods. The existing emissions inequalities are thus part of an objectionable inequality.

6 CONCLUDING REMARKS

There has been relatively little sustained analysis of climate change by political philosophers – though that is now, thankfully, changing. One important question that arises is how energy rights should be distributed. In this paper I have criticized several prevailing ways of thinking about this – some favored by political and business elites (grandfathering) and others favored by environmental activists (equal per capita emissions). Neither approach is, I have argued, adequate.

In the course of so doing, I have made a methodological claim and a substantive claim. The methodological claim is that it is a mistake to try to create a theory of justice that takes as its sole subject the fair distribution of carbon emissions. One should not have a theory of greenhouse gas justice (any more than a theory of aluminum justice or titanium justice). It is better to create a theory of justice that, *inter alia*, governs the emissions of greenhouse gases. In addition to this methodological point, I have endorsed an approach which is committed to meeting core needs and have drawn attention to two different ways of realizing this ideal. It may be that it, too, is subject to powerful objections, in which case, I call on others to develop it further or to show how an alternative is superior.

CHAPTER 5

Common atmospheric ownership and equal emissions entitlements

Darrel Moellendorf

In this chapter, I assess the view that every person shares in a common property right to the Earth's atmosphere.[1] This view is advocated by some environmental non-governmental organizations, which take it as the philosophical basis of the claim that every person has an equal entitlement to emit CO_2. For the sake of simplicity I shall limit my focus in this chapter to CO_2 emissions without assuming that CO_2 is the only important greenhouse gas. The common atmospheric property rights view has received little careful scrutiny in the emerging philosophical debate on climate change and justice.[2] I hope to contribute to remedying that.

In the next section, I discuss the relationship between the common atmospheric property rights view and three different traditions of liberalism that take rights seriously. I argue that the view is plausibly entailed by two versions of Lockean (libertarian) liberalism. Common property rights in the atmosphere have a weaker foundation in egalitarian liberalism despite the endorsement of an equal entitlement to emit CO_2 by some egalitarian liberals. I then argue

[1] A distant ancestor of this chapter was written under contract with the Robert Schalkenback Foundation (RSF). I would like to thank the Robert Schalkenback Foundation for permitting me to publish a substantially revised descendant of a distant ancestor written under contract to them. Thanks also to Denis G. Arnold, Luc Bovens, William Pierce and Lea Ypi for comments along the way. Early versions were presented to the Society of Philosophy and Public Affairs in conjunction with a meeting of the Pacific Division of the American Philosophical Association, at the Association for Practical and Professional Ethics and to the project Justitia Amplificata, Johann Goethe Universität. Thanks to the organizers and participants of these events for the opportunity to discuss the paper.
[2] The Centre for Science and the Environment (see www.cseindia.org/) and the Global Commons Institute (see www.gci.org.uk/) support an equal entitlement to emit greenhouse gases. The view is defended in Anil Agarwal and Sunita Narain, *Global Warming in an Unequal World: The Case for Environmental Colonialism* (New Delhi: Centre for Science and the Environment, 1991) and in Tom Athanasiou and Paul Baer, *Dead Heat: Global Justice and Global Warming* (New York: Seven Stories Press, 2002). The philosophical attention that the view has received has been relatively scant. But see, for example, Peter Singer, *One World: The Ethics of Globalization* (New Haven, CT: Yale University Press, 2002), 27–34; Martino Traxler, "Fair Chore Division for Climate Change," *Social Theory and Practice* 28 (2002): 121; and Stephen M. Gardiner, "Ethics and Global Climate Change," *Ethics* 114 (2004): 578–583. These discussions focus nearly entirely on the implications of common atmospheric ownership for matters of global justice, not intergenerational justice.

that the common atmospheric property rights claim is *prima facie* consistent with egalitarian liberalism.

The second section takes up CO_2 emissions and intergenerational justice. It argues that avoiding intergenerational injustice, as understood according to the common atmospheric property rights view, requires either institutions that enforce emissions limits at a rate which is less than what the climate system can absorb without creating perturbations in the climate system or institutions that make appropriate compensatory intergenerational transfers for the failure to limit emissions sufficiently. Section 2 also contains refutations of three arguments that the rights of future persons are incapable of supporting a duty of present persons.

In the third section I discuss the implications of the common atmospheric property rights view for global justice. I argue that it requires an equal per capita emissions entitlement. I discuss several objections to this entitlement, including those raised by Simon Caney in Chapter 4. I argue that the most important objection to equal emissions entitlements is that the principle of equal per capita emissions is inconsistent with the morally important aim of human development.

In the final section I argue that, despite the *prima facie* considerations to the contrary canvassed in Section 1, the common atmospheric property rights view is incompatible with a plausible interpretation of what egalitarian liberalism requires by way of both global and intergenerational justice. Assessment of the merits of the common atmospheric rights view then rests on matters of fundamental political philosophical principle. I close by arguing that to the extent that the two libertarian versions of liberalism are committed to the common atmospheric property rights view, they are committed, as a matter of intergenerational justice, to climate change mitigation that is significantly more demanding than liberal egalitarianism is.

The view that emerges from this chapter is that philosophical disagreement over common atmospheric ownership and an equal per capita CO_2 emissions entitlement runs deep. I cannot hope to adjudicate the deep disagreement within liberalism here. However, the inconsistency of the equal per capita CO_2 emissions entitlement claim with the aim of human development and the especially demanding mitigation requirements of the common atmospheric rights view are strong reasons against common atmospheric ownership.

I COMMON ATMOSPHERIC PROPERTY RIGHTS

Here I consider the relationship between the claim that persons have an equal share in the common property right to the atmosphere and three different traditions of liberal political philosophy that take rights seriously: the orthodox interpretation of the Lockean tradition, which conditionally permits private appropriation of natural resources by assigning individuals a natural right to private property only if a proviso is satisfied; left libertarianism, which denies the justice of private appropriations of natural resources; and egalitarian liberalism, which typically abjures claims of natural property rights but which requires that social and political institutions not create inequality on the basis of morally arbitrary differences between persons.

Orthodox Lockean accounts of the justified private appropriation of natural resources typically assume that original ownership of natural resources vests in humanity. In John Locke's words, "God ... hath given the World to Men in common."[3] These accounts allow that individuals have a natural private property right to the part of the natural world that they have appropriated only if they have satisfied certain conditions. Although the details vary, typically these conditions require that similar value is available for others to appropriate privately. For Locke this takes the form of the well-known proviso that there be "enough, and as good left for others."[4] According to Robert Nozick's variant, the proviso requires that "the position of others no longer at liberty to use the thing ... not be worsened."[5] Even if differently demanding, provisos such as these have a comparable function. They constrain private appropriation of the originally commonly held resource to circumstances in which similar value is available to others.

One of the many things that the atmosphere is good for is the disposal of CO_2 emitted from energy production and use. Polluting activity in the form of emitting CO_2 is an appropriation of atmospheric capacity to absorb CO_2. This capacity is part of the natural carbon cycle in which carbon is removed from stores on or below the Earth's surface through atmospheric absorption, but eventually recycled to the Earth's surface into the oceans and plants. Because of the greenhouse properties of CO_2, the carbon cycle plays a vital role in maintaining climate equilibrium. When humans sufficiently accelerate the removal of carbon from below the Earth's surface by means of fossil fuel burning, or when they decrease the rate of recycling back to the Earth through deforestation, climate perturbations occur.

Equal shares in a common atmospheric property right may permit persons limited use of the atmosphere's absorptive capacity for the disposal of CO_2. As long as one does not emit more than one's share of the threshold

---

[3] John Locke, *Two Treatises of Government*, bk. II (Cambridge University Press, 1970), 304.
[4] Ibid., 309.   [5] Robert Nozick, *Anarchy, State and Utopia* (New York: Basic Books, 1974), 178.

beyond which climate perturbations arise, neither Locke's nor Nozick's version of the proviso seems to be violated. Locke holds, "For he that leaves as much as another can make use of, does as good as take nothing at all."[6] The relative scarcity of the atmosphere's capacity to absorb CO_2 without climate perturbations establishes a limit on justified private appropriation by means of CO_2 emissions. If a person emits more than the amount that every other can emit without climate perturbations, the proviso has been violated since this leaves smaller shares for everyone else unless climatic perturbations are to occur. According to these Lockean accounts, emissions beyond one's share of the threshold violate the natural property rights of humanity, whose members own the atmosphere in common.[7]

This view might be challenged as being out of keeping with the orthodox Lockean tradition on several grounds. First, perhaps the reasoning involves a misapplication of the proviso since polluting the air is dissimilar to enclosing a piece of land insofar as the former does not exclude access to an identifiable, particular piece or quantity of property. Although it is the case that polluting does not deny access to a particular piece of property, polluting in excess of one's share of the capacity of the atmosphere to absorb CO_2 without creating climate perturbations and enclosing land that denies equal value for others are nonetheless relevantly similar insofar as both lessen the value of the remaining resources that others might appropriate equally. There are differences between the two actions. Fences can be easily torn down but, due to the long residence of CO_2 in the atmosphere, emitting CO_2 in excess of one's share of the atmosphere's capacity to absorb diminishes the remaining resource for users long into the future. Moreover, enclosing land does not necessarily create harms other than the loss of the remaining value for others, but polluting can be seriously harmful to many human interests, arguably constitutive of human well-being. Neither of these differences, however, diminishes the moral force of the claim that emitting CO_2 that exceeds one's share of the threshold is a violation of the Lockean proviso. Rather, both seem to amplify the moral wrong of such emissions.

Second, one might invoke Locke's discussion of the introduction of money as compensation for the lack of access to natural resources that have been privatized. Since CO_2 emissions facilitate economic growth, which increases the global per capita GDP, perhaps there is no injustice on Lockean grounds in emitting more than one's share. The problem with this argument is that we can have no confidence that the economic growth that CO_2 emissions facilitate will sufficiently compensate many people

---

[6] Locke, *Two Treatises*, 309.   [7] For a similar account of this, see also Singer, *One World*, 27–34.

who will be affected by the climatic consequences of the emissions. There is likely to be regional variation in the effects of climate change. The United Nations Human Development Programme's (UNDP) 2007–2008 *Human Development Report* warns that climate change induced reversals in human development in many parts of the world.[8] Droughts and heatwaves seem likely to threaten the revenues and food security of dry land farmers in sub-Saharan Africa. Accelerated glacial melting in the Himalayas could initially produce flooding in northern China, India, and Pakistan, but eventually reduce the flow of water available to major river systems for irrigation. Similar melting in Latin America could threaten water supplies for drinking, irrigation, and hydroelectricity generation. Rising sea levels might inundate low-lying areas of Bangladesh, Egypt, and Vietnam. Third, one might argue that the Lockean tradition allows for a variety of property rights encroachments such as easement, adverse possession, and the doctrine of laches. Perhaps these are grounds upon which CO_2 emissions in excess of one's share of the climatic system's capacity to absorb without creating climate perturbations can be justified even if they restrict the property rights of others.[9] There are, however, two problems with appealing to any of these doctrines in the present case. The first is the great harm to the poor and vulnerable – as summarized above – that reliance on such doctrines would seem to justify, which harm runs foul of typical proviso limitations. The second is that these doctrines apply, in any case, only after justified private appropriation; they expand or limit individual property claims within a regime of private property. They are not principles used to justify private appropriation of a shared resource *per se*.

Thus, the claim that each person has an equal share of a common property right to the Earth's atmosphere that is violated by emitting CO_2 in excess of one's share as constrained by avoiding climate perturbations seems entailed by the general normative commitments regarding property rights of the orthodox Lockean tradition.

Left libertarian versions of the Lockean tradition typically hold that private appropriation of natural resources violates the natural rights of persons to common ownership. One way to understand this is to take left libertarians as differing from orthodox Lockeans on the proper interpretation of the moral demands of the Lockean proviso.[10] Left libertarians

---

[8] United Nations Development Programme, *Human Development Report 2007–2008*, 27–30.
[9] Thanks to Luc Bovens for making this point to me.
[10] For a more general rejection of the orthodox Lockean proviso in favor of an egalitarian one, see Michael Otsuka, *Libertarianism Without Inequality* (Oxford University Press, 2003), 22–29.

following Henry George argue that the strictures of the proviso require that all of humanity be compensated for the exclusive use of land or natural resources generally, although not for the value added by the labor that improves the natural resources."[11] The compensation should be paid in the form of a tax on the value of the resource. Hence, humanity maintains a property right against the occupiers of land, which right entitles humanity to the value of the natural resource. Assuming this background of commitments regarding ownership of natural resources in general, the particular claim that each person is co-owner of the Earth's atmosphere seems to follow. Those who emit in excess of their share of the allowed total, as defined by the constraint against climate perturbations, are privately appropriating a resource – the atmosphere's absorptive capacity – and they owe a debt of compensation to the rightful owners.

Egalitarian liberal political philosophy does not, as a rule, include a commitment to natural property rights. John Rawls's view on property rights is illustrative. He rejects a basic right to private property in natural resources on grounds that it is "not necessary for the adequate development and full exercise of the moral powers [of equal citizens], and so . . . [is] not an essential social basis of self-respect."[12] Whether individuals can have a private ownership right to the means of production, including natural resources, or whether society is the rightful owner does not depend on natural rights to property, but is contingent on what, given the circumstances, would best serve the development and exercise of the moral powers of persons. So, a common property right to the Earth's atmosphere is not blocked by a prior commitment to natural rights to private property.

Why might common property rights in the Earth's atmosphere seem plausible on egalitarian liberal grounds? A positive case for common atmospheric property rights can perhaps be constructed from a concern to neutralize the influence of morally arbitrary conditions and circumstances on persons. Just as many egalitarian liberals judge that institutions should not allow for a person's race, ethnicity, sex, gender, sexual orientation, religious commitments (or lack thereof), and parental wealth to be sources of significant social advantage, in accounts of global justice, some egalitarian liberals have argued that the relationship between a state and its natural resource base is morally arbitrary. This claim is sometimes used in an argument that the role of just global institutions is to constrain the impact of these arbitrary differences by globally redistributing wealth based on

---

[11] See Henry George, *Progress and Poverty* (New York: Robert Schalkenbach Foundation, 2003).
[12] John Rawls, *Justice as Fairness: A Restatement* (Cambridge, MA: Harvard University Press, 2001), 114.

natural resource control and use.[13] This amounts to a non-natural rights basis for the claim that all persons have a share in the common ownership of the natural resources of the Earth. Since the Earth's atmosphere is a natural resource, which has the capacity to absorb CO_2, the general claim to property rights to natural resources is applicable to the atmosphere.[14]

The common atmospheric property rights view contains at least *prima facie* plausibility because there are *pro tanto* arguments for it from within three well-respected liberal accounts of justice. I shall next consider an application of the common atmospheric rights approach to intergenerational justice. A proper account of the intergenerational demands on climate change policy – as I hope will become clear presently – is central to any full account of the justice of climate change.

2 INTERGENERATIONAL JUSTICE AND EMISSIONS

Intergenerational justice looms large in discussions of climate change largely because of two important time lags in the climate system. One is the lag between reducing CO_2 emissions and stabilizing CO_2 concentrations. About 20 percent of the CO_2 emitted remains in the atmosphere for millennia.[15] This results in a long delay between reducing emissions and reducing concentrations. The other is between stabilizing CO_2 concentrations and stabilizing the climate system. Warming could continue for 1,000 years after emissions cease totally.[16] Sea levels are expected to continue to rise into the twenty-third century.[17] Hence, people who are not yet born, indeed who will not be born for a hundred years or more, will experience some of the most important effects of our current energy policies.

How it will affect other generations [handwritten marginal note]

[13] See, for example, Charles R. Beitz, *Political Theory and International Relations* (Princeton University Press, 1979), 136–143; and Brian Barry, "Humanity and Justice in Global Perspective," in *Nomos XXIV, Ethics, Economics, and the Law*, edited by J. Roland Penncock and John W. Chapman (New York: New York University Press, 1982), 219–252.

[14] See also Brian Barry, *Why Social Justice Matters* (London: Polity Press, 2005), 264; and Steve Vanderheiden, *Atmospheric Justice: A Political Theory of Climate Change* (New York: Oxford University Press, 2008), 99–109.

[15] G. A. Meehl *et al.*, "Global Climate Projections," in S. Solomon *et al.*, eds., *Climate Change 2007: The Physical Science Basis. Contribution of Working Group I to the Fourth Assessment Report of the Intergovernmental Panel on Climate Change* (New York: Cambridge University Press, 2007), 824. Available online at www.ipcc.ch/pdf/assessment-report/ar4/wg1/ar4-wg1-chapter10.pdf.

[16] Susan Solomon *et al.*, "Irreversible Climate Change Due to Carbon Dioxide Emissions," *Proceedings of the National Academy of Sciences of the United States of America* 106 (2009): 1704–1709. Available online at www.pnas.org/content/early/2009/01/28/0812721106.full.pdf+html.

[17] Intergovernmental Panel on Climate Change, "Summary for Policymakers," in S. Solomon et al., eds., *Climate Change 2007*, 17. Available online at www.ipcc.ch/pdf/assessment-report/ar4/wg1/ar4-wg1-spm.pdf.

The threshold for a generation's CO_2 emissions according to the common atmospheric property rights view is collective emissions not greater than those which would result in CO_2 concentrations that would create significant climate perturbations for subsequent generations. Staying below that threshold ensures that there is no loss of value for future persons as the result of the depletion of the common resource. According to this view, should there be such a loss of value, members of the earlier generation are exceeding their share of the common resource. This would be analogous to a community using water from a communally owned spring-fed lake at a rate faster than is replenished by the spring. Such consumption takes from subsequent users and depletes the intergenerational commons.

Emitting beyond the threshold violates the property rights of future persons and produces both direct and indirect costs. The direct costs are those that must be borne to adapt to climate change. The indirect costs are those necessary to compensate future persons for the lower threshold of emissions that they will be required to conform to in order not to create further future perturbations. It would in principle be permissible to emit above the threshold if the emitters also passed on savings that fully compensated members of future generations for their direct and indirect costs. Compensatory revenue could be generated either by a public sale of entitlements for emissions that exceed the threshold of climate perturbations or by a scheme of taxing emissions that exceed the threshold. In either case, a public fund is established to compensate future persons for their costs. There might be good reasons to prefer one or the other of these institutional arrangements, but both seem able to satisfy the demand of compensation required by intergenerational justice.

A general feature of the conclusion drawn from this section is that the view that persons have equal shares of a common atmospheric property rights places members of earlier generations under duties to persons comprising future generations because of the property rights of the latter. Several philosophical arguments challenge the coherence of the claim that the rights of future persons can entail duties for present persons. These must be addressed if the atmospheric rights view is to maintain plausibility.

One argument holds that the rights of future persons cannot place present persons under duties because future persons do not yet exist and persons cannot have rights if they don't exist.[18] We do not, however, need to assume that future persons have rights now in order to maintain that we have duties

---

[18] See Wilfred Beckerman and Joanna Pasek, *Justice, Posterity, and the Environment* (Oxford University Press, 2001), 15–23.

to them. We may have duties not to knowingly bring into existence relationships that entail duties that we cannot fulfill. For example, imagine you are considering marriage; in particular you are considering assuming the duties that spouses have in virtue of being spouses. You are, *ex hypothesis*, not yet married. You do not have spousal duties to your possible future spouse. But it does not follow that you have no duties in virtue of the possible future spousal relation. If you were to marry intending not to fulfill your spousal duties, you would be acting wrongly even though you have not violated your spousal duties. More importantly, if you now plan the marriage with the same intention you are also acting wrongly. The moral duties that you would have in the future constrain what you may do now.

This marriage example illustrates that one may act wrongly by acting in ways that will bring about relations in the future that generate moral duties that one intends not to fulfill. This idea can be applied to future generations. If we knowingly act in ways that bring it about that persons come to exist who are owed duties by us that cannot be fulfilled because of our actions, then we are acting wrongly. So, it is not necessarily incoherent to claim that we have a present duty not to act in ways that will knowingly bring about situations in which the rights of persons will not or cannot be respected.[19]

A second argument derives from what Derek Parfit calls *the non-identity problem*.[20] The problem arises on the very plausible assumption that any energy policy we now pursue will play an important role in affecting who will exist several generations hence. This is because of the pervasive effects of our energy policy on our lives. Couples might or might not meet depending upon the transportation that they use to go to work and social events; particular evenings might be more or less romantic depending upon the room temperature; climatic disruptions will cause people to flee or to move in pursuit of economic opportunities, and thus to meet people whom they would not have met if they had not moved; and so on. This fact about energy policy coupled with the moral claim that our action harms a person only if it renders the person worse off than she would have been had we not acted produces the non-identity problem. For, if a person would not have lived if we had pursued an alternative energy policy, our policy makes the person worse off than the alternative policy would have only in the extreme case in which it would have been better for the person not to have existed at all. Short of the person's living conditions being that extremely

---

[19] This idea is developed by James Woodward, "The Non-Identity Problem," *Ethics* 96 (1986): 804–831.
[20] Cf. Derek Parfit, *Reasons and Persons* (Oxford University Press, 1987), ch. 16.

bad, our policy does not harm the person. The energy policy cannot then be condemned on the moral ground that it harms persons who will exist in the future.

The common atmospheric property rights view is able to bypass this problem since it is concerned with the moral wrong of violating the property rights of future generations, not harming them. Generally, these are conceptually distinct matters. This is illustrated by James Woodward's example of a person denied a plane ticket on racist grounds, only to be thereby unwittingly spared from a fatal plane crash.[21] The person's right to purchase a ticket has been violated, but the person has not been rendered worse off as a result. If one wants to insist that violating a person's rights to common ownership of the atmosphere is harming her, this can be accepted but only insofar as harm is reconceived as not being limited to rendering a person worse off. In that case, however, since the claim that a person's property rights have been violated does not entail the claim that she has been made worse off, the former claim is not subject to the non-identity problem.

Parfit, however, contends that a focus on the rights of future persons does not provide a viable way around the non-identity problem. Future persons alive in part due to an earlier generation's CO_2 emissions, which have created climate perturbations, could simply choose to waive some of their rights.[22] In the case of the rights view under discussion, the waiver might cover their rights to common ownership of the atmosphere. Presumably, future persons might consider doing so after realizing that they would not have existed if the earlier generation had significantly different energy practices. By so waiving their rights, the actions of members of the earlier generation are rendered permissible, just as a defendant who waives his right to trial by pleading guilty is not wronged by not being tried. In both cases it is not merely that the persons do not complain. By waiving their rights they have no moral grounds for complaint.

However, to argue that appealing to the rights of future persons will not avoid the non-identity problem, it is not enough to claim that future persons *might* waive their rights since, of course, they *might not*. In that case, a just energy policy, like a just trial policy, would have to conform to requirements ensuring that those who did not waive their rights did not have their rights violated. Parfit's argument requires more. It must rely on there being some sort of irrationality on the part of future persons who do not waive their rights. It would have to be the case that the fact they would not be alive but for our energy policies (assuming the conditions of life are

---

[21] Woodward, "The Non-Identity Problem," 810–811. [22] Parfit, *Reasons and Persons*, 364.

such that their lives are worth living) rationally commits them to waiving their rights. Since rationally they must waive their rights, there were no rights violations by the energy policy that contributed to their existence but permitted CO_2 emissions in excess of a person's equal per capita share. According to this line of argument, then, one cannot appeal to rights to circumvent the non-identity problem. The equal share of common atmospheric property rights view, then, is rejected not on normative grounds, but on grounds that it simply makes no sense to invoke the rights of future persons.

It is implausible, however, that the only rational course is to waive one's rights in the case discussed above. It is possible that a person has not been rendered worse off as the result of a rights violation. When a person has not been rendered worse off as the result of a rights violation, the claim that it would be irrational for her not to waive her rights follows only if it would be irrational of her to press a complaint about the rights violation. Consider again Woodward's example of the racist refusal to sell a ticket. The waiver in that case would be retrospective since the customer was not asked upon purchase to waive her right to purchase the ticket. But unlike a waiver in advance when given a choice, it would not be irrational for the customer to maintain later that her rights had been violated, even though she owes her existence to the violation. Likewise in the energy policy case. Even if we assume that existence is a great good, it is implausible that a person would *necessarily* be irrational not to waive her otherwise violated right because the action that apparently violated it also brought her into existence. So, the common atmospheric rights view, then, is not undermined by the possibility that in the future people might choose to waive their rights to an equal share of the atmosphere. And Parfit's argument against invoking rights is not convincing.

A third argument, also deriving from Parfit, contends that we cannot violate the rights of persons who would not exist but for the actions that allegedly violate their rights since in that case it is impossible to fulfill their rights.[23] If rights cannot be fulfilled, they cannot be violated. The strength of the inference rests on its apparent invocation of the familiar principle, ought-implies-can, or more precisely its contrapositive, cannot-implies-no-duty. But this principle does not in fact apply in this case and the inference is invalid. Consider a case in which we have a duty not to violate a person's right, but the right can be violated only if we bring it about that the person exists. If we do not bring it about that the person exists, we have not violated his right and therefore have not done what we ought not to have done. In this case, we do wrong by bringing the person into existence, but do no wrong by not bringing him into existence. This applies directly to the common atmospheric rights claim. The claim holds that we ought to avoid violating the property rights of possible future persons by not adopting certain energy policies, which will bring it about that people exist who do not have an equal share of the atmosphere.[24] So, the argument that the property rights of future persons cannot place present persons under an obligation is unconvincing.

3 GLOBAL JUSTICE AND EMISSIONS

In this section I discuss the implications of common atmospheric rights for global justice by considering how duties of intergenerational justice to future generations should be distributed intragenerationally. This is a matter of global justice since emissions can originate in any geographic location and diffuse uniformly in the atmosphere. The emitters and those affected by the emissions can live halfway around the world from each other. Hence, adequately addressing the matter requires an effective global regulatory framework.

makes
assumption

I begin by assuming that there is a limit on overall level of emissions required by intergenerational justice. Since the hypothesis under discussion takes all persons to share in a common atmospheric property right, the entitlement to emit should be equal. The entitlement can be distributed on an equal per capita basis by dividing the overall intergenerationally acceptable emissions level by the global population. According to this view, equal emissions entitlements are derived from the morally prior common atmospheric property rights; equal emissions entitlements are not then morally basic rights.

Climate change negotiations under the United Nations Framework Convention on Climate Change (UNFCCC) treaty involve states as parties to the treaty. So, as a matter of realism, advocates of the common atmospheric property rights view might seek application of the view through the mediation of state institutions. Such an application would result in each country being assigned an overall emissions entitlement in proportion to its population as indexed to the same particular year. As other writers have observed, making the entitlement a function of the population in a year not too far in the future avoids providing states with an incentive to develop a population policy that encourages population growth in order to increase

---

[24] See also Woodward, "The Non-Identity Problem," 815.

their emissions entitlements.[25] This international scheme of emissions entitlements differs from the scheme that the Kyoto Protocol establishes in two important ways. First, the assignments granted to states are in conformity with the underlying principle of the common atmospheric property rights of all persons and are not the result of the capacity of the state to elicit concessions in a negotiation process.[26] Thus, in addition to conforming to the demands of equal rights, the scheme also possesses the virtue of transparency. Second, the system of entitlement is fully comprehensive and is not limited only to historical gross emitters, in particular not only to the states that industrialized first.

However, equalizing emissions entitlements need not necessarily work through state mediation. Individuals could be given emissions accounts. Some might favor this approach for moral reasons along the lines that Caney cites in Chapter 4 of this book: It would reduce the incentives to corruption that are produced by what Thomas Pogge calls "the resource privilege."[27] If the entitlement to emit can be liquidated by state leaders on an international market, this might encourage corrupt ascension to power in the pursuit of wealth. Establishing individual emissions accounts would also prevent state leaders from effecting a maldistribution of the entitlements internally.[28] These moral considerations favor individual assignment, but political realism suggests that negotiators representing the interests of states are likely to ensure that states are assigned the entitlement to emit. Regardless of whether the equal emissions entitlement is assigned to states on a per capita basis or to individuals, the overall initial assignment of emissions permitted within a particular state is not affected.

The justification of an equal entitlement to emit on grounds of a common atmospheric property right is importantly different from basing it on other important human interests in, say, keeping warm, cooking, traveling, communicating, and engaging in recreational activities. Caney convincingly argues that basing equal entitlements on such important human interests undermines the importance of equalizing emissions entitlements in isolation

---

[25] See, for example, Dale Jamieson, "Climate Change and Global Environmental Justice," in Clark A. Miller and Paul N. Edwards, eds., *Changing the Atmosphere: Expert Knowledge and Environmental Governance* (Cambridge University Press, 2001), 287–307; and Singer, *One World*, 36.

[26] Under the Kyoto Protocol, Australia is allowed an 8 percent increase above its 1990 emissions, whereas the EU is required to reduce by 8 percent. Figures available at the United Nations Framework Convention on Climate Change web page: http://unfccc.int/kyoto_protocol/background/items/3145.php.

[27] See Thomas Pogge, *Global Poverty and Human Rights* (London: Polity Press, 2002), 113.

[28] Simon Caney canvasses reasons such as these in Chapter 4 of this volume, "Climate Change, Energy Rights, and Equality," pp. 77–103.

from the distribution of the enjoyment of these other interests.[29] But if the entitlement to emit has its basis in morally prior atmospheric property rights, then its distribution can be considered independently of these other interests. Based on common atmospheric property rights, the distribution of the entitlement to emit need not be at all sensitive to the distribution of the human interests canvassed above.

Caney presses three other important objections against an equal entitle-ment to emit for all persons: Because the entitlement is a version of resourcism – an account of justice that takes distributive justice to concern the distribution of resources – it fetishizes the resource distributed; it is needs insensitive; and it is implausibly indifferent to past emissions.[30] I do not have the space to give these criticisms the hearing that they deserve, but I suspect that advocates (or at least some of them) of the version of equal entitlements to emit that I discuss here can construct plausible responses along the following lines. The charge of resource fetishism does not seem applicable. The present defense of equal entitlements to emit does not assume the independent value of a resource. Rather, it is based on the common property right in the atmosphere. It is not a version of resourcism, but a rights-based view. The charge of needs insensitivity seems correct, but its critical force will depend upon the extent to which needs should trump rights. Many egalitarian liberals will find the objection pressed by Caney relevant, but it is not likely to disturb libertarians who think that the justice of distributions depends on whether they respect property rights, not whether they meet needs. With respect to the third objection, historical emissions in the now developed world could possibly affect responsibility for financing compliance with equal entitlements to emit and thereby intro-duce sensitivity to the past. The idea would be that insofar as the relative wealth enjoyed by persons in the industrialized world is due to high past emissions that exceeded the intergenerational threshold, even if present persons in the industrialized world are not at fault for this past injustice they are not entitled to the wealth that it produced either. Some of that wealth might be used to subsidize the costs of conforming to an equal emissions entitlement regime in the developing world.

There is another, more serious weakness in the global application of an equal entitlement to emit CO_2. This is based on the constraints that an equal emissions entitlement puts on the ambitions of developing and underdeveloped countries to pursue human development. This pursuit is recognized in the right to sustainable development in Article 3, paragraph 4

---

[29] Ibid.   [30] Ibid.

of the UNFCCC.[31] I have discussed this problem at length elsewhere; so I only summarize it here.[32]

There are two main factors that, in combination, are likely to reduce the equal per capita emissions entitlement below what is necessary for states to mount an effective economic development policy. One is the extent of global emissions reductions required to satisfy the demands of intergenerational justice; the other is population growth. The common atmospheric property rights view requires that members of an earlier generation emit CO_2 below a threshold that would cause significant and uncompensated climate perturbations. What precisely this would require is unclear. But consider the widely held view that warming should be kept below 2°C, a view adopted by in the Copenhagen Accord. The Intergovernmental Panel on Climate Changes (IPCC) projects significant climate perturbations even below 2°C. A 1°C temperature increase is projected to put up to 30 percent of existing animal and plant life at increased risk of extinction and to result in increased coral bleaching. A 2°C increase is expected to increase damage from floods and storms and to produce extensive bleaching of most coral reefs.[33] So, the 2°C limit is certainly too high for a standard that only seeks to prevent climate perturbations. But perhaps these could all be compensated for so that the 2°C threshold would satisfy the requirement not to create uncompensated perturbations.[34] Let's examine the implications for development on the charitable assumption that 2°C warming is consistent with the demands of intergenerational justice according to the common atmospheric property rights view.

According to the IPCC, keeping warming within 2°C requires that by 2050, global emissions be 50–85 percent below 2000 levels.[35] Average global per capita emissions in 2000 were 3.92 metric tons (mt) CO_2 for a population that was just over 6 billion.[36] If we assume that population growth yields a global population of 9 billion in 2050, the per capita emission entitlement of a 50–85 percent overall reduction would be in the range of

[31] The United Nations Framework Convention on Climate Change is available online at http://unfccc. int/resource/docs/convkp/conveng.pdf.

[32] See my "Treaty Norms and Climate Change," *Ethics and International Affairs* 23 (2009): 247–265. Available online at www.cceia.org/resources/journal/23_3/features/001.

[33] Intergovernmental Panel on Climate Change, *Climate Change 2007: Synthesis Report: Summary for Policymakers*, 10. Available online at www.ipcc.ch/pdf/assessment-report/ar4/syr/ar4_syr_spm.pdf.

[34] The compensation would have to be based upon *ex ante* calculations of costs, and this presents problems due to uncertainty about sea-level rise due to uncertainty about ice-sheet collapse.

[35] IPCC, *Climate Change 2007*, 20–21.

[36] See the United States Energy Information Administration web page, www.eia.doe.gov/.

0.4–1.33 mt CO_2. We do not know exactly how much CO_2 must be emitted to achieve human development but there is very good reason to believe that it is much higher than this range. Of the countries listed in the category of having very high human development by UNDP's 2009 Human Development Index (HDI), the country with the lowest per capita emissions is Portugal, whose emissions in 2008 were 5.4 mt CO_2.[37] Of the 91 countries in the top half of the HDI only two – Albania and Peru – have per capita emissions within the 0.4–1.33 mt CO_2 range. Hence, we have very good reasons to believe that utilizing existing energy technology will require emissions to be significantly higher than this range. So, requiring developing countries to conform to an equal per capita emissions regime within the overall emissions requirements set by the common atmospheric rights approach to intergenerational justice would likely be inconsistent with permitting human development. The evidence above suggests that allowing underdeveloped and developing countries to make significant progress in human development requires allowing them to emit more than would be allowed on an equal per capita emissions scheme consistent with the restrictions of intergenerational justice. If the demands of intergenerational justice are to be met – that is, if additional climate perturbations are not to occur – the highly developed countries (which happen to have more resources to devote to renewable energy) will have to emit less than their equal per capita allotments to compensate for the increased emissions in underdeveloped and developing countries.

A trading scheme could provide some relief from the constraints on human development that an equal per capita emission regime would impose. Through the purchase of additional emissions entitlements, states, or their populations, could increase the volume of allowed emissions. But this would raise the cost of human development and could result in delayed or forestalled human development. Use of renewable energy in developing and underdeveloped states would allow for the pursuit of human development that is less reliant on increasing CO_2 emissions. But over the medium term, at least, such energy sources will be more costly than fossil fuels. So, a distribution of emissions entitlements that requires poor countries to use such technology in order to keep within their emission limits will have development-retarding effects. Given UNFCCC's recognition of the right to sustainable development, the legitimacy of an equal emissions entitlements

[37] 2009 Human Development Index in United Nations Human Development Programme, *Human Development 2009: Overcoming Barriers: Human Mobility and Development* (Basingstoke: Palgrave Macmillan, 2009), 171–174.

regime in conformity with the intergenerational demands of equal atmospheric ownership is then doubtful. Although the discrepancy between the emissions that human development seems to require and those that an equal per capita emissions regime in conformity with intergenerational justice will allow poor countries need not affect the libertarian endorsement of the regime, it saddles libertarianism with a liability that might weaken its general attractiveness. The threat to human development posed by the regime, however, provides liberal egalitarians with a good reason to question the compatibility of common atmospheric property rights with their basic commitment to equality.

4 EGALITARIAN LIBERALISM AND COMMON ATMOSPHERIC OWNERSHIP RECONSIDERED

A commitment to common atmospheric property rights seems to require significant climate change mitigation due the demands of intergenerational justice. Insofar as libertarianism (in both its right- or left-wing versions) entails the commitment to equal atmospheric property rights, libertarians ought to be committed to significant climate change mitigation. This is a surprising result because the political expression of libertarianism – at least in the USA – does not typically include strong support for climate mitigation proposals.[38] The reasons for this discrepancy could be manifold, but one may be a general suspicion that regulating the market infringes property rights. In the case of climate change such suspicions could lead libertarians astray. Libertarians might have good reasons to favor free markets in goods that may be privately owned, but if the atmosphere is not such a good (for the reasons discussed in Section 2), then the reasons favoring market distributions do not obtain. Unless the market is regulated on behalf of persons comprising later generations, the common natural right in the atmosphere will be undermined.

For egalitarian liberals the plausibility of the common atmospheric property rights view depends in large part on whether a scheme of common property rights in the atmosphere achieves the sort of distributive equality that is most important. The argument of the last section that equal emission entitlements within the constraints of a 2°C warming limit would probably frustrate human development is reason to believe that on grounds of global justice common atmospheric property rights should be rejected by

---

[38] I could find no document on the Cato Institute's web page that supports significantly mitigating climate change. See www.cato.org/.

egalitarian liberals as the wrong kind of equality. Since egalitarian liberals
have no basic commitment to equal natural property rights, and to the
extent that egalitarian liberals are committed to achieving social conditions
in which inequalities in wealth, resources, well-being, capabilities, or opportunities are appropriately constrained, an account of the distribution of
entitlements to emit CO_2 that would consign billions of people to abject
poverty is objectionable on grounds of global distributive justice.[39]

Less obvious are the egalitarian liberal reasons to doubt common atmospheric property rights on grounds of intergenerational justice. I assumed in
Section 2 that one reason to endorse common atmospheric property rights
on egalitarian liberal grounds derives from the morally arbitrary character of
one's generational membership. One need not take the point of justice to be
the correction of natural arbitrary differences, as some luck egalitarians do,
to be drawn to this line of reasoning.[40] One might take the less Promethean
view that it is unjust for social institutions to endow arbitrary natural
differences with moral significance.[41] But if egalitarian liberals generally
have reason to reject social inequalities deriving from arbitrary natural
differences, they need not necessarily be committed to the approach of
atmospheric property rights as the best way to mitigate such inequalities.
Liberal egalitarians are little concerned about property rights for their own
sake, but typically only in relation to how systems of property rights secure
basic liberties, appropriately constrain social inequalities, and achieve efficiency in production. So, even if one's morally arbitrary generation of birth
affects one's access to use the atmosphere, this is not necessarily morally
relevant. It becomes relevant only with respect to its relationship to how
institutions secure liberty, social equality, and efficiency.

Much of the talk in policy circles about the long-term effects of climate
change concerns its costs and the extent to which these costs will be passed
on to future persons, and especially the extent to which the costs will be
heaped on poor people. A focus on common atmospheric property rights
is not directly relevant to these concerns. I have argued elsewhere for a
version of intergenerational egalitarianism that directly captures concern
about costs and should therefore be more attractive on egalitarian liberal
grounds.[42] The idea is that impartial regard for persons across generations

[39] See my *Global Inequality Matters* (Basingstoke: Palgrave Macmillan, 2009). See especially chs. 1–4.
[40] An important early source of luck egalitarianism is Ronald Dworkin, *The Sovereign Virtue* (Cambridge, MA: Harvard University Press, 2000), ch. 2.
[41] See John Rawls, *A Theory of Justice*, rev. edn. (Cambridge, MA: Harvard University Press, 1999), 87.
[42] See my "Justice and the Intergenerational Assignment of the Costs of Climate Change," *Journal of Social Philosophy* 40 (2009): 204–224.

requires that the percentage of climate-change-related costs of energy production and use be equally distributed across generations. Roughly, *ceritus paribus*, as a percentage of the GDP, the per capita mitigation costs of an earlier generation should equal the per capita adaptation costs of a later generation. Since I have argued for this view at length elsewhere, I will not defend it in any detail here; but I note two attractive features: It allows liberal egalitarians to engage directly with the discussions concerning the future costs of climate change and it is sensitive to the moral arbitrariness of generational membership.

So despite the *prima facie* plausibility on egalitarian liberal grounds of the common atmospheric rights approach that I argued for in Section 2, and despite the fact that some egalitarian liberal philosophers have defended equal emissions entitlements on luck egalitarian grounds, further attention to the argument provides good reasons of both global and intergenerational justice for egalitarian liberals to reject common atmospheric property rights and equal emissions entitlements. The unusual agreement between libertarians and egalitarian liberals is more apparent than real.

The mitigation requirements of equalizing atmospheric property rights are significantly more demanding than those of equalizing the proportion of intergenerational costs of CO_2 emissions. The requirement of equalizing property rights in the atmosphere limits CO_2 emissions of persons in a prior generation to levels no greater than what would create climate perturbations for later generations. In other words, the prior generation is required to assume the full (climate-change-related) costs of emitting CO_2 so that there should be no uncompensated adaptive costs for subsequent generations. The mitigation requirements of the equal proportional costs approach are fewer. Members of the prior generation should not pass on greater adaptation costs (as a percentage of the GDP) to members of the subsequent generation than the costs (as a percentage of the GDP) that members of the prior generation paid in mitigation. Costs are shared intergenerationally, rather than fully assumed by the earlier generation.

The requirement to internalize the full cost of CO_2 emissions has all the intuitive plausibility of the general requirement to internalize negative externalities. The plausibility, however, weakens upon further consideration. For the externalities passed on to future generations by a prior generation's energy production and consumption are not wholly negative. The capital investments and the availability and quality of consumption goods made possible by energy production and consumption are benefits that can also be passed on. A principle of justice that requires members of the earlier generation to assume all of the costs, even though it passes on some of the benefits, assigns considerable advantages to persons born in later generations. Insofar as egalitarians are committed to limiting the social effects of the arbitrariness of generational membership, the common atmospheric rights approach is inferior to the equal intergenerational cost-sharing approach.

I'll close by drawing a surprising inference from the thread of discussion running through several sections. I have considered the reasons that liberals of various stripes might have to support common property rights in the atmosphere. The natural property rights positions discussed in Section 2 are often subsumed under libertarianism in its right- or left-wing varieties. On the grounds of natural property rights libertarians seem committed to common atmospheric property rights. Egalitarian liberals have reasons of global and intergenerational justice to reject this commitment. It now appears that the principle of common atmospheric property rights requires an approach to mitigating climate change that is more demanding than a principle of equalizing intergenerational proportional costs, which principle egalitarian liberals have reason to prefer. Hence, libertarianism appears to be committed to greater anthropogenic climate change mitigation than egalitarian liberalism is.

CHAPTER 6

A Lockean defense of grandfathering emission rights

Luc Bovens

A core issue in the debate over what constitutes a fair response to climate change is the appropriate allocation of emission rights between the developed and the developing world. Various parties have defended equal emission rights per capita on grounds of equity. The atmosphere belongs to us all and everyone should be allocated an equal share.[1] Others have defended higher emission rights per capita for developing countries on grounds of historical accountability. Developed countries are largely responsible for the threat of climate change due to their past emissions and, since they currently continue to enjoy the benefits thereof, they should be willing to accept lower emission targets.[2]

However, in reality we see that developed countries currently have much higher emission rates per capita and will continue to have higher rates than developing countries for some time to come. There is talk of "grandfathering" – setting emission targets for developed countries in line with their present or past emission levels. What, if anything, can be said in defense of grandfathering?

In Chapter 4 of this book Caney discusses grandfathering and puts the matter very bluntly: "No moral and political philosopher (to my knowledge) defends grandfathering, presumably assuming that it is unjust."[3] Grandfathering can at best be defended by means of pragmatic arguments. In *Realpolitik* we need to make some concessions in order to get all the

---

[1] See Darrel Moellendorf, "Treaty Norms and Climate Change Mitigation," *Ethics and International Affairs* 23 (2009), 264, n. 32, and Simon Caney, "Justice and the Distribution of Greenhouse Gas Emissions," *Journal of Global Ethics* 5 (2009), 130–3, for an overview of philosophers and institutions defending this position.

[2] E.g., Henry Shue, "Global Environment and International Equity," *International Affairs* 75 (1999), 531–45, Eric Neumayer, "In Defence of Historical Accountability for Greenhouse Gas Emissions," *Ecological Economics* 33 (2000), 185–92, and Axel Gosseries, "Historical Emissions and Free-Riding," *Ethical Perspectives* 11 (2003), 36–60. For an overview of the debate, see Caney "Justice and the Distribution of Greenhouse Gas Emissions," 133–5.

[3] Caney, "Justice and the Distribution of Greenhouse Gas Emissions," 128.

parties on board. But this is like making concessions in negotiations with the Mafia. Nobody deems such concessions to be fair, but it sure beats the blood bath that may come about due to the lack of an agreement. Similarly, any agreement that would give developed countries more emission rights than developing countries would not be fair, but it sure beats the ice caps melting. This of course is not much in the way of a moral argument.

Neumayer does point in the direction of a moral argument for grand-fathering: "It is sometimes suggested in the spirit of Locke and Nozick that a long history of emission rights might have established the right for devel-oped countries to prolong current emission levels into the future and that such 'squatters' rights' can be derived from the common law doctrine of 'adverse possession' (e.g. Young and Wolf 1991 [*sic*])."[4] However, he also thinks that there is not much of a moral argument here, because "even Nozick (1974, p. 175) ... acknowledged an appropriation of property rights can only be regarded as just if 'the situation of others is not worsened', which is clearly not the case with global warming."[5]

Young and Wolf actually do not mention squatters' rights or adverse possession, but they do write that there is a theory in support of grand-fathering emission rights that "considers current emissions as a claim established by usage and custom"[6] and this is similar to existing policies of assigning fishing rights based on current catch levels. Sterner and Muller refer to a principle of "prior appropriation" (giving rights to first users) that can provide a "rights-based perspective" in support of grand-fathering emission rights.[7]

Raymond provides the most extensive discussion of a Lockean justifica-tion of grandfathering in allocating claims to common pool resources. He argues that this kind of justification is present in allocating grazing rights but absent in the allocation of emission rights for greenhouse gas (GHG) emissions. He provides a purely positive account of why this is the case, citing five reasons (without any pretense that these are *good* reasons). Compare the usage of land with the usage of the atmosphere in GHG emissions. The usage of land is *tangible* – i.e., our labor affects a token plot

---

[4] Neumayer, "In Defence of Historical Accountability for Greenhouse Gas Emissions," 188. The reference is to H. Peyton Young and Amanda Wolf, "Global Warming Negotiations: Does Fairness Matter?," *Brookings Review* 10 (Spring 1992), 46–51. (This article is indexed by EBSCOhost under H. Peyton Young and Amanda Wolfberg, "Global Warming Negotiations.")

[5] Neumayer, "In Defence of Historical Accountability for Greenhouse Gas Emissions," 188. The reference is to Robert Nozick, *Anarchy, State, and Utopia* (Oxford: Blackwell, 1974).

[6] Young and Wolf, "Global Warming Negotiations: Does Fairness Matter?," 49.

[7] Thomas Sterner and Adrian Muller, "Output and Abatement Effects of Allocation Readjustment in Permit Trade," *Climatic Change* 86 (2008), 35–6.

of land – and *beneficial,* i.e. productive.[8] In contrast, in GHG emissions, (i) we do not affect a token quadrant of the atmosphere and (ii) the emissions are just a by-product of the wealth-generating process. Furthermore, GHG emissions by developed countries (iii) have limited (if any) beneficial effects on developing countries and (iv) have long-lasting negative effects on the atmosphere. And finally, (v) inequalities in the usage of the atmospheric absorption capacities match economic inequalities in today's world.[9]

Authors who mention a moral argument for grandfathering emission rights have done so either in passing or with the aim to reject it. I will argue that we can make at least a sustained, yet qualified, moral argument in support of grandfathering emission rights on Lockean grounds. I will consider what the scope and limits are of such an argument and what place it should have in setting carbon emission targets for countries at different levels of development.

I COPENHAGEN AND EQUAL EMISSION RIGHTS

Resistance to grandfathering on egalitarian grounds was very much present in the COP15 in Copenhagen. In the early days of the COP15 in Copenhagen, a document dubbed "the Danish Text" was leaked to the *Guardian.*[10] The Danish government had prepared this text jointly with other developed countries as a discussion text, which, once leaked, incited a huge outcry among developing countries. What were the contested issues in this document?

There are two issues that concern us here. First, the Kyoto protocol required developed countries to cut emissions, but not developing countries. The Danish text, on the other hand, imposes constraints on emissions of emerging economies which would be monitored by the international community. Second, the Danish text imposes a 50 percent *global* emission reduction from 1990 levels to 2050 and an 80 percent reduction for *developed* countries from 1990 levels to 2050. These requirements, together with population level forecasts, make it possible to calculate the projected emissions per capita that are required from *developing* countries.

---

[8] Leigh Raymond, *Private Rights in Public Resources – Equity and Property Allocation in Market-Based Environmental Policy* (Washington, DC: Resources for the Future, 2003), 53.
[9] Ibid., 167–8.
[10] John Vidal, "Copenhagen Climate Summit in Disarray After 'Danish Text' Leak," *Guardian,* December 8, 2009. www.guardian.co.uk/environment/2009/dec/08/copenhagen-climate-summit-disarray-danish-text, and "Draft Copenhagen Climate Change Agreement – the Danish Text," *Guardian,* December 8, 2009. www.guardian.co.uk/environment/2009/dec/08/copenhagen-climate-change.

The *Guardian* reports that such calculations were carried out in a "confidential analysis of the text by developing countries," yielding projected emission rates for developed and developing countries at a ratio of roughly 2:1.[11] The calculations themselves are not uncontroversial. But let us bracket this issue. What concerns us here is that both (i) the imposition of emission cuts on a subset of developing countries, viz. on emerging economies, and (ii) projections of unequal emission targets by 2050, are deemed offensive and unfair by developing countries. From the point of view of the developing countries, as long as there is no convergence to equal emissions per capita, the obligation is on the side of developed countries to cut back emissions and any action by developing countries should be voluntary, since it is over and above the call of duty. And furthermore, developing countries expect that convergence be achieved much earlier than 2050.

In the end, the COP15 produced the "Copenhagen Accord."[12] With this Accord, both developed and developing countries (excluding Least Developed Countries and Small Islands Developing States) have taken on responsibility for setting emission-cut targets for 2020 – though no specific targets were actually set at the meeting. Still, the phrasing is subtly different for developed and developing countries, suggesting a dissimilar type or level of obligation[13]: developed countries "*commit* to implement mitigation actions" (sect. 4, emphasis added) whereas developing countries "*will* implement mitigation actions" (sect. 5; emphasis added). Furthermore, developed countries commit to "predictable and adequate funding" (sect. 8) to developing countries approaching $30bn per year by 2010–12 and increasing towards $100bn per year by 2020.

With the Copenhagen Accord we are moving away from the binary position that treats mitigation efforts of developing countries as voluntary, and the efforts of developed countries as obligatory. But the language still suggests that, as long as we have no convergence toward equal emissions per capita, the level of obligation on developed countries is greater than on developing countries. And in exchange for developing countries obliging themselves to undertake mitigation efforts, the developed world has to increase its level of financial support for any mitigation undertaken by the developing world.

---

[11] Vidal, "Copenhagen Climate Summit in Disarray After 'Danish Text' Leak."

[12] UNFCCC, "Draft Decision CP/.15. Proposal by the President. Copenhagen Accord," December 18, 2009. http://unfccc.int/resource/docs/2009/cop15/eng/l07.pdf.

[13] "Copenhagen Debriefing. An Analysis of COP15 for Long-Term Cooperation," *Climatico*, January 10, 2010, 22. www.climaticoanalysis.org/wp-content/uploads/2010/01/post-cop15-report52.pdf.

At the same time the Copenhagen Accord does move toward the task of setting portfolios of emission reduction targets for countries at various levels of economic development through prolonged negotiations. Alongside concerns for equality and responsibility for past pollution, a concern to respect investments will carry some weight in these negotiations. So there is an urgent need to understand the moral weight of this concern. Nothing is gained by dismissing policies that result in unequal emission rights in the near future as mere *Realpolitik* – as some form of expediency in accommodating unduly recalcitrant parties in negotiation that has no moral grounds. A proper understanding of the moral argument for grandfathering is constructive in setting emission-reduction-targets portfolios for countries at various levels of economic development that are both realistic and morally justifiable.

2 THE LOCKEAN ARGUMENT: FROM PRIVATE PROPERTY IN LAND TO EMISSION RIGHTS

Here is a popularized version of the Lockean argument[14] in defense of private property rights with respect to land. Let there be a commons that is genuinely unmanaged and unproductive. Some people decide to fence in part of the commons to work the land. Suppose that every such act of homesteading is such that some are better off and nobody is worse off, where such welfare evaluations are understood in terms of reasonable preferences. This is the Lockean enough-and-as-good condition which Nozick dubs "the Lockean Proviso."[15] Now some may decide to homestead larger plots, some smaller plots, all dependent on their needs and aspirations in life. Some people may choose not to homestead, since they would not derive any joy from such enterprise and they prefer to work for wages by selling their labor to homesteaders. But nobody is allowed to homestead a plot of land that is larger than what he or she can reasonably put to good use. That is the Lockean no-waste condition. Let us suppose that this homesteading constrained by both Lockean conditions goes on for a while. At some point it becomes clear that further homesteading would no longer satisfy the enough-and-as-good condition. The practice of homesteading is then stopped. The outcome of this procedure is that some people own smaller plots of land, some own larger plots of land, and some own no land whatsoever. But this does not make the procedure or the resulting allocation unfair.

---

[14] John Locke, "Chap. V. Of Property," in *The Second Treatise of Civil Government* (1690). www.constitution.org/jl/2ndtreat.htm.
[15] Nozick, *Anarchy, State, and Utopia*, 175.

We now extend this Lockean argument for the allocation of land to the allocation of the atmospheric absorptive capacity. Before industrialization, the atmosphere was a relatively unproductive commons. (Certainly it allowed us to breathe, but it is capable of doing so much more without interfering with our capability to breathe.) The atmosphere was capable of absorbing a certain amount of GHGs without adverse consequences, but there was, as of then, no technology emitting worrisome amounts of GHGs as by-products. Then we made technological advances – entrepreneurs came along and started using portions of this atmospheric absorptive capacity. Some used large portions, others used small portions – all depending on their needs and capacities. Initially this was done within the constraints of the Lockean enough-and-as-good and no-waste conditions: Many benefited, nobody was made worse off, and all usage was productive usage.

At some point we came to realize that the atmospheric absorption capacity was running out – any expansion beyond present usage would impose harm, violating the enough-and-as-good condition. So we closed the commons. We were not to expand beyond present usage. Just as land usage (through homesteading) established claim rights over land, usage of atmospheric absorption capacity established claim rights over atmospheric absorption capacity. Once the commons was closed, we could trade these claim rights, but we could not simply increase them by starting to use another part of the commons, be it the commons of land or the commons of atmospheric absorption capacity.

Past usage establishes differential claim rights to present and future usage of the atmospheric absorption capacity; that is, to differential claim rights to emit GHGs. Emitters can all continue to emit at their past levels. Any changes in emission rights must come through trade. People (companies, countries, etc.) certainly have unequal emission rights. But why would this be unfair while we had no objection to an allocation procedure of land that yielded unequal property rights?

As it stands, this argument is problematic if not outright laughable. And yet, while one might question whether the Lockean argument for property rights in land is the whole story, it would be hard to deny that it has at least some appeal. Why would the same argument not have any appeal for emission rights? How is it that property rights in land are so different from emission rights? Certainly there are differences, but do any of these differences provide good reason to retain the right-libertarian intuition for *property rights in land*, yet not retain it for *emission rights*?

3 GHG EMISSION RIGHTS VERSUS
PROPERTY RIGHTS IN LAND

I will consider three salient differences between the usage of land and the usage of the atmosphere and argue that none of these differences blocks my transposition of the Lockean argument to emission rights.[16]

3.1 Private goods versus common pool resources

Land and the atmosphere are resource systems. What we consume is some portion of a particular capacity of the resource system. In the case of land, we consume a portion of the produce-yielding capacity of the land. In the case of the atmosphere, we consume a portion of the absorptive capacity of the atmosphere – i.e., the capacity of the atmosphere to neutralize GHGs over time so that they do not have any detrimental effect on the climate. Now land is a *private good*; i.e., it satisfies the conditions of rivalry and excludability. As to rivalry, my consuming a portion of the produce-yielding capacity of the land subtracts from your opportunity to consume any such portion. As to excludability, barbed wire may exclude you from consuming any portion of the produce-yielding capacity of the land. Now in the case of the atmosphere, rivalry holds but not excludability. As to rivalry, relative to the constraint of global warming, my consuming a portion of the absorption capacity of the atmosphere reduces your opportunity to use any such portion. But there is no barbed wire. I cannot exclude you from setting up a business consuming additional units of absorptive capacity of the atmosphere. This, according to the orthodoxy, makes land into a private good and the atmosphere into a common pool resource.

Note that this does not say anything about how these goods should be governed. The term "private good" is deceptive in this respect. No claim has been made that such goods should be privately owned. To say that something is a private good is to say no more than that it is characterized by rivalry and excludability. To say that something is a common pool resource is to say no more than that it is characterized by rivalry and non-excludability.

Of course excludability is a matter of degree. It may be more or less difficult to exclude others from consuming. A stealth bomber reduces the capacity to set up a polluting company – once it is located, it can be

---

[16] My discussion draws on Elinor Ostrom, *Governing the Commons – the Evolution of Institutions for Collective Action* (Cambridge University Press, 1990), 1–58.

taken out. And before the invention of barbed wire it may have been more difficult to exclude people from trespassing on land. So there is a sliding scale from private goods to common pool resources. But still, land is on the side of private goods and the atmosphere on the side of common pool resources of this scale.

Does this block the analogy? Well, let us move to the best-known common pool resource, say, a lake that has a certain fish-yielding capacity. I can't stop you from putting another boat on the lake (non-excludability) but there is a threat of overfishing and exhaustion of fish stocks (rivalry). Let us see whether we can tell the same story about the lake as we told about land. The lake was originally an unproductive commons. Similar to farmers homesteading smaller or larger plots of land, fishers invested in fishing rods, trawlers, or whole fleets (depending on need and entrepreneurial spirit) and they thereby came to use different-sized portions of the fish-yielding capacity of the lake. This was done respecting the Lockean conditions – leaving enough and as good for others and making sure that every fish caught is put to good use. The fishers thereby come to acquire claim rights in these fish-yielding capacities of the lake.

How should we give shape to these claim rights? In the case of land, we do so by partitioning sections of land and assigning property rights to them. This is effective, since plots of land tend to have fixed produce-yielding capacities and it encourages good stewardship of the land. Similarly, we may assign different-sized sections of the lake to various people. But alternatively, we may let fishers roam freely over the whole lake but impose quotas on how much they are allowed to catch. These quotas are set relative to past usage, which in turn is determined by the size of their investments. Now, in the case of the atmosphere, we cannot assign segments of the atmosphere to give shape to these claim rights. The only thing that we can do is to impose quotas relative to past use determined by investments.

Why is there this difference? In the case of land, the segment of the commons that one is working roughly determines the portion of the produce-yielding capacity of the commons that one is using, assuming good stewardship. This is also somewhat the case for the lake, though less so – fish move around and last year's good spot may no longer be a good spot this year. In this case, it may be better to assign quotas rather than segments to capture these claim rights. In the case of the atmosphere, this relation is absent – the segment of the atmosphere that one is "working" does not determine the portion of the absorption capacity of the atmosphere that one is using. Pittsburgh, PA, "works" a very small segment of the atmosphere (simply by being adjacent to it), yet uses a huge portion of the atmospheric absorption capacity. So the weaker the correlation between the size of a segment in a resource system and the productive capacity of that segment, the more fitting it is to express claim rights in terms of quotas on resource usage rather than as property rights over segments of that resource system. However, this does not undercut the Lockean argument. It only means that claim rights will not be translated into property rights over segments of the resource system but rather will be expressed in terms of quotas on permissible resource usage.

3.2 Long-standing violations of the enough-and-as-good condition

So far, we have considered cases in which we were vigilant and identified the exact point at which the enough-and-as-good condition was violated. But this did not happen in the case of GHG emissions. We are long past the point at which the atmosphere could comfortably absorb GHG emissions without there being any tangible effects on the environment or on the well-being of third parties. Due initially to a lack of the requisite scientific knowledge and later to the lack of political will, appropriations of the atmospheric absorption capacity have gone far beyond what is permissible on the enough-and-as-good condition.

Was there a time, say in the early days of the Industrial Revolution, when such appropriations did pass the enough-and-as-good condition? Well certainly the first steam engine in England did little harm – nobody in Tuvalu was worse off because of *that* little puff of GHGs. And furthermore, the Industrial Revolution also benefited countries that were not themselves involved in emitting GHGs. There was a sharp drop in poverty indicators not only in industrializing countries but also in countries in which industrialization started much later.[17] I do not wish to downplay the horrors of colonialism and its connection to industrialization. For the purposes of our discussion, all that needs to be established is that, feasibly, there was at one point a period of time during which the appropriation of the atmospheric absorption capacity via industrialization satisfied the enough-and-as-good condition – i.e., a period during which negative externalities were not yet present (at least in the sense of posing a threat of climate change) and the

---

[17] E.g., consider the drop in deaths of infants under 1 year old per 1000 live deaths for select countries in Africa and Asia in the nineteenth and twentieth centuries in Brian R. Mitchell, *International Historical Statistics – Africa, Asia and Oceania, 1750–2000* (London: Macmillan, 2003), 80–83. Of course it is an open question whether these early infant death rates would have dropped in the absence of the industrialization in the West.

overall effects of the industrialization on non-industrializing nations were non-negative.

So when did this time of unproblematic appropriations of atmospheric absorption capacity end? I do not know. Note that it ended earlier than the time when we *found out* about the threat of climate change due to excessive GHG emissions. At that time, one might argue, the inaction of developed countries due to the lack of political will became culpable (as opposed to illicit but non-culpable). Before that point in time, there was no culpability, since we simply did not *know* that appropriations of the atmospheric absorption capacity were wrong on grounds of violations of the enough-and-as-good condition. But we are not interested here in when such appropriations became culpable, but rather when they became illicit, independent of what we knew to be the case. When was it the case that, from the perspective of an omniscient being, it was time to start worrying about the negative externality of the threat of climate change caused by industrialization? I do not know, but I submit that it was at a point in time when today's inequalities had roughly taken shape, bracketing development in some recently emerging economies.

Now let us return to land appropriations and fishing rights. Suppose that we were to face the same problem of a late discovery of the fact that the enough-and-as-good condition had been violated. For instance, suppose that we cultivate orchards (of different sizes) through homesteading and then realize that these orchards are drawing on a common water source that cannot support fruit farming of such intensity. Or suppose that we only realize that we have permitted too many vessels to enter the lake when fish stocks are already in serious jeopardy. In each case, we need to cut back – but how should we cut back? Do we say that everyone in the vicinity – fruit-farmers or not, fishers or not – should now have equal access to the fruit-yielding capacity of the land or fish-yielding capacity of the lake and hence that larger operations should drastically downscale? I do not think so. We would, at least to some extent, respect differential investments made, especially the investments made at the time when these were morally unproblematic (in the sense of being licit, not in the sense of being non-culpable). For instance, with fish stocks dwindling, the EU does not assign fishing quotas to the member states so that the allocated catch per capita is equalized. Rather, quotas are set with a sensitivity to the relative dependencies of national economies on fishery.[18]

---

[18] Christian Lequesne, "Fisheries Policy – Letting the Little Ones Go?," in *Policy-Making in the European Union*, ed. William Wallace and Mark A. Pollack (Oxford University Press, 2005), 366.

When we catch violations of the enough-and-as-good condition too late, we bring in multiple considerations to rectify the situation. We may demand disproportionate sacrifices from those who are well off and hence more able to scale back. But at the same time, we may also turn away newcomers or target recent expansions. However, the argument that all who live in the vicinity should now have equal rights to the land or the lake carries little weight. When we did catch the violation of the enough-and-as-good in a timely fashion, such an appeal to equality had little weight. If we fail to catch it in time, matters become more complicated. But it is far from obvious that an appeal to equality should all of a sudden become the sole principle of decision making.

If this would be our policy in the case of farming and fishing, why would we act any differently in the case of industries emitting GHGs? Developed countries should be able to demand that, in deliberations, some respect be paid to their appropriations of the atmospheric absorption capacity that predate the cutoff point at which the enough-and-as-good condition was first violated. When violations have been ongoing, this is not the sole principle, since we also need to impose rectification on illicit appropriations past this cutoff point. And granted, these are for a large part due to growth in developed countries (but also to the GHG-intensive development of emerging economies). That *some* respect be paid to differential investments made during the time when there were no violations of the enough-and-as-good condition is common in such policy decisions. This, I take it, is the moral ground for grandfathering in setting caps on emission rights.

3.3 The structure of the harm infliction

Locke's example of respecting the enough-and-as-good condition is one person drinking from a river without reducing another person's chance to drink.[19] So a violation would be a case in which upstream people take so much water that the supply of water to the downstream people is reduced (without offsetting gains in well-being from other sources). Or, think of a case in which the upstream people catch so much fish that the opportunity to catch fish for the downstream people is reduced. In these cases the constraint on one's actions comes from the harm that would be caused by reducing other people's opportunities to perform actions *of the same kind.*

However, this is not how the structure of the harm operates in the case of GHG emissions. If I emit excessively, then a third party will become

---

[19] Locke, "Chap V: Of Property," Par. 33.

harmed *in a very different way*. For example, Tuvalu will be flooded and its inhabitants will have to move. If we collectively consider this to be the kind of harm that we ought not to inflict, then it is the case that my excessive GHG emissions stand in the way of your emitting GHGs with the same intensity. So, in the case of the upstream and downstream fishers, the harm caused by upstream overfishing is that it reduces the opportunities of the downstream fishers. In the case of GHG emissions, the harm caused by extensive GHG emissions affects third parties and has nothing to do in the first instance with the opportunities of others to emit GHGs. It is only relative to the fact that we wish to avoid the harm to third parties in that excessive GHG emissions reduce the opportunities of others to emit GHGs.

It is easy to import this restriction into the original problem of appropriating land from the commons. Suppose that there is land in abundance for farming, but even limited farming affects the much needed recreational opportunities of urban consumers in the neighboring metropolis. So now the enough-and-as-good condition also kicks in because of harm to third parties. Suppose that we catch the effects on urban consumers in time and we block any new acquisitions or expansions of existing farming operations. Would we not simply respect existing farms as they are, assuming that the acquisition process was fair? Would the closure of the commons due to third-party harm provide grounds to strive for land reform on egalitarian grounds? I do not see why this would be so.

It is not an objection to our analogy that the typical harm structure in the case of land is different from the harm structure in the case of the atmosphere. The reason is that if we impose the third-party harm structure onto land appropriation from the commons, then we could still run the standard Lockean argument. So the difference in harm structure does not block the analogy.

In conclusion, none of the distinctions outlined above between the commons of land and the global atmospheric commons make for a moral difference. The Lockean argument that can provide for a justification of unequal landownership due to differential appropriations through homesteading retains its relevance for the global atmospheric commons. For the commons of land, earlier appropriations and good stewardship within the Lockean constraints establish future claims and undoing these through egalitarian land reforms would be an injustice. Similarly, in the global atmospheric commons, certain earlier appropriations of the atmospheric absorption capacity establish future claims. An appeal to grandfathering aims to respect these claims. A radical egalitarian reform of emission rights without any concern for historically established claims is no less problematic than egalitarian land reforms without any concern for historically established claims.

4 INTERNAL AND EXTERNAL OBJECTIONS TO
LOCKEAN EMISSION RIGHTS

But clearly it would be bordering on moral madness to tell India and the USA that, since their GHG emissions per capita were, say, 1:100, at the time that climate change posed no threat, we will now fix the ratio of their future emission rights per capita at 1:100. So what can be said to modify this claim? For an answer to this question, we need to delve into critiques of Lockean thought. I distinguish between a critique that is external to Lockean thought and a critique that is internal to Lockean thought.

The critique that is *external* to Lockean thought echoes Nagel's response to Nozick – historical arguments that rest on appropriations from the commons are just one concern in determining what constitutes a fair division of land today. Other concerns should carry weight as well.[20] Humanitarian concerns can be voiced – for example, the concern that nobody should be so disenfranchised so as to fall below a minimally decent standard of living. Egalitarian concerns can be voiced – in particular latecomers or future generations will object that they never had the opportunity to homestead land and are disenfranchised now due to no fault of their own. Utilitarian concerns can be voiced – namely, we wish to avoid allocations of property rights that are hugely suboptimal. These, as well as other concerns, should certainly be taken into consideration in the fair allocation of property rights today. But nonetheless, Lockean concerns should carry *some* weight at least in planning for earlier stages. Setting policy requires a careful balancing of all these concerns with particular sensitivities to the case at hand. And there is no algorithm that covers all cases. Similarly, in determining a fair allocation of emission rights, historical emission patterns of GHG should carry some weight. But they should be balanced against other moral concerns – concerns that make historical appropriations less than sacrosanct and that typically moderate existing inequalities.

The critique that is *internal* to Lockean thought centers on the question of whether we should understand the Lockean conditions as constraining only the initial acquisition of the land, or the continued ownership of the land. To address this question, it is useful to reflect on Nozick's intriguing

---

[20] Thomas Nagel, "Libertarianism Without Foundations. Book Review of R. Nozick, *Anarchy, State, and Utopia,*" *Yale Law Journal* 85 (1975), 146.

observations on the legitimacy of continued well ownership under conditions of desertification.

5 NOZICK'S WELL

Nozick discusses a case in which there are limitations on one's property rights due to a change in circumstances. Suppose that a number of people have drilled wells (or, in Nozick's terms, water holes). The enough-and-as-good condition was satisfied, they have appropriated these wells and are selling the water in a competitive market. (Nozick does not include Locke's no-waste condition.) Now conditions change and all wells run dry except for one. The owner of this well now has a monopoly position and can extract monopoly prices. Nozick suggests that this might be permissible if the situation came about due to this person's good stewardship (and, presumably, the poor stewardship of others) rather than just luck. But it is not permissible if it came about due to desertification and the simple good luck that this person owns a well in the only location where there is still water to be tapped.[21] What is going on here?

Nozick has little to say about why he holds these intuitions. In this section, I will assess whether we can give some kind of justification for Nozick's intuitions on the basis of the Lockean tools at our disposal. In the next section, I will then consider whether any of the insights gained from reflecting on Nozick's well may be useful in reaching a less extreme Lockean position on the allocation of emission rights.

Nozick's well suggests that the enough-and-as-good condition does not apply only at the point of the initial acquisition. But should we then just apply it continuously – i.e., private property of a resource is only justified if it is Pareto superior to the return of the resource to the commons? This, I think, would make a travesty of the institution of property. Suppose that I appropriated a piece of land that was an eyesore to the neighboring community. My appropriation was Pareto superior at the time – my intention was to create a beautiful orchard and everyone would benefit from this. But now, once the work is done, my continued ownership of the orchard may not be Pareto superior any longer. If the community is minimally responsible, many may benefit from a return of the orchard to the commons and its dedication to parkland. One does not need to be a diehard libertarian to agree that the exercise of eminent domain would not be acceptable in this case.

---

[21] Nozick, *Anarchy, State, and Utopia*, 180.

And yet we do examine whether appropriations continue to be justified as circumstances change. It is not sufficient that the initial acquisition satisfies the Lockean conditions. So what else is required for continued ownership in changing circumstances? Nozick's example suggests two such circumstances:

(i) Extracting monopoly rents on water would threaten people's subsistence. The government can revoke property rights when people's subsistence is being threatened. This is reminiscent of Hume's point that during famines it is permissible for the government to open up the granaries and divide the goods equitably – the property rights of the granary owner simply cease to exist.[22]

(ii) Monopolies create inefficiencies and the government can revoke property rights that, due to changing circumstances, have come to block the operation of the free market.

To distinguish these cases we could construct the following tests. Suppose that there are still multiple wells and a free market for water, but still, due to changing circumstances, the resource has become scarce and a hike in prices threatens the livelihood of the villagers. Could we then return the wells to the commons? If so, it is (i) that matters and not (ii). Suppose, on the other hand, that we are talking not about water but about a luxury good like diamonds. There used to be multiple mines, but due to changing circumstances, one mine has remained open and now has a monopoly. Nobody's subsistence is threatened, but the owner of the mine does extract monopoly prices. Could we then return the diamond mine to the commons? If so, it is (ii) that matters and not (i).

However, neither of these answers could provide the whole story, since neither accounts for the difference that luck versus good stewardship makes. An account of what Nozick is after should incorporate this difference as well. So how can we do that?

Think of the no-waste condition. There are two dimensions to this condition. First, I should not plant and harvest more than I can consume. Second, I should not homestead a piece of land that is bigger than what I am capable of working or willing to work (or to manage). The second dimension is quite interesting, because it does base my ownership on a willingness to work the land; i.e., on good stewardship. One should extend this aspect of the no-waste condition to continued ownership – the benefits of my continued ownership must be deserved by a continued willingness to work

---

[22] D. Hume, *An Enquiry Concerning the Principles of Morals* (1889), Sect. 3, Part 1. www.anselm.edu/homepage/dbanach/hume-enquiry%20concerning%20morals.htm#sec3.

the land. And this is all the more so if these benefits become excessive and at the expense of the well-being of others.

So here may be the moral. Continued ownership, unlike initial acquisition, does not require *strict* applications of the no-waste condition and the enough-and-as-good condition. I can let my land lie fallow or leave my house unoccupied for short periods of time. I can hang on to my land or my house even though my continued ownership is not Pareto superior to its return to the commons. But if it is the case that there are *huge* gains to be made from a return to the commons (and hence that the Pareto condition on my continued ownership is massively violated), then I may lose my property rights. If it is the case that the stewardship of my property is *seriously* lacking, then I may lose my property rights. In other words, serious violations of the enough-and-as-good and no-waste conditions may jeopardize my *continued ownership*.

Furthermore, there is an interaction effect. When there are serious violations of the enough-and-as-good condition, minor violations of the no-waste condition may tip the scale and cause a revocation of property rights. This is what we see with squatters' rights. If housing needs become acute, it may become more important for owners to establish continued usage in order not to lose property rights to squatters. With acute housing needs, the enough-and-as-good condition becomes more pressing. And when the enough-and-as-good condition is pressing, even a slight waste (short-term non-occupancy) may jeopardize ownership.

And the opposite holds true as well. When there are serious violations of the no-waste condition, a minor violation of the enough-and-as-good condition may tip the scale and cause a revocation of property rights. For example, suppose I seriously "waste" my land resources by being absent for a prolonged period of time. People start crossing my land to take a shortcut which provides them with a relatively minor benefit. If I were to exercise my property rights and hinder them from trespassing, then this would constitute a minor cost to them. But due to my absence, I fail to notice their trespassing. Then this may lead to a loss of property rights when the trespassers acquire an easement on my property through adverse possession, even if what is gained thereby for them is just a minor improvement in their situation, viz. the opportunity to take a shortcut.

Nozick's well is not accounted for simply by an appeal to blocking monopolies or fending off threats to subsistence. The logic is more complex. Certainly, monopolies or subsistence threats due to evolving patterns of ownership may constitute a violation of the enough-and-as-good condition. But there is also an interaction effect with the no-waste condition. If I am the only one exercising good stewardship, then clearly the resource was not wasted on me. I did not take more than I could manage – as a matter of fact I managed the resource extremely well in comparison to others. So the no-waste condition – under a particular interpretation – is strongly respected. And if the no-waste condition is strongly respected, then the violation of the enough-and-as-good condition by itself is not enough to revoke my property rights.

6 IMPOSING LIMITATIONS ON LOCKEAN EMISSION RIGHTS

Let us now compare a case in which a strict regime of Lockean claim rights does have intuitive appeal to a strict regime of Lockean emission rights. The purpose of this exercise is to understand how it is that the internal and external critiques of Lockean claim rights do impose restrictions on their implementation. I am substituting a boating example for the earlier fishing examples, because it makes for a starker contrast with GHG emissions.

(i) In 1800, there was a lake that lay in the commons. Except for some routine tasks (bathing, washing) it was barely used for anything. Recreational boating started taking off, and over the years, some people have added larger and smaller boats to the lake. All was well until, say, around 1960, there was a threat of overuse. Additional boats would be unpleasant to present users (as well as other recreational users of the lake). All boat owners were granted licenses for their respective vessels (specifying sizes) and no further licenses were to be granted. So investments in recreation were respected and nobody was required to sell their boat. Newcomers or incumbents wishing to upgrade their boats can buy permits from present owners. Some trading has happened, but waiting times are long and, of course, many of these boat licenses can still be traced to families whose history in the region goes back for centuries.

(ii) In 1800, the atmosphere lay in the commons. Except for some routine tasks (e.g., breathing), it was barely used for anything. Industry started taking off and, over the years, some countries have started using this atmosphere as a sink for GHGs – some to a larger extent, some to a smaller extent. All was well until, say, around 1960, there was a threat of overuse. (The 1960 figure is entirely fictional.) Suppose that, contrary to fact, we recognized this fact off the bat. Increased usage of the atmosphere as a sink in this manner would set us on the path to climate change. All users were given quotas corresponding to their respective usage levels. So investments were respected and nobody was required to sell their companies. Newcomers or incumbents who wish to extend companies may buy permits from present GHG emitters. Some trading has happened, but of course, many of the presently industrialized countries are the countries that had the benefit of early entrance.

It strikes me that there is very little wrong with the case of recreational boating. This seems like a reasonable way to run such a common pool resource. As a newcomer to the region, I may find it somewhat upsetting that it is so difficult to obtain a license. One might want to tweak the policy somewhat so that newcomers who are persistent and show determination do have a chance to join the Marina Bay Club. But the basic idea of the policy is morally sound.

But if there is not much wrong with this regime, then what would have been wrong with handing out emission licenses to the various countries of the world in 1960 at the levels of GHG emission at the time? Clearly, if we had had the knowledge and the nerve to do this, then we would have a world today not threatened by global warming but probably even more unequal in industrialization levels than what we witness in the real world. In short, nobody would have stood for *that*. Such a regime would have blocked the rise of emerging economies. It would not just be grandfathering, which implies some time horizons, but would provide a license for continued inequalities (in the absence of the unlikely event of developing countries buying their way into emission licenses).

So why is it that sauce for the goose is not sauce for the gander? What makes the gander so different?

Following our *external* critique of Nozick, one might say that in the case of boating, there are very few conflicting claims. There is no issue of respecting subsistence needs in this case and egalitarian ideals with respect to luxury goods just do not carry much weight.

One can also provide an *internal* critique. Such a critique shows that not even a sophisticated Lockean could insist on emission rights that are strictly determined by historical practice. The clue lies in the continued enough-and-as-good condition. In the case of boating, some people do miss out because they cannot obtain a license, but the loss is quite minimal. It does not threaten their livelihood, they can try to buy a license, there are other hobbies to practice, and there are other lakes to drive to. But in the case of industrialization, countries who do not have emission licenses miss out radically in all aspects of life. How would newcomer countries gather the cash to buy emission rights? The lack of industrialization within their borders keeps them in dire poverty. There is little else to do and there are no other places to go. So emission quotas that are fixed by early industrialization division keys would violate the continued enough-and-as-good condition to such an extent that a correction is clearly needed – just as a correction was needed for Nozick's well owner who did not respect the continued enough-and-as-good condition.

Does this mean that we need to move as swiftly as possible to the equivalent of a radical egalitarian land reform – i.e., to equal emission rights per capita? I do not think so. Equal emission rights per capita is simply not the proper starting point for the allocation of common pool resources *in medias res* – as little as it is for the private good of land. If we had caught the onset of violations on time, then the proper *starting point* would be the existing allocation at that time. We then move away from this starting point because it strongly violates the continued enough-and-as-good condition – and we would move away with much more haste and determination than in the boating case.

What we learned from Nozick's well is that property rights can be revoked if there are serious violations of the enough-and-as-good conditions and I cannot justify my advantage by attributing it to my good stewardship. Now many people would be condemned to abject poverty if we were to continue with quotas set by the actual historical appropriations of atmospheric absorption capacity. This would be a serious violation of the enough-and-as-good condition. And an appeal to good stewardship would only go so far. The owner of the well that survived desertification through the owner's hard work might appeal to this. But could the Malibu-surfing heir to the well? Good stewardship wears off fast as we pass down the generations. So similarly, on grounds of the serious violation of the enough-and-as-good condition, we would wish to scale back developed countries' historical claim rights to the atmospheric absorption capacity. Initially we might wish to scale back conservatively in order to respect investments and good stewardship. But also the appeal to do it conservatively wears off as we make projections for future generations.

The no-waste condition is relevant to determine future emission rights and cuts in two ways. Relative to consumption patterns in the developing world, much of developed-world consumption of the atmospheric absorption capacity is inefficient and frivolous. For example, three of Socolow and

Pacala's 15 "wedges" to cut global emissions aim at reducing end-user efficiency and conservation.[23] On such grounds, the developed world loses emission rights due to failing the continued no-waste condition. On the other hand, the developing world violates the no-waste condition by the use of dirty industries and hence its poor performance in GHG emissions per unit of GDP. So they should commit to technological improvements in order to gain the emission rights that the developed world loses. The developed world violates the no-waste condition through inefficient and frivolous consumption, the developing world through irresponsible production. Portfolios of commitments to mitigation should be sensitive to different requirements generated by the no-waste condition for different countries, depending on how they may be liable to violate it.

Furthermore, one should not forget that developed countries do carry responsibility for expanding their emissions *past* the time that the commons were closed. Appeals to rectification are justified for excessive emissions by developed countries that occurred after the cutoff point when the enough-and-as-good condition on initial appropriations was violated. This is a legitimate appeal to the *polluter-pays* principle and can be invoked to argue for financial support to developing countries for mitigation and adaptation efforts.

7 CONCLUSION

My approach to GHG emission rights leads to a distribution of emission rights that will gradually become more and more egalitarian. But it does not get us to this point by preaching a strong egalitarianism complemented by *Realpolitik*-style concessions to grandfathering devoid of any moral justification.

What I defend is a regime in which relative emission rights are negotiated by carefully balancing

(i) a concern for respecting differential investments, as determined by the pre-proviso-violation distribution of the GHG-absorption-capacity resource;

(ii) a concern for rectification on grounds of the *polluter-pays* principle, considering the illicit post-proviso-violation pollution levels of developed countries;

(iii) egalitarian concerns and a concern to raise developing countries above the subsistence level, on grounds of our external critique of Locke;

---

[23] Robert H. Socolow and Stephen W. Pacala, "A Plan to Keep Carbon in Check," *Scientific American* 295 (September 2006), 54.

(iv) a concern that there is enough-and-as-good of the GHG-absorption-capacity resource left to support developing countries in their economic development, respecting the continued enough-and-as-good condition in our internal critique of Locke;

(v) a concern to reduce waste in both consumption and production, respecting the continued no-waste condition in our internal critique of Locke.

In practice, this will lead to a regime with steadily converging but initially unashamedly unequal emission rights and with developed countries contributing financially to adaptation and mitigation through investment and technology transfer in developing countries.

What is not called for is a regime in which the obligation to reduce GHG emissions affects only developed countries, in which they are branded as scoundrels for every inch that they deviate from equal emission rights per capita, and in which they are forced to foot the climate-change bill single-handedly as if developing countries are owed Versailles-style wartime reparations. Such an attitude is both unwarranted and unhelpful in climate change negotiations that aim to yield feasible and morally justifiable solutions.[24]

[24] I am grateful to Denis G. Arnold, Karin Edvardsson Björnberg, Simon Dietz, Alice Obrecht, Leigh Raymond, Laura Smead, Peter Vallentyne, and Alex Voorhoeve for discussion and comments. My research was supported by the Grantham Research Institute on Climate Change and the Environment, and the Centre for Climate Change Economics and Policy, which is funded by the Economic and Social Research Council, UK.

CHAPTER 7

Parenting the planet

Sarah Krakoff

In our time, for the first time, one human species can be envisaged, with one common technology on one globe and some surrounding "outer space." The nature of history is about to change. It cannot continue to be the record of high accomplishments in dominant civilizations, and of their disappearance and replacement. Joint survival demands that man visualize new ethical alternatives for newly developing as well as over-developed systems and identities. A more universal standard of perfection will mediate more realistically between man's inner and outer worlds than did the compromises resulting from the reign of moral absolutes; it will acknowledge the responsibility of each individual for the potentialities of all generations and of all generations for each individual, and this in a more informed manner than has been possible in past systems of ethics.[1]

– Erik Erikson

INTRODUCTION

The Earth is under our thumb. Global warming is the latest example of how human activity has reached every nook and cranny of the Earth's natural systems, but it is not the only one. The effects on the ozone layer, the collapse of fisheries throughout the world, and the accelerated species extinction rate, among many other phenomena, indicate the planetary scope of human impacts. As Nobel Prize winner Paul Crutzen has put it, we have entered the "Anthropocene," the era of ubiquitous human influence on the Earth's geological systems.[2] Robert H. Socolow similarly suggested that today we might think of ourselves as "planetarians," due to our wide-ranging impacts and, arguably, correspondingly broad responsibilities.[3] This stage,

[1] Erik H. Erikson and Joan M. Erikson, *The Life Cycle Completed* (New York: W. W. Norton & Co., 1997), p. 157.
[2] Paul Crutzen and Eugene F. Stoermer, "The Anthropocene," *Global Change Newsletter* 41 (2000), 17–18.
[3] Robert H. Socolow and Mary R. English, "Living Ethically in a Greenhouse," Chapter 8 in this volume.

the Anthropocene, the Planetarian, or whatever label we choose to apply, provides the occasion to reconsider our relationship with the natural world. Just as importantly, it provides the occasion to dwell on what it means to be human and the legacy that we would like to leave behind. Despite the need for sophisticated technological solutions to address the many challenges of global climate change, ultimately our decisions will reflect our moral and ethical commitments to other humans and to the natural world, even if they will not reflect them perfectly.

Since the Industrial Revolution, progress has gone hand-in-hand with technological innovation. For roughly the past forty years (dating, some-what arbitrarily, from the first Earth Day in 1970), technology has, in significant measure, also allowed us to rein in some of the negative environmental consequences of industrialization. The Western developed world made substantial progress toward addressing, for example, air and water pollution through a mix of regulation and technology. Even in less obviously technology-dominant areas such as species preservation, the combination of scientific knowledge and human ingenuity resulted in important conservation victories, such as bringing the bald eagle, the California condor, and other less telegenic species back from the brink of extinction. These technological and scientific successes have been domi-nated by technological frames of thought, including welfare economics, market liberalism, and other rationalist/individualist approaches, that have monopolized politics and decision making in much of the Western devel-oped world. These frames have in common an outlook of perpetual growth that is dependent on unstated assumptions about boundless resources. The appeal of these frames is obvious. But the benefits of perpetuating a vision of the good life that is bound up in these frames may be receding.

What I want to explore in this chapter is the possibility of a conception of how we relate to the planet that might supplant the dominant frames with a timelier and perhaps more enduring vision of ourselves and our obligations. The seemingly intractable collective action features of climate change, which render it a commons problem of global and intergenerational proportions, make it all the more ripe for a conception of ourselves that does not depend predominately on individual rational self-interest to explain human motiva-tion. An appropriate, though admittedly not perfect, metaphor for this new stage is that of parenting. The features of global climate change, like the features of parenting in the ordinary sense, are such that daily and indefinite behavior change is called for, and personal and hedonistic desires may have to be set aside when necessary; and yet, with any luck, we will derive deep satisfaction and joy in our new role even if we glance with occasional longing at the more libertine phase we have left behind. And finally, parenting the planet will require us to accept that, even if, or perhaps *especially* if, we do the best we can and all goes as well as possible, we will never know the end of the story. To be a parent is inherently tragic in this sense, even while parenting also has the potential to magnify the best aspects of the human experience – love, joy, passion – all gained through the necessary loss of displacing the self as the center of the universe. Finally, parenting captures the paradox that taking on the challenge of displacing one's own needs requires the willingness to recognize that one has a monopoly on how to control and influence the needs of others. Like parents, we as a species are in the driver's seat. Whether we exercise our control and influence to allow the flourishing of others (including other human communities, other species, and future generations) or not is up to us.

Philosophers have provided an array of theories and arguments in support of an ethical relationship with the natural world. Some are grounded in utilitarian theories, some in deontological approaches, and some in theories of virtue or character. Indeed, there is a burgeoning literature in the third category.[4] The interest in virtue-based accounts of an ethical relationship with nature might reflect an emerging sense that in the Anthropocene, positive environmental outcomes depend, more than ever before, on the fulfillment of human moral potential. The renewed interest in virtue may also reflect an inchoate sense that positive environmental outcomes are more elusive than ever before, and that we need a reason to be good that does not depend on them. The idea developed in this chapter – that we need a conception of ourselves and our relationship with nature that is in step with the Anthropocene – is akin to the virtue-based approaches for these reasons. Unlike virtue theory, however, my point is not to provide a philosophically convincing basis for an environmental ethic. Rather, my goal is to provide an accurate and bracing description of the stage that human beings are in with respect to our dominance of the planet, and then to sketch the ethical implications and possibilities. Being a parent, after all, is not itself a virtue. It is just a description of a very distinctive relationship. To be a good parent requires, at the very least, recognition of the particular power and influence that one is capable of wielding. The road to virtue may follow from that recognition, or not. But it certainly will not follow without it.

---

[4] Recent works on virtue-based environmental ethics include Ronald Sandler, *Character and Environment* (New York: Columbia University Press, 2007), and Ronald Sandler and Philip Cafaro, *Environmental Virtue Ethics* (Oxford: Rowman & Littlefield, 2005).

I THE STAGES OF HUMAN PSYCHOLOGICAL
DEVELOPMENT AS A METAPHOR

The psychoanalyst Erik Erikson outlined eight stages of human psychological development, starting with infancy and ending with old age. Later, when Erikson's wife and collaborator Joan lived a bit beyond even "old age," she sketched a ninth stage, which she did not name. Whether in eight or nine stages, the Eriksons' central idea, refined and modified by others since, is that human psychology is not static, and that we resolve certain psychological conflicts in order to meet the challenges of each phase of life. Erikson's approach is not strictly biological; his theory does not depend on a handful of immutable characteristics that can be universally applied regardless of time or circumstance. Rather, he is careful to say, "Wherever we begin ... the central role that the stages are playing in our psychosocial theorizing will lead us ever deeper into the issues of *historical relativity*."⁵ In other words, the stages provide a useful framework for asking the right questions. Those questions include: What are the central conflicts that we face, and what virtues might be cultivated to address those conflicts? The answers to those questions, however, are not automatic or universal, but rooted in the norms and values of particular cultures.

Erikson's insights, like Freud's and Jung's, now may seem somewhat basic. Just as most people accept as a general matter the role of unconscious motivation, most people in our culture accept that the mind of a child is not the same as the mind of an adolescent, and that the developmental issues we face as very young adults are different from those we face in middle age. What I want to borrow from this now familiar framework is the idea that different circumstances have different essential conflicts, and therefore call for the cultivation of different virtues and different behaviors. In the Anthropocene, the implications of attitudes and actions towards the environment are quite different than in previous eras, when human activity was capable of only the most ephemeral effects on the world.

Conceptualizing our relationship with the natural world in stages also has the following two interrelated benefits. First, this conceptual framework allows us to think, perhaps more objectively and less judgmentally, about the ways in which laws about the environment both reflect and reinforce the essential virtues and conflicts of a particular time and place. Thinking about stages may help us to see how laws generated during a previous stage, even laws that are protective of the environment, no longer are sufficient because they do not facilitate resolution of the essential conflicts we face today. To use

⁵ Erikson and Erikson, *The Life Cycle Completed*, 12–13 (emphasis in original).

an analogy to the human stages, we do not apply the same legal standards to the behavior of children that we do to the behavior of adults. We adjust our regulatory schemes and our expectations for behavior based on a sense of developmental appropriateness. Related to this, but going beyond legal reform, using the metaphor of stages (hopefully loosely, and not so dogmatically that I attempt to explain everything in terms of an elaborate Eriksonian developmental chart) may liberate us to think about our current situation in ways that make us feel hopeful, engaged, and ready to do our best rather than depressed, apathetic, frightened, or, maybe worst of all, like pretending that nothing is going on that warrants rethinking. Consider, for example, how strange and frustrating it would be to grapple with the challenges of adolescence equipped only with the conceptual tools and vocabulary of a preschooler. Clarity about the challenges and conflicts one faces at least provides the possibility for meaningful engagement with them, whereas uncertainty results in disorientation, confusion, and the potential for actions that are futile given the circumstances. A teenager cannot navigate the complex social environment of junior high school with the "will you play with me?" approach that worked in elementary school, and we cannot solve global warming by relying on the pollution-control strategies of the past.

Second, conceptualizing our relationship with the nonhuman world in terms of stages sidesteps the problem of whether we are distinct from or a part of "nature," or "wilderness." Taking a meta-view of how we see ourselves and our role in different periods, the question instead becomes one of both history and values. Have we *acted* as if we are members of the natural world, or have we *acted* as if the natural world is an object that is distinct from us? And what do our different attitudes and behaviors mean about our values? If we are indeed in the Anthropocene era, then the focus should appropriately shift from what is or is not "natural" to what we value and why.

This is, perhaps, all an elaborate way of saying that it is good to know where you are. We have, I believe, not yet grappled with where we are as a species in relationship with the rest of the natural world. Despite having entered the Anthropocene, we have not embarked on a widespread project of reconsidering what this might mean in terms of our obligations to other species, future generations, or even the many human beings who are on the short end of our effects on natural systems.

2 SOME FACTS ABOUT GLOBAL WARMING

So where are we? The scientific facts about global warming are covered in earlier chapters and so I will not review them all here. Instead I will provide a very brief summary, highlighting the aspects of climate change that lend themselves to a shift to a parenting conception of our relationship with the planet.

In terms of what we know about global warming, we now have roughly two decades of accumulated scientific studies. These studies are regularly reviewed by the International Panel on Climate Change (IPCC). In January 2007, the IPPC issued its fourth set of assessment reports on global warming. The IPCC concluded that "warming of the climate system is unequivocal,"[6] and also expressed "very high confidence" that human emissions of carbon dioxide (hereafter CO_2) and other heat-trapping gases (methane, nitrous oxide, various hydrofluorocarbons, various perfluorocarbons, and sulfur hexafluoride) since 1750 have caused the Earth's surface temperature to rise.[7] During that time, CO_2, the most important of the anthropogenic greenhouse gases, increased from a pre-industrial level of roughly 280 parts per million (ppm) to 382 ppm in 2007.[8]

CO_2, methane, and the other heat-trapping gases work in the following way. The sun's energy passes through the atmosphere in short, powerful waves. The Earth reflects the sun's energy back, but in longer heat waves. These longer waves are too big, molecularly speaking, to make it back through the "blanket" of CO_2 and other heat-trapping gases. This blanket keeps the Earth warmer than it would be if the heat energy were simply reflected back into space. If there were no heat-trapping gases in the layers of our atmosphere, the Earth's average surface temperature would be about 5°F (−15°C), far too cold for most of us. So the Earth is habitable thanks to greenhouse gases, but other planets are uninhabitable because they have atmospheres that trap too much heat. Venus, for example, has an atmosphere composed of 98 percent CO_2 and its surface temperature is 891°F (477°C). According to the scientist and author Tim Flannery, "if even 1 percent of Earth's atmosphere" were CO_2, it would, all other things being equal, "bring the surface temperature of the planet to boiling point."[9] Like Goldilocks, we want our atmosphere to be "just right." For millennia, it has been. But today, CO_2 and other gases are at levels that are far higher than they have been since the dawn of life as we now know it. Therefore, we may

[6] Intergovernmental Panel on Climate Change, "Climate Change 2007, the Physical Science Basis, Summary for Policy Makers," 4 (2007), available at www.ipcc.ch. [Hereafter "IPCC PSB Summary 2007."]

[7] Ibid. at 3. The authors define "very high confidence" as at least a 90 percent chance. Ibid., n. 7.

[8] Ibid. at 2.

[9] Tim Flannery, *The Weather Makers: How Man Is Changing the Climate and What It Means for Life on Earth* (New York: Atlantic Monthly Press, 2005), 23–4.

be fast approaching the point where global average surface temperatures are becoming, like Papa Bear's bowl of porridge, too hot.

The theory of global warming is not new. It has been with us since the 1890s, when a Swedish chemist demonstrated that a decrease in atmospheric CO_2 could have brought about an ice age, and further speculated that increasing levels of CO_2 due to coal burning would have a future warming effect.[10] Global warming advanced beyond the theoretical in the late 1950s, when scientists began to document the concern that human activities, including significant increases in CO_2 emissions, might be changing the way the atmosphere traps heat. In 1958, the scientists Roger Revelle and Charles Keeling established a research station at the top of Mauna Loa in Hawaii, from which they launched weather balloons and measured the amount of CO_2 in the atmosphere.[11] The measurements revealed a striking trend of annual increases in CO_2 concentrations, which, coupled with the physics of how CO_2 and other greenhouse gases trap heat, provided factual support for the hypothesis that human emissions would cause increases in the Earth's temperature.

Since this time, more and more evidence has been filling in. In retrospect, the scientific story is one of a sound hypothesis evolving year by year into an increasingly solid, and now all but indisputable, reality. There have been, and continue to be, distracting sideshows about the precision with which we know things. For example, a *Newsweek* story provides a nice summary of how the "denial machine," which the author describes as a "well-coordinated, well-funded campaign by contrarian scientists, free-market think tanks and industry," created a "paralyzing fog of doubt around climate change," for nearly two decades.[12] These efforts were funded by the American Petroleum Institute, the Western Fuels Association, and ExxonMobil Corporation. The denial machine is still in operation, but we are largely over the narrative of scientific doubt, at least in most mainstream circles. Instead, the doubt has shifted from whether humans are causing climate change to whether it is worth it economically to do anything serious to curb emissions.[13] We will come back to this later.

---

[10] Ibid. at 39–42.

[11] Al Gore, *An Inconvenient Truth: The Planetary Emergency of Global Warming and What We Can Do About It* (New York: Rodale, 2006), 38–39; Flannery, *The Weather Makers*, 24–5.

[12] Sharon Begley, "Global Warming Is a Hoax*," *Newsweek* (August 2007).

[13] One prominent example is Bjørn Lomborg, *Cool It: The Skeptical Environmentalist's Guide to Global Warming* (New York: Random House, 2007) (advocating some measures to reduce carbon emissions, but claiming that spending too much displaces other social and environmental priorities).

3 A BRIEF TOUR THROUGH THE EFFECTS

Warming's effects are also well covered in other chapters in this volume, and so I will not dwell on them for long. There are two aspects of global warming's impacts that I want to highlight, however, because they are particularly relevant to the idea of shifting to a new stage in terms of our conception of planetary obligations. First, global warming's effects on other species are, in many instances, already quite clear. While many know about the risk to the polar bear from the dramatic decrease in Arctic sea ice,[14] I suspect few have heard of the gradual displacement of a furry creature closer to my home, the pika. Pika live in colonies at high altitude, typically in skree fields on mountainsides. They are social, and communicate with one another through high-pitched squeaks or whistles. Pika require chilly temperatures, but, unlike humans, they cannot respond to the increasing heat by turning on the air conditioning. Instead, they have been migrating further up-slope to capture the high altitude benefit of cooler air. Yet some pika populations in the area known as the Great Basin, between the Sierra Nevada and the Rocky Mountains, have already disappeared because, presumably, they ran out of up.[15] At this point, it is tempting to reel off the many other species that are already at risk due to climate change. But I want to keep my promise not to dwell too long on these effects. We can think of the pika as a stand-in for the many other species that will not be able to adapt. Their loss will not be felt in any immediate, daily way, but they join the list of casualties to a process that we have set in motion.

Second, the effects of global warming are being, and will continue to be, felt disproportionately by developing nations and by poor people in general. As the IPCC Fourth Assessment Report concludes, "Africa is one of the continents most vulnerable to climate variability and change because of multiple stresses and low adaptive capacity."[16] The IPCC and other sources report a similarly disparate vulnerability for virtually all underdeveloped and developing regions. For example, a report by the Natural Resources Law Center at the University of Colorado documents disparate effects on Native

---

[14] See Proposed Final Rule: Endangered and Threatened Wildlife and Plants; Determination of Threatened Status for the Polar Bear (*Ursus maritimus*) Throughout Its Range, 50 C.F.R. Part 17 (May 14, 2008).

[15] Erik A. Beever, Peter F. Brussard, and Joel Berger, "Patterns of Apparent Extirpation Among Isolated Populations of Pikas (*Ochotona princeps*) in the Great Basin," *Journal of Mammology* 84, no. 1 (2003), 37–54; Sean F. Morrison and David S. Hik, "Demographic Analysis of a Declining Pika *Ochotona collaris* Population: Linking Survival to Broad-Scale Climate Patterns via Spring Snowmelt Patterns," *Journal of Animal Ecology* 76, no. 5 (2007), 899–907.

[16] IPCC, "Climate Change 2007: Climate Change Impacts, Adaptation and Vulnerability," 8.

American communities within the USA, including Native villages in Alaska, tribes throughout the increasingly arid Southwest, the salmon tribes of the Pacific Northwest, and the two Florida tribes.[17] Indeed, some Alaskan Native villages are already being forced to relocate due to global warming. Decreased sea ice has allowed more waves to pound the shore, and higher surface temperatures have made the shoreline less stable, causing coastal villages literally to slip into the ocean. The Native village of Kivalina, for example, used to comprise 54 acres. Erosion of the shoreline has shrunk the village to 27 acres. In 2006, the US Army Corps of Engineers concluded that in ten years the village would be uninhabitable. Relocation plans have estimated the costs of removal to range from $95 million to $400 million. Kivalina has filed a lawsuit against ExxonMobil and other corporate defendants, alleging nuisance theories as well as conspiracy to conceal facts about global warming and to mislead the public about its causes and effects.[18] Raising an even broader array of climate change's effects on Native life and culture, the Inuit Circumpolar Conference filed a petition before the Inter-American Commission on Human Rights alleging various human rights violations by the USA.[19] As in the context of effects on non-human species, this section could be the beginning of a much longer recitation of the disparate effects on poor and indigenous communities throughout the world, but this handful of examples will serve here to make the point.

To summarize, global warming is already having negative effects on other species, and not just the telegenic polar bear. And global warming is also either already affecting or likely to affect poor people everywhere, and particularly poor people in poor regions of the Southern Hemisphere as well as indigenous communities whose ways of life are tied to place. The ethical dimensions of climate change thus include obligations to other species and justice to the world's poor and indigenous peoples. This does not necessarily distinguish climate change from other global environmental problems. But the collective action features of climate change, outlined below, heighten the necessity as well as the difficulty of a truly global response.

[17] Jonathan Hanna, "Native Communities and Climate Change: Protecting Tribal Resources as Part of National Climate Policy," Natural Resources Law Center Publications, University of Colorado Law School (2007), available at www.colorado.edu/law/centers/nrlc/publications.
[18] *Native Village of Kivalina* v. *ExxonMobil Corp.*, No. 08-1138 (N.D. CA, filed February 27, 2008).
[19] Petition to the Inter-American Commission on Human Rights Seeking Relief from Violations Resulting from Global Warming Caused by Acts and Omissions of the United States (2005) (hereafter "Inuit petition").

4 THE POTENTIALLY TRAGIC STRUCTURE OF GLOBAL WARMING: TEMPORAL LAGS AND SPATIAL DISPERSION

If the reader is getting that bleak feeling, unfortunately we have to go a little bit further down before we can start climbing up. We have to discuss why global warming is different from other environmental problems, even if, at least compared to some, it is different only in magnitude. These, the potentially tragic features of global warming, are that it is both a temporally lagged and spatially dispersed phenomenon.

4.1 Temporal lags

Global warming is a severely temporally lagged phenomenon because CO_2 stays in the atmosphere for hundreds of years, so most of the molecules added since the dawn of industrialization are still hanging around. As a practical matter, every molecule we add is one that is increasing the "thickness" of our atmospheric blanket because none are going away within a time-frame that matters. This results in a lag between emissions increases and the effects on warming. The effects from today's blanket will be felt throughout the rest of the century (meaning increased warming and so on) even if we were to stop all carbon emissions today. Likewise, we are now feeling the effects not of our own emissions, but those of our parents and grandparents. The problem compounds over time, because even if we begin to reduce emissions, we are reducing relative to a base with significant longevity. Reflecting this, a study by Susan Solomon and others found that changes in surface temperature, rainfall, and sea level are largely irreversible for more than 1,000 years after emissions are completely stopped.[20]

One challenge presented by the time lag is one of perception. It is understandable that we have a hard time experiencing today's daily activities as contributing to an increasingly intractable global problem when the effects of these normal, culturally reinforced activities (driving, heating our homes with fossil fuels, flying to visit relatives) will be felt decades from now. A related challenge is that it puts us in the position of setting targets for emissions based on predictions about the future rather than certainties about the here and now. For example, there is a worrisome range of scientific assessments regarding safe levels of global CO_2. James Hansen and several scientific coauthors are calling for reductions and

[20] Susan Solomon *et al.*, "Irreversible Climate Change Due to Carbon Dioxide Emissions, " *PNAS* 106 (February 10, 2009), 1704–9.

stabilization at 350 ppm of atmospheric CO_2 in order to avoid perpetual climate catastrophes.[21] Not very long ago, the working assumption was that stabilization somewhere between 450 and 550 ppm would suffice.[22] Today's levels, measured as recently as 2008, are already at around 384 ppm. Given current rates of emissions, we have a very small window (estimates vary and depend on trends in the next few years, but sometime between now and midcentury) to decrease and then zero out our emissions.[23] So, to summarize, the temporal features of global warming are such that (1) we feel the effects tomorrow of our actions today, (2) tomorrow will be with us for a long time, and (3) the best plan for tomorrow is to start dramatically reducing emissions today, even if we never achieve 100 percent consensus on the non-catastrophic level of emissions.

4.2 *Spatial dispersion*

Spatial dispersion is what makes global warming a problem requiring a truly global solution. The atmosphere is a global commons. No matter where in the world you are, your emissions contribute to the increasing insulating properties of the atmosphere. And the atmosphere is not and cannot be compartmentalized. So the fact that the USA has the highest historical greenhouse gas emissions does not mean that our atmosphere is "thicker" and our effects from global warming are proportionately higher than other countries. (In fact, in terms of effects, the contrary is true. Regions that have contributed the least are likely to feel the most severe effects.[24]) The spatial dispersion means that reductions in one part of the globe can be rendered meaningless by increases in another part of the globe. If the total parts per million of CO_2 continues to rise overall, it doesn't matter where the parts come from. The "commons" nature of global warming is what makes policymakers say things like "What about China and India? If they are not part of a global regime to reduce emissions, we may be tightening our carbon belts for nothing." Nobody wants to be a dupe. Rational choice and collective action theorists label this kind of commons problem a prisoner's

---

[21] Available at www.columbia.edu/~jeh1/2008/TargetCO2_20080407.pdf.

[22] Stephen Pacala and Robert Socolow, "Stabilization Wedges: Solving the Climate Problem for the Next 50 Years with Current Technologies," *Science* 305 (August 13, 2004), 968.

[23] James Hansen *et al.*, "Target Atmospheric CO₂: Where Should Humanity Aim?," *The Open Atmospheric Science Journal* 2 (2008), 217–31, www.columbia.edu/~jeh1/2008/TargetCO2_20080407.pdf.

[24] Sarah Krakoff, "Ethical Perspectives on Resources Law and Policy: Global Warming and Our Common Future" in *The Evolution of Natural Resources Law*, Larry Macdonald and Sarah Van de Wetering, eds. (New York: ABA Publications, 2009).

dilemma.[25] Each entity, acting in its "rational self-interest" has an incentive not to curb emissions even though the interests of all would be served if we would agree to reduce emissions.

The temporal and spatial dispersion together heighten the nature of this challenge. As Stephen Gardiner has put it, they create a true intergenerational collective action problem.[26] Each generation has an incentive, under rational choice assumptions, not to reduce emissions because the "burdens" of reduction will be felt now and the "benefits" of curbing emissions will be felt by subsequent generations. If we look around at our behavior over the last two decades, and even now as we dither about whether to do anything serious at the national or global level about mitigating emissions, we might find ourselves persuaded by the rational choice description: rational self-interest in our own well-being (which is heavily dependent on our carbon economy) has led us not to act for the benefit of future generations, other species, and less well-off human communities, even though the moral case for doing so is heightened by the fact that we are the generation that could make the biggest difference. As Gardiner has articulated, the "perfect moral storm" that makes global warming an acute yet elusive moral issue leads to behaviors characterized by "moral corruption," which include distraction, complacency, unreasonable doubt, selective attention, delusion, pandering, false witness, and hypocrisy.[27] Indeed, these behaviors have all been evident in the reaction to climate change at the national level. And yet, is the "we have done nothing," mother-of-all collective action descriptions accurate? Have "we" been doing nothing? And furthermore, is our conception of ourselves limited to a being that calculates benefits and costs and acts "rationally" on them?

5 ETHICS FOR A POTENTIALLY TRAGIC AGE

The remarkable thing is that despite the potentially tragic structure of global warming and the fact that warming's effects fall disproportionately on poor people, other species, and future generations, people all over the world, including the developed world, are trying to do something about it. Are they acting in a consequentialist and rationalist way, because they hope to

---

[25] Stephen Gardiner, "The Real Tragedy of the Commons," *Philosophy and Public Affairs* 30 (2001), 387.
[26] Stephen Gardiner, "A Perfect Moral Storm: Climate Change, Intergenerational Ethics and the Problem of Corruption," *Environmental Values* 15 (August 2006), 397.
[27] Ibid. at 407–8.

succeed? In part they must be. But they likely also know that they may not succeed, for the reasons just described, and furthermore that they will never know if in fact they do. So while participants in various local arrangements to reduce greenhouse gas emissions want their actions to be part of a larger and ultimately successful movement, their behavior also reflects other values and motivations. What follows is a very impressionistic survey of some of the things happening around the world. I want to suggest that what is going on in these communities is neither silly, idealistic delusion nor grim self-denial in sacrifice to a preservationist goal. I want to suggest that there are pockets of humanity fashioning an alternative subjective identity, an identity whose meaning derives from participating in the daily tasks of parenting the planet. This identity is in part an end in itself, a way of constructing a meaningful and even joyous life in the face of tragic circumstances, even while it also has both existential and even potentially consequentialist implications.

There are too many individual and local initiatives on climate change to provide a comprehensive account. Instead, I will provide a brief overview of activities at the subnational level, spending a little more time on particularly salient examples. The overview starts at the highest subnational level of coordination and moves down the scale from there.

5.1 US states and regions

At the state level, there is a great deal of activity.[28] Thirty-six states have completed "climate actions plans," which are initial documents laying out steps for reducing emissions and preparing for the already inevitable effects of warming. Forty-three states have greenhouse gas inventories, allowing them to track emissions. More impressively, twenty-one states have emissions reduction targets, and a number of those have an actual carbon cap and offset program for power plants. Twenty-nine states, plus the District of Columbia, have renewable portfolio standards, which require that a certain percentage of the state's electric needs come from renewable sources. In Colorado, the legislature recently raised the percentage of renewables in our renewable portfolio standard, requiring that we achieve 20 percent renewable sources by 2020.

California has been the leader in all of these efforts, and in 2006 enacted a law setting the goal of reducing the state's greenhouse gas emissions to 1990

[28] For an overview of state and regional initiatives and programs, see www.pewclimate.org/states-regions.

levels by 2020.[29] The legislation authorizes the California Air Resources
Board to adopt a market-based system to regulate greenhouse gases, and
mandates enforcement of emissions standards against regulated sources.[30]
In addition to action by individual states, there are several regional initiatives throughout the country, in which states (sometimes together with
Canadian provinces) have combined to address climate change and in some
cases are working toward regional cap-and-trade systems. The Regional
Greenhouse Gas Initiative, which was the first of the regional efforts,
includes ten states in its mandatory cap on emissions from the power sector.

5.2 Cities

Moving down the scale, cities have been very engaged with enacting climate
policies. In 1993, Portland, Oregon, became the first city to adopt a strategy
for reducing emissions of CO_2. In June 2005, Portland issued a "Progress
Report" which concluded that the city and surrounding county had reduced
per capita emissions by 12.5 percent since 1993.[31] Other cities including
Seattle, Washington, and Salt Lake City, Utah, have joined Portland in
establishing emissions reduction targets. To unite and further catalyze these
efforts, Mayor Greg Nickels of Seattle created the US Mayors Climate
Protection Agreement. The agreement urges federal and state governments
to enact policies that meet or surpass the Kyoto target of reducing global
warming pollution to 7 percent below 1990 levels by 2012 and also calls on
Congress to pass greenhouse gas reduction legislation. The agreement states
that signatory mayors will strive to meet or exceed the Kyoto targets within
their own communities by creating an inventory of emissions in their cities,
setting reduction targets, and increasing use of alternative energy sources.
More than 800 mayors have signed the agreement.[32] All this activity
demonstrates that governments at the local level are taking a leading role
in setting and meeting greenhouse gas emissions reduction goals. My own
hometown, Boulder, Colorado, has even managed to pass the country's first
relatively modest carbon tax. The tax is imposed on residential and commercial energy customers, and is collected by the local electric utility. The
charge is based on electricity use, and wind energy customers are exempt.

[29] California Global Warming Solutions Act, California Health and Safety Code, Div. 25.5 (signed into
law on September 27, 2006).
[30] Ibid.
[31] Portland Online, "A Progress Report on the City of Portland and Multnomah County Local Action
Plan on Global Warming," www.portlandonline.com.
[32] "The U.S. Conference of Mayors Climate Protection Page," www.usmayors.org.

The tax is low, costing the average household only $1.33 per month, and is designed to fund Boulder's Climate Action Plan, a multi-faceted effort to increase efficiency and transition to renewable energy sources in order to meet Boulder's goals of complying with Kyoto's emissions reduction targets.

5.3 Non-governmental community efforts

Beyond the governmental realm, there are also many examples of norms and practices emerging from within different local cultural communities. The following three examples highlight the impressive diversity among these efforts. First, the Inuit of the Arctic Circle, who have developed a culture over millennia of living low-carbon lives, now find that they must engage in international legal efforts to preserve that way of life. Second, a subsect of evangelical Christians has fashioned a religiously based movement to reduce emissions and care for the poor as well as other species. Third, neighbors in some countries are banding together to pledge to reduce their individual carbon footprints.

The Inuit community

It is understandable that the Inuit, residents of the Arctic Circle, where warming is twice that of the global average, are very engaged with climate change. The Inuit Circumpolar Conference is a non-governmental organization representing "approximately 150,000 Inuit of Alaska, Canada, Greenland and Chukotka (Russia)."[33] Sheila Watt-Cloutier, an Inuit and member of the conference, filed a human rights petition (hereafter the Inuit petition) against the USA in the Inter-American Court of Human Rights in 2005.[34]

The petition intersperses scientific data with direct observations by Inuit people, including clear and poignant descriptions of how natural cycles are changing far faster than human culture can adapt, and how those changes are causing the loss of key cultural practices.[35] It also describes in vivid detail the many signs that other species are struggling, and perhaps heading toward extinction.[36]

---

[33] Inuit petition. [34] Ibid., 9.

[35] Ibid., 48–9 (describing cultural loss, including passing on to future generations how to build an igloo).

[36] Ibid., 212 (describing effects on seal pups and polar bears); 47 (describing effects on caribou); 4–5 ("For Inuit, warming is likely to disrupt or even destroy their hunting and food sharing culture as reduced sea ice causes the animals on which they depend on to decline, become less accessible and possibly extinct.") (citation omitted).

The Inuit community, as viewed through the prism of their involvement with the Inuit petition and other efforts to participate in international climate discussions, embraces moral commitments to place and to future generations, and simultaneously engages with the highly technocratic and bureaucratic forms of science and international law. The Inuit, through these varied commitments and strategies, are attempting to preserve their place-dependent cultures while simultaneously integrating themselves into the global web of legal and technical relationships that will be required for them to succeed. Their efforts, although necessitated by dire circumstances, are therefore also imbued with hope and a sense of optimism.

The evangelical climate community
The culture of megachurches with Christian rock groups and Bible study classes may seem a far cry from the culture of seal and whale hunting, igloo building, and fine-tuned reading of ice and snow, but global warming highlights their similarities. Like the Inuit, the evangelical climate com- munity is engaged in a modern, multi-faceted campaign rooted in cultural and spiritual values. Those values extend care and obligation both forward in time, to the inheritors of the world that evangelicals believe that they are charged with "stewarding," and outward in space, to the residents of other nations who will feel most acutely global warming's effects.

While the evangelical climate community is less rooted in a particular geography than the Inuit communities, there are otherwise some interesting parallels. Like the Inuit, the evangelical impetus to address global warming grows out of a spiritual worldview. And also like the Inuit, the evangelical climate community embraces the science documenting global warming as well as the need for comprehensive legislative solutions. This is evident in the Evangelical Climate Initiative (ECI) statement, "Climate Change: An Evangelical Call to Action."[37] The statement describes the scientific con- sensus that global warming is real and human-induced, and argues that Christian moral convictions demand a response to the problem. The state- ment urges policymakers both to mitigate climate change and provide aid to the poor in order to help them adapt to the changes already underway.[38] Lastly, the statement calls on the USA to pass national legislation requiring reductions in CO_2 emissions through market-based mechanisms.[39] The evangelical climate movement has also attempted to draw attention to individ- ual activities that can reduce greenhouse gas emissions. In 2002, the group

---

[37] Evangelical Climate Initiative, "Climate Change: An Evangelical Call to Action."
[38] Ibid.   [39] Ibid.

formed a campaign entitled "What Would Jesus Drive" (WWJD), arguing that transportation choices are moral choices.[40]

CRAGS

The third and final community-based example is even further down the scale of size and organization. In the UK and the USA, small groups have formed whose members pledge to one another to live low-carbon lives. Carbon Rationing Action Groups, or CRAGS, as they are called, are communities that keep one another true to their principles by formulating a yearly limit of emissions for members and then meeting regularly to monitor one another. There are currently 160 people active in some 20 CRAGS across the UK, with another 13 in the formulation stage. These have been joined by two working CRAGS in Canada, three in the USA, and others in various start-up phases.[41]

The goal for the majority of CRAGS in the UK is to reduce personal carbon footprints by roughly 10 percent each year to achieve a reduction of 90 percent of current levels by 2030. To meet their goals, members are changing their daily habits, including using less light and different sources of fuel. According to a *New York Times* article, CRAG member "Jacqueline Sheedy has turned the former coal barge where she lives into a shrine to energy efficiency: She reads by candlelight in midwinter, converts the waste from her toilet into fertilizer, and hauls freshwater home on a trailer attached to her bicycle. Now Ms. Sheedy has set herself a new goal: to stop burning coal for heat and instead use wood from renewable resources."[42] Similar groups are forming all over, including in New York, Oregon, Maryland, and even Texas, which has the highest per capita emissions of any state in the USA.[43] In Boulder, Colorado, at least two CRAG-like groups have also formed, aided by the leadership and support of the City's Climate Action Program.

6 VIRTUES AND PRACTICES FOR A WARMING WORLD

From a rational choice/welfare economics perspective, all of these subglobal initiatives might be described as irrational. When we consider the spatial and intergenerational collective action features of global warming, people

---

[40] Discussion Resources and Fact Sheets, available at http://whatwouldjesusdrive.info/intro.php.
[41] See CRAG Homepage at www.carbonrationing.org.uk.
[42] James Kanter, "Members of New Group in Britain Aim to Offset Their Own Carbon Output," *New York Times*, October 21, 2007.
[43] US CRAG Groups, available at www.carbonrationing.org.uk/groups?country=us.

and communities are denying themselves a benefit in order to achieve a goal which may be undermined by increased carbon emissions by their neighbor the next house down, or the next city over, the next state over, and, even more threateningly, subsequent generations. Furthermore, conventional understandings of human empathy and planning – that near-empathy is more powerful than abstractions, and that the human time horizon does not reach much beyond two generations – are called into question by these local carbon mitigation activities.

What all of this activity might reflect is a shift in the way that we conceive of our role on the planet, and the identities that we are constructing to make our lives have meaning. In the face of a potentially tragic problem, we are not merely slotting in a strong preference for a better long-term outcome and acting on it, which could be a rational choice explanation for this behavior. Rather we are creating daily habits and rituals that make our lives feel good and meaningful, irrespective of whether we will succeed at stabilizing our greenhouse gas emissions at levels that could avoid severe and volatile outcomes. Consistent with virtue theory, we are fashioning an ethic that does not depend solely on a narrow version of rationality or consequences, but rather on an account of a fully realized human life. Moreover, we are acting ethically in the face of potential futility. Doing so may reflect a transition from a stage of control and dominance to one of care-taking and wisdom. Love, care, and wisdom are, according to Erikson, the central virtues of adulthood.[44] Analogously, virtue theorists have suggested that love, care, and wisdom, as well as a string of other virtues associated with communion, environmental activism, and sustainability, are central to good environmental character.[45] Perhaps the actions being taken across the globe are signs that we, as a species, are growing up.

The motivations to make this shift are as varied as the people and the communities that are making it. For some, like the Inuit, this conception of life as daily tending to the Earth has deep and ancient roots in culture and religion. As the late Justice William Brennan once put it: "[many indigenous] religions regard creation as an on-going process in which they are morally and religiously obligated to participate."[46] The Inuit have adopted this view of their role in the new challenge. For others, like the ECI and its members, the motivation is their own religious worldview. For all, there are, I suspect, some common threads. One recurring theme is that people want

---

[44] Erik H. Erikson, *Insight and Responsibility* (New York: W. W. Norton & Co., 1972), 115.
[45] Sandler, *Character and Environment*, 22.
[46] *Lyng v. Northwest Cemetery Ass'n*, 485 U.S. 439, 460 (1988), Brennan, J. dissenting.

to be seen by subsequent generations as people who tried their best. They do not want to leave a legacy of indifference behind, even if they also are aware that what they are trying to do may not be enough. For example, Thomas R. Carper, a US Senator from Delaware, was quoted as saying that he "did not want his children and grandchildren chastising him for inaction in decades to come. 'I don't want them to say, "What did you do about it? What did you do about it when you had an opportunity? Weren't you in the Senate?"' Carper said, adding that he hoped to tell them, 'I tried to move heaven and Earth to make sure we took a better course.'"[47] What is particularly telling about this quotation is that it assumes that the Senator has failed in his efforts. So his hope is not just that he succeeds politically. Surely he hopes that, but also that he is seen as someone who tried his hardest to avert harm to future generations, even if his work was unavailing.

Another theme is that daily engagement with something linked to a higher purpose is itself meaningful. One of the CRAG members put it this way, "I don't want our credits to be like taxes that we only think about once a year . . . I want them to be the lifeblood of the way we operate every day."[48] The Maryland Craggers state in their website that "we felt like we were part of something that mattered and could help other people."[49] Like parents, these politicians and ordinary folks are acting every day for the next generation. And like parents, CRAG members and others have found meaning and enjoyment in their new practices, even if occasionally they wish they could just turn up the darned heat. Also like parents, they will not know if their contributions, in the end, have had a happy conclusion. For parents in the usual sense, of course, do not hope to outlive their children, making the best end of the parenting story an inevitable unknown. (When parents in fact know the end, because their children have predeceased them, it is therefore an awful one.)

What all this may point to is that we are on the verge of a very different way to think about success, security, meaning, and happiness than the dominant ways of thinking about such things, at least in the Western developed world, for the last couple of centuries. And this all might be, not just a fine thing, but a wonderful, joyful thing. We may be developing a more dispersed and global sense of identity and obligation, and accompanying flexible attitudes and behaviors regarding how to measure contentment and satisfaction.

---

[47] Juliet Eilperin, "Lawmakers on Hill Seek Consensus on Warming," *Washington Post*, January 31, 2007.

[48] Kanter, "Members of New Group in Britain."   [49] www.carbonrationing.org.uk/maryland.

On a darker but not unrealistic note, these are the same attitudes that our children and grandchildren will need if our generation continues to fail to address climate change, and they are living in a world where formulating ethics in the face of natural resource devastation will be the presiding challenge.[50] That possibility is hardly indistinct, given the ongoing failure of the UN-sponsored process to formulate an enforceable multilateral climate treaty and the tenuousness of any substitute political arrangements between the USA, China, and India. Our political ingenuity seems to lag behind our technological prowess, leaving future generations and the natural world with the legacy of the Anthropocene as an atmospheric, geologic, and ecological matter, but not with respect to constructing an enduring and enforceable legal regime that can match these pervasive effects. The prospect of political failure at the global level serves only to heighten the importance of cultivating virtue locally. Without a personal and identity-based sense of why to care about nature, other species, or future generations, the very idea of an environmental ethic may simply slip away. Tragedy thus haunts environmental virtue as a theoretical and practical matter. Moreover, the tragedy is of human origin. The parenting metaphor captures this in a way that other virtue-oriented metaphors do not. There is nothing inherently dominating or tragic about friendship or stewardship. But, parenting, as a matter of description and potential, has these features. We might not like to think of our species as having overwhelming influence on the planet, and even if we recognize our control and influence, we might also prefer to believe that we have the intelligence and skill to fix everything. Yet climate change, the paradigmatic evidence of the Anthropocene, gives the lie to both. Parenting as a metaphor captures these aspects of control and tragedy, which is not to say that others may not do just as well. But a metaphor that lacks these features will leave us ill-equipped to confront current and future realities.

7 GROWING UP IS NOT THE ONLY OPTION

As suggested at the outset, the decisions we make at this crossroads in our relationship with the planet will depend on our values, our conception of

---

[50] Two fictional renderings of such a world include Cormac McCarthy, *The Road* (New York: Vintage International, 2006), and James Howard Kunstler, *World Made by Hand* (New York: Atlantic Monthly Press, 2008). Both books explore what life might look like after catastrophic global events that eliminate modern technology. In *The Road*, life is reduced to its barest elements, and nearly all animal and plant species have been eliminated, leaving humans to wander the scorched Earth in search of sustenance. Kunstler's version of the post-apocalypse is slightly more familiar. Local flora and fauna have survived in some regions, and though all infrastructure from the energy economy is gone, there are vestiges of technology and engineering from which to build premodern communities.

ourselves, and the role we would like to play. They also depend on our sense of the ontology of the planet (which itself is a matter of background values). What is the planet? If viewed as an endlessly malleable resource, which, when we apply our dazzling ingenuity to it, can yield ever increasing wealth for humans, then the choice to address climate change by reducing emissions is not at all obvious. This view of the planet, which is the one that underlies the logic both of consumer capitalism and its supporting academic disciplines, including welfare economics and rational choice theories generally,[51] lends itself to skepticism about whether it is worth it economically to change our consumption patterns and energy infrastructure. This skepticism has two components. First, we are likely to fail to rein in emissions enough to avoid dramatic effects from climate change anyway. Second, we may be able to engineer our way out of the gravest difficulty.[52]

If, on the other hand, the planet is viewed as a bounded system with resources that are by definition limited, and further that there is beauty, meaning, and value in the way the Earth's flora and fauna (including the human fauna) interact, then putting that same dazzling human ingenuity to work to place humanity within, rather than above, the rest of the planet has greater appeal. At bottom, many of the disputes about how much mitigation (i.e., emissions reduction) to do to address climate change, versus how much to rely on geo-engineering, sequestration, and other technology-only fixes, come down to this difference in views about the world. Accompanying these different senses of the Earth's ontology are different outlooks on the human experience. If the Earth is a small, limited system that landed, for whatever set of reasons (and senses of this vary depending on religious and metaphysical orientation), in our hands, then part of being human is to care for it. If the Earth is, like Mary Poppins' magical bag, a source of endless material for satisfaction of human needs, then the predominant human mission is to exercise our powers to extract use out of that material.

Global warming may nudge more people to see the Earth as a bounded system, but it is not automatic that it will do so. The facts about global

---

[51] For a good summary and critique of rational choice theories generally, see Michael Taylor, *Rationality and the Ideology of Disconnection* (New York: Cambridge University Press, 2006). For a similarly critical assessment of the modern discipline of economics in particular, see Stephen A. Marglin, *The Dismal Science: How Thinking Like a Scientist Undermines Community* (Cambridge, MA: Harvard University Press, 2008).

[52] For different versions of this view, see Alan Carlin, "Why a Different Approach Is Required if Global Climate Change Is to Be Controlled Efficiently or Even at All," *William & Mary Environmental Law & Policy Review* 685 (2008), 32 (arguing that geo-engineering solutions are far preferable to emissions reduction approaches); and Lomborg, *Cool It* (arguing that the costs of mitigation are too high, and that adaptation and poverty relief should instead be the dominant policies).

warming, like the facts about almost anything, can play a role in shaping values and worldviews, but they can also be slotted into pre-existing frames.[53] Further, there is a point at which the facts simply run out. Will geo-engineering solutions, such as spraying particulate matter into the stratosphere or seeding the ocean with iron to grow more algae,[54] work seamlessly, with no down-side effects on ecosystems or even human health? We may have factually based reasons to have serious doubts, but we do not know for certain, which is why we rely on background values to adopt rules of decision. For those in the "Earth as bounded system" camp, the rules that tend to rein in actions with potentially harmful, even if uncertain, effects on natural systems are preferred. For those in the "Earth as malleable resource" account, some version of cost-benefit analysis, with discounts for future generations, is the preferred rule. The rules we adopt are based ultimately on our sense of ourselves and how we want to spend our time here on earth, which brings us back to the metaphor of developmental stages. According to Erikson's account, the central virtues of childhood include hope, will, purpose, and competence.[55] These virtues seem eerily resonant with the view of the Earth as an infinitely malleable resource: "In the individual here and now [an exclusive condition of hopefulness] would mean a *maladaptive optimism*. For true hope leads inexorably into conflicts between the rapidly developing self-will and the will of others from which the rudiments of will must emerge."[56] Will, purpose, and competence, like hope, provide the necessary groundwork for mastering the world. What is lacking in these virtues, however, is a quality of sustaining love or obligation, as well as a sense of balancing one's own assertions of will with the world's many limitations. Those virtues arise later, first in adolescence, with fidelity, and then in adulthood, when, if the cultural and ideological framework is there to reinforce them, the virtues of care, love, and wisdom come to the fore.[57] The adult virtues build on each other, adding dimensions of obligation and widening concern.[58] Love "pervades the intimacy of individuals and is thus the basis of ethical concern," but can also become "a joint selfishness in the service of some territoriality, be it bed or home, village or

---

[53] There is a large and growing body of literature on this, including Dan Kahan and Donald Braman, "Cultural Cognition and Public Policy," *Yale Law & Policy Review* 149 (2006), 24.

[54] Sid Perkins, "Scientists Work to Put Greenhouse Gas in Its Place," *Science News* 173, no. 16 (May 10, 2008), www.sciencenews.org/view/feature/id/31431 (describing ocean seeding and sequestration possibilities); Philip J. Rasch *et al.*, "Exploring the Geoengineering of Climate Using Stratospheric Sulfate Aerosols: The Role of Particle Size," *Geophysical Research Letters* 35, LO2809 (2008) (exploring spraying aerosols into the stratosphere to block incoming heat radiation).

[55] Erikson, *Insight and Responsibility*, 115.   [56] Ibid., 118 (emphasis added).

[57] Ibid., 131–4.   [58] Ibid.

country."[59] Care, then, extends beyond this potential for mutual narcissism. According to Erikson,

Care is a quality essential for psychosocial evolution, for we are the teaching species. Only man ... can and must extend his solicitude over the long, parallel and overlapping childhoods of numerous offspring united in households and communities. As he transmits the rudiments of hope, will, purpose, and competence ... he conveys a logic much beyond the literal meaning of the words he teaches, and he gradually outlines a particular world image and style of fellowship. All of this is necessary to complete in man the analogy to the basic, ethological situation between parent animal and young animal ... Once we have grasped this interlocking of human life stages, we understand that adult man is so constituted as to *need to be needed* lest he suffer the mental deformation of self-absorption, in which he becomes his own infant and pet.[60]

Displacement of the self by taking on the care of others is the key to moving beyond the "mental deformation of self-absorption." To carry this to the species and planetary level, perhaps now is the moment when humanity must choose either to remain static, suspended in a state of self-love, or to move on to a stage of caring for the planet we live on and the many lives, human and otherwise, that it sustains. Finally, of wisdom, Erikson has this to say:

For if there is any responsibility in the cycle of life it must be that one generation owes the next that strength by which it can come to face ultimate concerns in its own way – unmarred by debilitating poverty or by the neurotic concerns caused by emotional exploitation.[61]

Wisdom, which is the knowledge accumulated over a lifetime of reflecting on the human condition, enables the broadest view of human responsibility. Wisdom entails an acceptance of one's own decline and mortality while seeing the needs of the coming generations.[62] "Wisdom, then, is detached concern with life itself, in the face of death itself."[63] Wisdom, like climate change, has a tragic structure. It may take a virtue wrought from tragic circumstances to match a structurally tragic commons problem.

These virtues – love, care, and wisdom – are the grown-up ones. They are the virtues necessary to accept a role of care-taking, of reducing our own demands on the Earth in order to cultivate the conditions for *all* human communities, in company with other species, to make their way.

We may not be entering a stage in which these virtues predominate. We may, as a world community, be too entrenched, too dispersed, and too

[59] Ibid., 130.   [60] Ibid., 131.   [61] Ibid., 133.   [62] Ibid.   [63] Ibid.

path-dependent on the dominant technological frames to make the leap. Whether, as a whole, we are making that transition or not, certain factions of humanity, including but not limited to the ones described above, already are. And it may well be that these factions become the subcultures from which the rest of the world will have to learn, when and if we reach the bleaker stages of global warming that many scientists predict in the absence of immediate and radical action.[64]

CONCLUSION

Vice President Al Gore has made much of the fact that the Chinese word for *crisis* comprises two characters, the first the symbol for "danger" and the second the symbol for "opportunity."[65] In the film and book, *An Inconvenient Truth*, Gore catalogues the dangers of the "climate crisis," and also outlines the two kinds of opportunity it presents, both practical and moral. As to the latter, Gore writes that what we are presented with is the opportunity "of knowing: *a generational mission*; the exhilaration of a compelling *moral purpose . . . the opportunity to rise.*"[66] In an interesting parallel, Erikson theorizes that the move from one stage of human development to another is prompted by a crisis of personality. According to Erikson,

> Crisis once meant a turning point for better or for worse, a crucial period in which a decisive turn *one way or another* is unavoidable . . . Such crises occur in man's total development sometimes more noisily, as it were, when new instinctual needs meet abrupt prohibitions, sometimes more quietly when new capacities yearn to match new opportunities, and when new aspirations make it more obvious how limited one (as yet) is. [67]

We appear to be experiencing a crisis in this Eriksonian sense of an unavoidable turn. We have put the conditions in motion to alter the atmosphere of the planet in dramatic and pervasive ways. Whether we choose to respond by attempting to match our moral and emotional capacities to these circumstances or not, our actions will affect the fate of the entire Earth and its future. Another way in which Erikson's version of crisis seems apt is that, in the course of a human life cycle, crises are normal; just part of growing up. It takes a crisis to prompt the transition to the next

---

[64] For a survey of the data on the likelihood of cataclysmic effects, see Michael C. MacCracken *et al.* (eds.), *Sudden and Disruptive Climate Change* (London: Earthscan, 2008).

[65] See Gore, *An Inconvenient Truth*, 10.   [66] Ibid., 11.

[67] Erikson, *Insight and Responsibility*, 139 (emphasis in original).

stage. It is normal, in other words, to confront the limits of a previous frame. Seeing it this way might dampen the fear of judgment that creeps into many discussions about obligations to less well-off human communities and other species. Were we bad and wrong to have emitted greenhouse gases in the way that we did? The question is somewhat irrelevant to figuring out how to react today. The normality of the Eriksonian version of crisis is helpful in one final way. If crises are normal, then we must learn to live with them for the long haul. The global climate crisis, rather than being experienced as a sudden conflagration, will be felt in myriad and dispersed ways, often not viscerally traceable to our emissions patterns. Despite this, the crisis calls for us to respond, one way or another, to the astounding ways in which we have put ourselves in the position, wittingly or not, of parenting the planet.

CHAPTER 8

Living ethically in a greenhouse

Robert H. Socolow and Mary R. English

It was made clear at the December 2009 conference on climate change in Copenhagen (Conference of the Parties 15) that the nations of the world are only beginning to concede that they face a common threat. It was widely reported that there was a deep divide at Copenhagen between delegates from "developed" countries and delegates from "developing" countries, and that the depth of the anger of the delegates from developing countries surprised the delegates from developed countries. Should the anger have been surprising? Not only had some of the developed countries – most notably, the USA – failed to take significant steps prior to the meeting to reduce the impacts of their economies on the climate. In addition, the developed countries had come to the meeting to revise the global structure of climate change mitigation such that all countries (or at least all of the major economies) would share the task. This arrangement, all conceded, entailed a sharp departure from the previous structure, in place since the 1992 United Nations Framework Convention on Climate Change, which dealt with equity across nations by dividing the world into two groups of countries with "common but differentiated responsibilities." Only the group of "Annex 1" countries (approximately, the countries of the Organization for Economic Cooperation and Development plus Russia) was obligated to make legally binding mitigation commitments.

Our chapter identifies a critical requirement for progress: the widespread development of moral imagination, in order for many more individuals to develop a planetary identity that augments their other loyalties. We defend a fresh formulation of equitable allocation of responsibility. We argue for the merits of accounting rules that focus on the world's individuals first and its countries second. Such an accounting would treat equally all individuals whose contributions to global emissions are the same, irrespective of whether they live in the USA or in Bangladesh. This accounting would therefore reflect individual lifestyles, as well as the institutions in each country that mediate lifestyles to create environmental impacts.

The next few decades are a crucial time to develop common values and aspirations through dialog. There is a need, for example, to discuss the desirability of a totally managed planet with many species of plants and animals found only in botanical gardens and zoos, versus a world with greater randomness and wildness. Philosophers have a major role here. Their professional assignment has long been to think about and help others think about what it means to be human. Our chapter argues that they now have an additional task: to help us think about what we as human beings should strive to accomplish during the millennia that lie ahead.

[margin note: Philosoper's]

We are mindful that most of our analysis is predicated on the future bringing only modest changes in the globally dominant conceptualization of the good life. Given such a premise, the global targets endorsed at Copenhagen will be very hard to reach. Therefore, our chapter necessarily takes a positive view of the promise of technology to lead the way to an environmentally safer world. We argue for a nuanced view of technology that presumes that the implementation of every option can be done badly or well.

[margin note: looks at promise of technology]

Returning to our original point, attaining the ultimate goal of long-term CO_2 stabilization will require not only a technological but also a moral transformation: one that, we argue, necessitates cultivating a planetary identity using the tool of moral imagination. This moral transformation can and should be fostered now. Realistically, however, it will be slower to take root than a technological transformation. Both the immediate technological transformation and the fundamental moral transformation are essential.

[margin note: Do you believe we need both a fundamental moral transformation or technological transformation? Jamieson argues moral]

STRUCTURE OF OUR ARGUMENT

The global climate change problem intersects ethics in countless places. Many of these intersections are well mapped, because they have been encountered with other environmental issues. Here, we seek to identify nine major intersections of climate change and ethics. The first four concern the planet. The remaining five concern humans, individually and collectively.

From the point of view of the planet:

1. The problem for the Earth is making room for all of us.
2. The Earth has provided us coastlines for cities and weather for agriculture, but it has had many coastlines and all sorts of weather, and they will change again.
3. The Earth's climate is being changed by us at an unprecedented rate.
4. Solutions that can limit climate change are not innocuous. *[margin note: ← define]*

From the point of view of humans:

5. Each person's "share" of future CO_2 emissions must be small.

6 Regardless of where they live, the world's individuals with the largest emissions must reduce their emissions to much lower levels.

7 Global climate change raises both the opportunity and the necessity of forging a "planetary" identity, using the tool of moral imagination.

8 Global climate change also raises the opportunity and necessity of cultivating *prospicience*.

9 We should focus on managing the climate change problem, not on "solving" it.

Which is the most important?

I THE PROBLEM FOR THE EARTH IS MAKING ROOM FOR ALL OF US

The Earth is small, relative to the demands we put upon it. Our demands result from values of self-realization and equal opportunity coupled with consumer values. Many of these values have become widespread only in the past century. In short, global climate change is a consequence of the success of the modern agenda. The good life is nearly universally defined as exuberant consumerism; in other words, as the self-gratifying acquisition of plentiful and varied goods, services, and experiences. This conception of the good life can and should be challenged – see Section 7 below – but for now, it is a major force behind the rapid increase in the emissions of CO_2 and other greenhouse gases. *Discuss the good life our tendencies*

Population

The problem of global climate change is exacerbated by global population growth. The collision of economic and population growth with environmental limits was prominently foretold in the 1970s,[1] but we collectively chose to disparage and shun the messenger. As a result, the need for action is even more urgent than if we had started to act thirty years ago. In 1960, when the possible effects of CO_2 were first being recognized, the global population was 3 billion. By 1980, the global population had grown to nearly 4.5 billion, and in early 2008 it was nearly 6.7 billion. Looking forward, however, rapid upward or downward changes in global population are not expected. The climb in global population is now less steep, because the transition to replacement-level birthrates is largely accomplished in many parts of the world. As for a future decline, perhaps our descendants will find a graceful way to reduce the global population without war or pestilence – say, down to the 1950 level of 2.5 billion by 2200. If that happens, it will become

---

[1] Donella H. Meadows, Jorgen Randers, Dennis L. Meadows, and William W. Behrens III, *The Limits to Growth: A Report for the Club of Rome's Project on the Predicament of Mankind* (New York: Potomac Associates, 1972).

easier for humans to fit on the planet. The time scale, however, is longer than
the one we are considering here.

As Socrates famously said of morality, "We are discussing no small matter,
but how we ought to live." In the following sections, we concentrate on
carbon-intensive human practices but do not deal with the issue of human
reproduction. Nevertheless, that too is a question of how we ought to live.

2 THE EARTH HAS PROVIDED US COASTLINES FOR CITIES AND WEATHER FOR AGRICULTURE, BUT IT HAS HAD MANY COASTLINES AND ALL SORTS OF WEATHER, AND THEY WILL CHANGE AGAIN

Historically, we shaped our civilizations around particular environmental
circumstances: We built our cities near rivers and coasts, and we planted our
crops where the rain fell. These practices were logical at the time, but they
are not fortuitous given global climate change. When the earth emerged
from the last ice age, the sea level rose nearly 100 meters, but sea level has
changed very little during the period of human settlement. A rise of another
5 meters could result if either the ice sheet on Greenland or the ice sheet
in West Antarctica were to slide entirely into the sea.

The consequences of sea-level rise for two parts of the world, one rich and
one poor, are seen in Figures 1 and 2. Already severely affected by storm
surges, Bangladesh is one of the nations most vulnerable to sea-level rise. It
is also one of the world's poorest nations.[2]

Climate change can mean disruption for farmers as a result of an over-
abundance of water in some places and drought in others. Effects can be
subtle. In Australia, for example, a persistent drought has led rice farmers to
sell some of their land and water rights to wine producers, who can produce
a crop of equal or greater monetary value from each acre with one-third
as much water. The Deniliquin mill, the largest rice mill in the Southern
Hemisphere, once processed enough rice to meet the needs of 20 million
people but is now mothballed.[3]

With severe and persistent flooding, drought, and other environmental
conditions triggered by climate change, will resettlement be necessary? If so,
there will be massive human costs, largely endured by some of the world's
poorest people.

---

[2] UNEP/GRID-Arendal, "Potential Impacts of Climate Change," *Vital Climate Graphics*, online at
www.grida.no/climate/vital/33.htm.
[3] Keith Bradsher, "A Drought in Australia, A Global Shortage of Rice," *New York Times*, April 17, 2008.

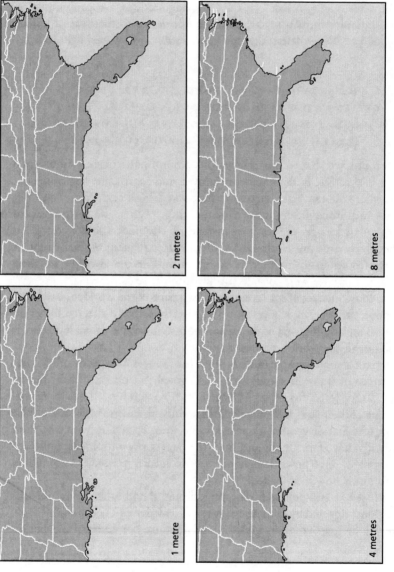

Figure 1 New coastlines for the southeastern Gulf states in the USA resulting from sea-level rises of 1, 2, 4, and 8 meters.
Source: T. Knutson, Geophysical Fluid Dynamics Laboratory, National Oceanic and Atmospheric Administration.

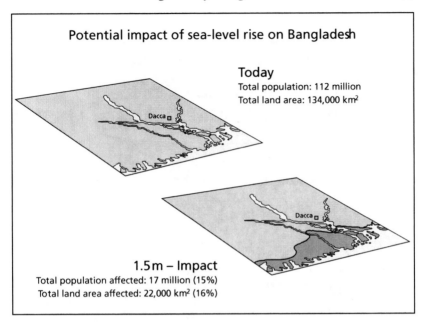

Figure 2 Effect on Bangladesh of a 1.5-meter sea-level rise.

3 THE EARTH'S CLIMATE IS BEING CHANGED BY US AT AN UNPRECEDENTED RATE

The climate change problem can be conceptualized by representing the CO_2 in the atmosphere as a bathtub (Figure 3). Here, we discuss only CO_2 – the largest contributor to greenhouse gas emissions – while noting that other gases, especially methane, also are important to the total picture.

As shown in Figure 3, the quantity of CO_2 in the atmosphere today is approximately 3,000 billion metric tons, which equals a concentration of 390 parts per million (ppm). Adding 7.7 billion metric tons of CO_2 (in which there are 2.1 billion metric tons of carbon) raises the concentration of CO_2 in the atmosphere about one part per million. In the 250 years since the "preindustrial" (i.e., pre-1750) era, we have added as much CO_2 to the atmosphere as was added during the 20,000 years of emergence from the depth of the last ice age. With only 1,400 billion more metric tons of CO_2 in the atmosphere, we will have doubled the concentration of CO_2 relative to preindustrial levels.

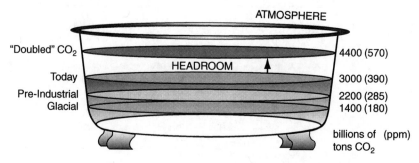

Figure 3 Past, present, and potential future levels of CO_2 in the atmosphere.

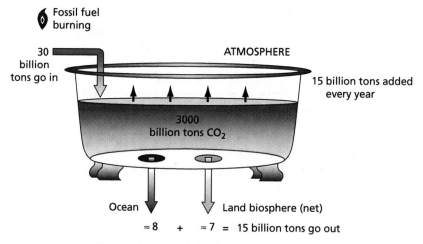

Figure 4 Net annual global increase in CO_2 today.

The bathtub has two drains, however (Figure 4). While 30 billion metric tons of CO_2 are added from burning fossil fuels annually (about 4.5 metric tons per capita globally, at the current population size), it is estimated that today the ocean drain and the land drain together remove about 15 billion metric tons of CO_2 annually. Much less clear is how the two drains (technically, these drains are called "sinks") divide the job. (Note that the symbol ≈ in Figure 4 means "approximately"; the apportionment of the total removal of 15 billion metric tons of CO_2 per year between the land and the ocean sinks is still uncertain.) The ocean becomes more acidic as a result of the CO_2 entering at the ocean surface.

Ocean becoms more acidic

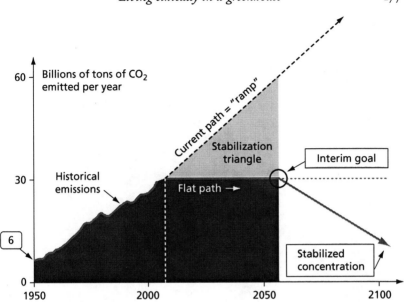

Figure 5 The stabilization triangle.

Why the land today is absorbing CO_2 is not well understood: evidently, in spite of human-caused deforestation, the forests and other plant life of the planet are gaining carbon; we know that this has not always been so – even in the past century.[4]

We have not always emitted CO_2 at the current global rate of 30 billion metric tons per year. As recently as 1950, the total global CO_2 emissions rate was about 6 billion metric tons annually, or approximately 2.4 metric tons per capita at the 1950 global population of roughly 2.5 billion. To stabilize CO_2 emissions at a level just below the 4,400 line in Figure 3, we need to hold total global CO_2 emissions at no more than 30 billion metric tons annually over the next 50 years as an interim goal and then drop to no more than 10 billion metric tons annually after another 50 years. The first 50 years of this job is captured by the "stabilization triangle" (Figure 5).

We have delayed long enough. If we delay yet longer, what will the consequences be, and on whom will they fall most heavily?

[4] Robert H. Socolow and Sau-Hai Lam, "Good Enough Tools for Global Warming Policy Making," *Philosophical Transactions of the Royal Society*, 365 (2007), pp. 897–934.

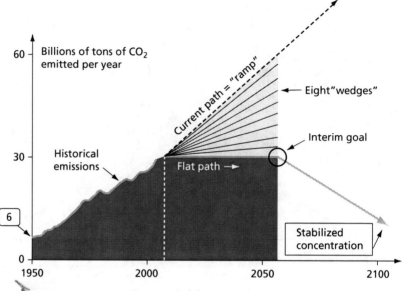

Figure 6 Stabilization wedges.

4 SOLUTIONS THAT CAN LIMIT CLIMATE CHANGE ARE NOT INNOCUOUS

Despite the urgency of the problem, it can be managed. The world today has a terribly inefficient energy system; most of the "physical plant" that we will have 50 years from now has not yet been built; CO_2 emissions have just begun to be priced. The need for CO_2 emission reductions is pressing, but opportunities abound.

Elsewhere,[5] it has been pointed out that the problem of managing climate change can be decomposed into "wedge strategies" using current technologies. Taken together, these wedges enable addressing the climate change problem – not forever, but for the next 50 years (Figure 6).

A "wedge" is a strategy – already commercially available – to reduce CO_2 emissions by 4 billion metric tons per year in 50 years, for a total reduction of 100 billion metric tons over the 50-year period (Figure 7).

---

[5] Stephen W. Pacala and Robert H. Socolow, "Stabilization Wedges: Solving the Climate Problem for the Next 50 Years with Current Technologies," *Science*, 305 (2004), pp. 968–972. (Note that in this chapter, a wedge is slightly smaller: One billion metric tons of *carbon* per year not emitted in 50 years. One billion metric tons of carbon is 3.67 billion metric tons of CO_2.) Also see Robert H. Socolow and Stephen W. Pacala, "A Plan to Keep Carbon in Check," *Scientific American*, 295, no. 3 (2006), pp. 50–57.

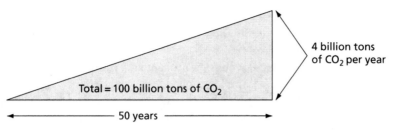

Figure 7 A wedge.

The stabilization triangle can be filled with eight wedges in such categories as:

- efficient use of fuel through, e.g., hybrid vehicles, mass transit, rail freight, and reductions in vehicle miles traveled
- substituting accessibility for mobility through, e.g., video-conferencing and at-home delivery of goods and services
- efficient use of electricity through, e.g., improved motors and lighting
- capture and reuse of wasted energy through, e.g., heat recycling and cogeneration of electricity
- "decarbonized" electricity from, e.g., nuclear power, wind turbines, and coal with CO_2 capture and storage
- "decarbonized" fuel – e.g., biofuels and geothermal heating and cooling
- methane management.

To these pragmatic technologies that are well understood we may be able to add other, more radical climate change management measures through "earth engineering" technologies. The two most discussed examples are (1) tuning the atmospheric CO_2 concentration by capturing large amounts of CO_2 directly from the air and storing it underground, and (2) compensating for global warming with global "dimming" by placing reflective particles in the upper atmosphere.[6] We may also be able to benefit from the arrival of commercially viable nuclear fusion and other options still much in the research stage today.

Neither the current technologies for managing climate change nor the prospective, more sweeping technologies are innocuous.[7] It is a misconception

[6] J. J. Blackstock, D. S. Battisti, K. Caldeira, D. M. Eardley, J. I. Katz, D. W. Keith, A. A. N. Patrinos, D. P. Schrag, R. H. Socolow, and S. E. Koonin, *Climate Engineering Responses to Climate Emergencies* (Santa Barbara, CA: Novim, 2009) (archived online at: http://arxiv.org/pdf/0907.5140).

[7] National Research Council, *Hidden Costs of Energy: Unpriced Consequences of Energy Production and Use* (Washington, DC: National Academies Press, 2009).

to think that the technological alternatives we can draw upon are without flaws and that only a combination of conspiracy and inertia prevents us from sailing into calm waters.

On the contrary, every alternative is problematic, from seemingly benign wind energy,[8] to deeply complex biofuel,[9] to geopolitically challenging nuclear power;[10] Energy conservation may mean regimentation; nuclear power is haunted by questions about waste management and nuclear proliferation; coal – often called "clean" when emissions at power plants are controlled – still can produce detrimental effects on miners, surface water and ground water, and the land mined; and renewable energy at large scale can make immoderate demands on land. Implementation in all instances can be done badly or well. The cure can all too readily become worse than the disease. But it does not have to.

"Solution science" – that is, the study of the environmental and social costs and benefits of stabilization strategies – is emerging. Through this study, which is necessarily interdisciplinary, solutions can be examined for flaws as well as benefits. Ultimately, however, no single solution will suffice. Portfolios of solutions will be needed. And ultimately, the choices about which solutions to pursue and which to forego will be political and ethical choices, based on individual and collective values. We individually and collectively will need to answer the question: What criteria should we use to compare the disruption triggered by various solutions with the disruption triggered by climate change?

5 EACH PERSON'S "SHARE" OF FUTURE CO$_2$ EMISSIONS MUST BE SMALL

To achieve *long-term* stabilization, roughly 10 billion metric tons of CO$_2$ emissions per year can be emitted: that is, one-third of today's global CO$_2$ emissions rate (Figure 8). If the flat path and follow-on descending path in Figure 5 were followed, this rate would be reached in 2100. Comparing

[8] National Research Council, *Environmental Impacts of Wind-Energy Projects* (Washington, DC: National Academies Press, 2007).

[9] D. Tilman, Robert H. Socolow, J. A. Foley, J. Hill, Eric Larson, L. R. Lynd, Stephen W. Pacala, J. Reilly, Timothy Searchinger, C. Sommerville, and Robert H. Williams, "Beneficial Biofuels: The Food, Energy, and Environment Trilemma," *Science*, 325, no. 5938 (2009), pp. 270–271. See also Robert H. Socolow, J. A. Foley, J. Hill, Eric Larson, L. R. Lynd, Stephen W. Pacala, J. Reilly, Timothy Searchinger, C. Sommerville, and Robert H. Williams, "Response to Letters to the Editor," *Science*, 326 (2009), p. 1344.

[10] Robert H. Socolow and Alexander Glaser, "Balancing Risks: Nuclear Energy and Climate Change," *Dædalus*, 138, no. 4 (2009), pp. 31–44.

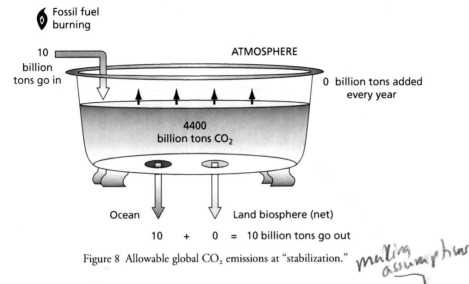

Figure 8 Allowable global CO_2 emissions at "stabilization." *making assumptions*

Figures 4 and 8, we see that we get this answer by assuming (arbitrarily) that at that future time the terrestrial biosphere will be carbon neutral and the ocean uptake of CO_2 will be slightly larger than today.

Dividing target global CO_2 emissions by the growing world population reveals that "stabilization" – that is, a total CO_2 emissions rate that will not contribute to further global climate change – is at a very low level of individual emissions. It is not sufficient simply to limit emissions in the prosperous parts of the world. If per capita emissions from the less prosperous nations of the world – for example, non-members of the Organization for Economic Co-operation and Development (OECD) – were allowed to fully "catch up," the planet would be overwhelmed by CO_2 emissions. Indeed, total emissions in non-OECD countries already are roughly half of the global total (Figure 9). *undeveloped countries* *Do you think it's right to tell them they can't?*

In a climate-stabilized world, the CO_2 emissions per capita would be equal to those of people today whom no one would call well-off. What are the implications of this ineluctable fact for both the less prosperous and the more prosperous people of the world?

One implication is that to reach such a low level of global emissions, the world will have to implement low-carbon technologies widely and aggressively. As this happens, everyone's CO_2 emissions will fall, even if they are doing exactly what they have done before: lighting their bedrooms, for example. Another implication is that, as seen in the next section, there is room for the world's least prosperous people to increase their fossil fuel

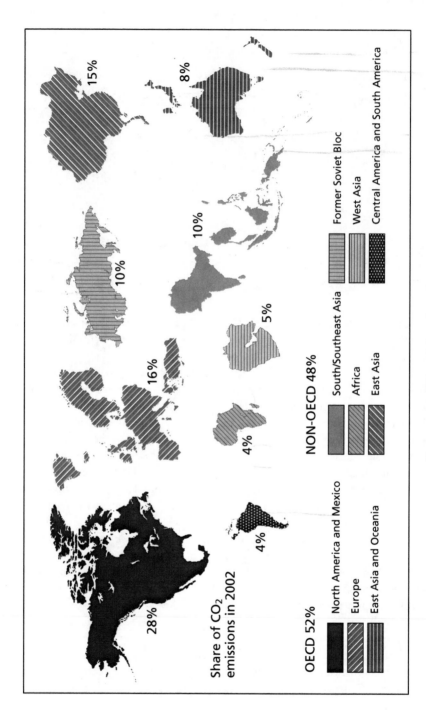

Share of CO$_2$ emissions in 2002

OECD 52%
- North America and Mexico — 28%
- Europe — 16%
- East Asia and Oceania — 10%

NON-OECD 48%
- South/Southeast Asia — 10%
- Africa — 4%
- East Asia — 15%
- Former Soviet Bloc — 10%
- West Asia — 5%
- Central America and South America — 4%
- 8%

Figure 9 2002 CO$_2$ emissions, by region.

emission rates now, even while the required decarbonization of the global economy is getting under way. The poor can be allowed – indeed, enabled – to use diesel engines for village-scale power; liquid propane gas for cooking; grid-connected electricity for lights, refrigerators, and cell-phone charging; and gasoline for motorbikes. A third implication, also discussed in the next section, is that the world's individuals with the largest CO_2 emissions must drastically cut their emissions.

6 REGARDLESS OF WHERE THEY LIVE, THE WORLD'S INDIVIDUALS WITH THE LARGEST EMISSIONS MUST REDUCE THEIR EMISSIONS TO MUCH LOWER LEVELS

As of 2003, the mean per capita CO_2 emissions level was 4.1 metric tons per year, but the distribution of emissions among people was extremely skewed. Only 27 percent of the world's population, or about 1.7 billion people, were above the mean, and they emitted 79 percent of the total global CO_2 emissions. Some of these 1.7 billion people are very rich; others, barely not poor. Moreover, only about 54 percent (about 900 million) lived in the OECD nations. The others (about 800 million) lived in developing nations. For 2030, according to a model embedding assumptions about trends in population and consumption, the individual CO_2 emissions of 2.8 billion people will exceed the 2003 mean, and 62 percent (more than 1.7 billion) of them will live in non-OECD nations.[11]

Evidently, the world needs a new conceptualization of global burden-sharing that takes into account the poor in rich countries and the rich in poor countries. The "greenhouse development rights" (GDR) framework of Baer, *et al.*[12] makes a similar point. The basic ethical implications are clear: Distributive justice in CO_2 emissions is not only about rich and poor nations but also about rich and poor individuals. "Common but differentiated responsibilities," since first invoked in 1992 in the United Nations Framework Convention on Climate Change, has been understood to refer to nations. This view has led to stalemate.

What sort of allocation scheme across the world's countries might emerge if one begins with a focus on the emissions of individuals? Figure 10 shows one

[11] Shoibal Chakravarty, Ananth Chikkatur, Heleen de Coninck, Stephen Pacala, Robert Socolow, and Massimo Tavoni, "Sharing Global CO_2 Emission Reductions Among One Billion High Emitters," *Proceedings of the National Academy of Sciences of the USA*, 106, no. 29 (2009), pp. 11884–11888 (doi:10.1073/pnas.0905232106).

[12] Paul Baer, Tom Athanasiou, Sivan Kartha, and Eric Kemp-Benedict, *The Greenhouse Development Rights Framework: The Right to Develop in a Climate Constrained World*, rev. 2nd edn (Berlin: Heinrich Böll Foundation, Christian Aid, EcoEquity, and the Stockholm Environment Institute, 2008).

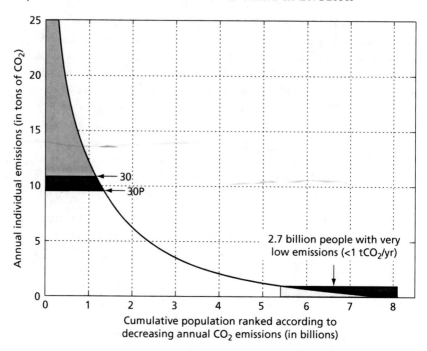

Figure 10 Projected distribution of global emissions across the
world's individuals in 2030 and associated numerical constructions
relevant to a proposed scheme for equitable distribution across countries.

Source: S. Chakravarty *et al.*, "Sharing Global CO_2 Emission
Reductions Among One Billion High Emitters," *Proceedings
of the National Academy of Sciences of the USA*, 106: 11884–11888.

approach, where the world's total emissions are allocated across all of the
world's individuals, who are lined up left to right, with the highest-emitting
individuals on the left. The curve shows projected global emissions in 2030 and
was created with the help of recent national income distributions, assumptions
about future regional emissions growth, and a few additional assumptions.[13]
The global population in 2030 is assumed to be 8.1 billion people, and the total
annual emissions of CO_2 in 2030 are assumed to 43 billion metric tons, based
on 2007 Energy Information Administration information.

Imagine that the world's nations look at this projected level of emissions
and decide, collectively, that it is too large. Instead, say, the world decides
that it would be more prudent for 2030 emissions to be only 30 billion

[13] Chakravarty *et al.*, "Sharing Global CO_2 Emission Reductions."

metric tons of CO_2; i.e., 13 billion metric tons of CO_2 less, or roughly the same as today. (This is just an example; in Copenhagen in December 2009 much of the world was discussing, implicitly, a somewhat tougher target.) The horizontal line marked "30" corresponds to this choice. The gray area at the upper left bounded at the bottom by "30" corresponds to the needed reduction of 13 billion metric tons of CO_2. It results when the 1.1 billion individuals with the highest emissions all cap their 2030 emissions at 10.8 metric tons of CO_2. Each country looks into the emissions of those of its inhabitants who are included in these 1.1 billion people and works out its share of the emissions in the gray area, which corresponds to that nation's allocation of the global emissions reduction. (In this specific analysis, the emissions of 270 million individuals living in the USA and 300 million individuals living in China exceed the cap. Because China's high-emitting individuals emit less, however, the magnitude of China's obligation for emissions reduction in 2030 is smaller.) The scheme leaves it to every country to choose how it achieves its targeted reduction of emissions.

Shifting the bottom boundary at the left downward to "30P" to include the black area immediately below the gray area corresponds to keeping the global emissions target unchanged but modifying the scheme so as to pay special attention to the world's very poor. These people closely correspond to the 2.7 billion lowest-emitting individuals represented in the light gray area at the extreme right. The extra emissions reduction allows the emissions of these individuals to be raised to 1 metric ton of CO_2 per year. The two black regions on the graph have the same area. Taking global poverty into account for this particular analysis, the global target of no more than 30 billion metric tons of CO_2 emissions for 2030 is achieved when the individual ceiling is 9.6 metric tons per year. Before the world acts, 1.3 billion people have emissions above that level and 2.7 billion people have emissions below 1 metric ton of CO_2 per year.

Conceivably, a conceptualization that treats all of the world's high emitters alike and all of its low emitters alike, no matter where they live, may be able to break the international logjam and free diplomacy to invent new solutions.

[handwritten: Do you think it would be good to do this?]

7 GLOBAL CLIMATE CHANGE RAISES BOTH THE OPPORTUNITY AND THE NECESSITY OF FORGING A "PLANETARY" IDENTITY, USING THE TOOL OF MORAL IMAGINATION

We each have multiple social identities as individuals, family members, members of a community or tribe, and members of nested political entities

[handwritten margin notes: Allows / raise / for / poor]

such as the village, the province, and the nation. Do our individual values and collective norms change as large numbers of us feel that, in addition, we owe allegiance to the planet and all of its people?

A planetary identity decreases the likelihood that I will see remote humans as alien, not worthy of moral consideration. It also increases the likelihood that I will be interested in the survival of other species sharing the planet with me. The emergence of planetarians may be a silver lining in the dark cloud of global climate change. To make the radical changes needed in our CO_2 emission practices, a planetary identity may be necessary. To develop this identity, however, moral imagination will be required. We must individually and collectively overcome our predisposition to see moral issues through narrow lenses.

Moral imagination has been called the ability to discover and evaluate possibilities that are not merely determined by a particular circumstance, with that circumstance's operative mental models, nor framed by a set of rules or rule-governed concerns.[14] It also has been described as the ability to form mental constructions of what is not real to oneself, permitting one to create possible worlds that are either morally better or worse than the world as we find it.[15]

As noted by Arnold and Hartman,[16] moral imagination has its roots in the work of philosophers several centuries ago: in particular, that of David Hume (*A Treatise of Human Nature*) and Immanuel Kant (*Critique of Pure Reason*). In addition, the early twentieth-century work of John Dewey, a philosopher, psychologist, and educator who advocated pragmatism, has contributed to contemporary concepts of moral imagination.[17] For example, John Dewey said of imagination that "only imaginative vision elicits the possibilities that are interwoven within the texture of the actual."[18]

In the past 20 years, moral imagination as a tool for ethical inquiry has been developed by philosophers such as John Kekes,[19] who noted that moral imagination can have not only an exploratory but also a corrective

---

[14] Patricia H. Werhane, *Moral Imagination and Management Decision-Making* (New York: Oxford University Press, 1999).

[15] Denis G. Arnold and Laura P. Hartman, "Moral Imagination and the Future of Sweatshops," *Business and Society Review*, 108 (2003), pp. 425–461.

[16] *Ibid.*

[17] Steven Fesmire, *John Dewey and Moral Imagination: Pragmatism in Ethics* (Indianapolis, IN: Indiana University Press, 2003).

[18] *Ibid.*, p. 68 (quoting Dewey in *Art as Experience*, 1934).

[19] John Kekes, "Moral Imagination, Freedom, and the Humanities," *American Philosophical Quarterly*, 28, no. 2 (April 1991), pp. 101–111.

function; and Mark Johnson,[20] who noted that moral reasoning is basically imaginative, because it uses imaginative structures such as images, narratives, and metaphors. Over the past dozen years, business ethicists in particular (e.g., Patricia Werhane, Denis G. Arnold) have developed and applied the tool of moral imagination.

Werhane has identified several requisites for moral imagination:
- being self-reflective about both oneself and one's situation
- disengaging from one's situation and being aware of the mental model or "script" dominating that situation
- imagining new possibilities outside the prevailing mental model
- evaluating from a moral point of view both the original situation and its dominant mental models and the new possibilities one has imagined.[21]

Moral imagination differs from conventional moral reasoning in two important ways. On the one hand, moral imagination avoids the rigidity that can come from relying mainly on abstract rules such as principles of distributive justice (e.g., the principle of equal shares, the principle of "from each according to his ability, to each according to his need," or the difference principle of John Rawls[22]) or abstract procedures such as cost/benefit analysis. On the other hand, moral imagination avoids the narrowness and, arguably, the flaccidness and moral relativism that can characterize purely situated moral reasoning. As Arnold and Hartman note,[23] citing Kekes: "Without the exercise of moral imagination, cultural myopia, ideology, and limited experience can individually and collectively constrain one's moral outlook."

Moral imagination is necessary but not sufficient for decision-making.[24] Moral imagination gives us freedom from our daily mental models, but we also need to be able to give good reasons for our judgments, if only to communicate them to others. According to Werhane, "a well-functioning moral imagination needs moral reasoning skills to amplify and justify intuitions, abstract and amplify what we learn from stories and cases, linking these together in what one hopes is a coherent and relevant point of view."[25]

---

[20] Mark Johnson, *Moral Imagination: Implications of Cognitive Science for Ethics* (University of Chicago Press, 1993).

[21] Werhane, *Moral Imagination*. Also see Patricia H. Werhane, "Moral Imagination and Systems Thinking," *Journal of Business Ethics* , 38 (2002), pp. 33–42.

[22] John Rawls, *A Theory of Justice* (Cambridge, MA: Harvard University Press, 1971).

[23] Arnold and Hartman, "Moral Imagination."

[24] Werhane, *Moral Imagination*. Also see Patricia H. Werhane, "A Place for Philosophers in Applied Ethics and the Role of Moral Reasoning in Moral Imagination: A Response to Richard Rorty," *Business Ethics Quarterly*, 16, no. 3 (2006), pp. 401–408.

[25] Werhane, "A Place for Philosophers," p. 405.

As with traditional moral reasoning, moral imagination and judgment were initially conceived as operating mainly at the individual level but can operate at collective levels as well. Drawing on systems thinking and the work of Henk van Luijk, Werhane[26] has extended the concept of moral imagination to the organizational and systems levels. She notes that each system or subsystem is goal-oriented, and that this goal-orientation – together with the structure and interrelationships shaped by the goals – accounts for the system's normative dimensions. "On every level, the way we frame the goals, the procedures and what networks we take into account makes a difference in what we discover and what we neglect."[27] Systems, like individuals, use mental models. As with an individual's mental models, a system's mental models can be challenged by moral imagination – in particular, by dispassionate self-evaluation.

The concept of moral imagination and judgment can be applied to the problem of global climate change at the levels of individuals, organizations and institutions, and political economies. Global climate change poses the challenge of balancing a critical but diffuse goal – the reduction of greenhouse gases – with more immediate and thus seemingly more vital goals. For example, an immediate goal for both individuals and collectivities is well-being. Typically, as discussed in Section 1, modern conceptions of well-being are translated into consumerism – i.e., the acquisition of abundant and varied goods, services, and experiences. Using the tool of moral imagination, however, well-being can be coupled with virtue and uncoupled from purely material prosperity. The "good life" (or *eudaimonia*) can take on a meaning that includes material simplicity as well as caring about and for others.

Global climate change also poses the challenge of balancing a temporally and spatially vast problem with pressing responsibilities to oneself and proximate others. Through moral imagination and judgment, however, we can individually and collectively realize that these pressing quotidian responsibilities can and should share the stage with our planetary responsibilities. One set of responsibilities does not trump the other. Because immediate responsibilities are tangible while planetary responsibilities are dauntingly huge, it is tempting to duck, ostrich-like, when confronted with the latter. One possible answer is to recognize that no single person or collectivity can "solve" the global climate change problem, but, as discussed in Section 8, all need to cultivate prospicience, and, as discussed in Section 9, all can contribute to step-wise management of the global climate change problem.

[26] Werhane, "Moral Imagination and Systems Thinking."   [27] Ibid., p. 36.

8 GLOBAL CLIMATE CHANGE ALSO RAISES THE OPPORTUNITY AND NECESSITY OF CULTIVATING *PROSPICIENCE*

Prospicience is defined in the *Oxford English Dictionary* as "the action of looking forward, foresight." Derived from the Latin *prospicientia*, the word is rarely used today. It can take on new meaning, however, to describe a new, much-needed intellectual domain. Prospicience can be thought of as "the art and science of looking ahead." In the past few decades we have become aware of our deep past: the history of our Universe, our Earth, and life at the genomic level. Can we achieve a comparably sophisticated sense of the Earth and human civilization at various future times?

For example, national populations that climb to a constant level and are stable ever after do not seem likely. Nor does a world in which we have all become middle class, live in peace and tend our gardens, and have smoothly functioning national and global institutions. Instead, directly ahead is an era of countries with falling as well as rising populations, entrenched poverty in specific places, shrinking wilderness, and conflict over access to water, food, minerals, and fuels. Many of us have long assumed that our children and grandchildren will be richer than we are, but environmental concerns force us to examine this assumption. In the face of global climate change, resource scarcity, and an increasing world population, what issues are likely to arise, when, with what options?

Prospicience can help us sort out our individual and collective goals and responsibilities for distinct time frames: for example, the next 5 years versus the next 50 years versus the next 500 years. The debate over carbon capture and storage, like the debate over radioactive waste storage, raises questions about what we can feasibly do now or in the near future, in light of what we know and don't know about the distant future. As discussed in Section 9, we must act in the face of necessarily partial knowledge.

9 WE SHOULD FOCUS ON MANAGING THE CLIMATE CHANGE PROBLEM, NOT ON "SOLVING" IT

In Section 3, we identified a path of constant global emissions at today's rate, 30 billion metric tons of CO_2 per year, for the next half-century (Figure 5). In Section 6, we further identified a scheme that would result in national allocations to achieve this goal, based on a cap on the world's high emitters and a floor on the world's low emitters (Figure 10). Figure 5, furthermore, shows a world that achieves a longer-term balance of CO_2 flows into and out of the atmosphere, but after not 50 but 100 years, when the emissions rate falls to 10 billion metric tons per year.

Our primary focus needs to be on what *we* should be doing *now and in the next few decades.* We have to slow the supertanker before its course can be reversed. This will require technologies available today, but today's technologies will not in themselves suffice. Using them, we must act now, but we also must lay the foundation for the future: for both better technologies and better-informed scientists and citizens. Rather than pursuing the quixotic goal of CO_2 stabilization as soon as possible, with minimal concern for the deficiencies of mitigation strategies, we should pursue a path of step-wise decision-making that leads us with all deliberate speed toward aggressive, globally coordinated CO_2 emissions reductions. The distinction between "as soon as possible" and "with all deliberate speed" may seem trivial, but it is not. "As soon as possible" suggests not only urgency but also hastiness. "With all deliberate speed" suggests not only urgency but also deliberation.

In our deliberations, we should honor the intelligence of those coming after us, even as we try to minimize the burdens of our faulty decisions. There are no once-and-for-all solutions: The well-meant but misguided 1982 Nuclear Waste Policy Act has taught us that. Predicated on the idea that we should shoulder the entire responsibility for our own nuclear waste and not impose this burden on future generations, the 1982 Act instead has resulted in endless delays in arriving at even temporary consolidated storage for spent fuel and high-level radioactive waste. Surely we can and should do better than that.

Over the past 25 years, philosophers, economists, and others have debated our responsibilities to future generations.[28] They ask: Are those responsibilities equal to our responsibilities toward today's generations? In contrast, do we have no responsibilities to future generations? Or is the middle road — that of diminishing responsibilities to future generations as they become more remote in time — the morally right view? Here, we side with the middle road, but mainly for pragmatic rather than ethical reasons. In practice, it seems most plausible to be capable of taking responsibility

[28] Ernest Partridge (ed.), *Responsibilities to Future Generations* (Amherst, NY: Prometheus Books, 1981); Derek Parfit, "Future Generations: Further Problems," *Philosophy and Public Affairs*, 11, no. 2 (1982), pp. 113–172; G. Bruntland (ed.), *Our Common Future: The World Commission on Environment and Development* (Oxford University Press, 1987); Robert M. Solow, "Sustainability: An Economist's Perspective," the Eighteenth J. Seward Johnson Lecture to the Marine Policy Center, Woods Hole Oceanographic Institution, at Woods Hole, Massachusetts on June 14, 1991. Reprinted in Robert N. Stavins (ed.), *Economics of the Environment*, 4th edn (New York: W. W. Norton, 2000), pp. 131–138 (available online at www.owlnet.rice.edu/~econ480/notes/sustainability.pdf).

for and planning for "the rolling present," as some have dubbed it.[29] Using this concept, our main responsibility to future generations is to provide the next one or two generations with the skills, resources, and opportunities they will need to cope with the problems we have left behind. These generations, in turn, have a similar responsibility toward the two succeeding generations, and so forth.

There is much hubris in believing that our generation can provide a final "answer" that spares all future generations. Instead, we should more modestly seek improvements within our capabilities, acknowledging that those who come after us will, if we carry out our responsibilities today, have greater knowledge – and, one hopes, at least as much wisdom. We are thus linked, generation to generation. We are the captain of the supertanker for a short time; then others succeed us. Our undoing – and theirs – comes if we think we can chart the course for them.

CONCLUSION

Despite the importance just noted of not seeking "final answers," we must remember that we want climate change management strategies to work. It is not enough to identify what's wrong with a strategy when it first is proposed. We then must ask: Are there changes that would make this strategy acceptable? How might we get there from here? Can we improve this strategy by addressing its technical defects, environmental risks, governance issues, etc.?

We may decide that a particular strategy for climate change management is simply too unpalatable to be adopted. But given the gravity of the climate change problem, we – both the professional communities and the public – cannot allow ourselves to be excessively squeamish about imperfect strategies. The more dire the consequences of environmental stress from climate change, the less we can allow ourselves to flatly reject strategies such as nuclear power or below-ground CO_2 storage. To achieve environmental soundness and adequate prosperity for all will require not only a technological overhaul but also an overhaul of our individual and collective thinking.

---

[29] Bayard L. Catron, Lawrence G. Boyer, Jennifer Grund, and John Hartung, "The Problem of Intergenerational Equity: Balancing Risks, Costs, and Benefits Fairly Across Generations," in C. Richard Cothern (ed.), *Handbook for Environmental Risk Decision Making* (New York: CRC Press, 1996), pp. 131–148.

CHAPTER 9

Beyond business as usual: alternative wedges to avoid catastrophic climate change and create sustainable societies

Philip Cafaro

There is a curious disconnect in climate change discourse, between explanations of the causes of global climate change (GCC) and discussions of possible solutions. On the one hand, it is widely acknowledged that the primary causes of climate change are unremitting economic and demographic growth. As the Fourth Assessment Report from the Intergovernmental Panel on Climate Change (IPCC) succinctly puts it: "GDP/per capita and population growth were the main drivers of the increase in global emissions during the last three decades of the 20th century ... At the global scale, declining carbon and energy intensities have been unable to offset income effects and population growth and, consequently, carbon emissions have risen."[1] On the other hand, most proposals for climate change mitigation take growth for granted and focus on technical means of reducing greenhouse gas emissions.

Climate scientists speak of the "Kaya identity": the four primary factors which determine overall greenhouse gas emissions. They are economic growth/per capita, population, energy used to generate each unit of GDP, and greenhouse gases generated per unit of energy. Over the past three and a half decades, improvements in energy and carbon efficiency have been overwhelmed by increases in population and wealth. Here are the numbers, again according to the IPCC: "The global average growth rate of CO_2 emissions between 1970 and 2004 of 1.9% per year is the result of the following annual growth rates:

population + 1.6%,
GDP/per capita + 1.8%,

---

[1] Intergovernmental Panel on Climate Change (IPCC), *Climate Change 2007: Mitigation* (2007), Technical Summary, p. 107.

energy-intensity (total primary energy supply (TPES) per unit of GDP) – 1.2%, and carbon-intensity (CO_2 emissions per unit of TPES) – 0.2%."[2]

Crucially, the IPCC's projections for the next several decades see a continuation of these trends (Figure 1). More people living more affluently means that under "business as usual," despite expected technical efficiency improvements, greenhouse gas emissions will increase between 25 and 90 percent by 2030, relative to 2000 (note the net change increases projected in Figure 1).[3] If we allow this to occur, it will almost surely lock in global temperature increases of more than 2°C over pre-industrial levels, exceeding the threshold beyond which scientists speak of potentially catastrophic climate change. Following this path would represent a moral catastrophe as well: the selfish over-appropriation and degradation of key environmental services by the current generation to the detriment of future ones, by rich people to the detriment of the poor, and by human beings to the great detriment of the rest of the living world.[4]

A reasonable person reading the IPCC report and subsequent scientific literature on climate change would likely conclude that we are bumping up against physical and ecological limits. Given the dangers of catastrophic GCC, a prudent and moral response might be: "Wow! This is going to be hard. We need to start working on this problem with all the tools at our disposal. Increasing energy and carbon efficiency, to be sure. But also decreasing the pursuit of affluence and overall consumption; and stabilizing or reducing human populations. Maybe in the future we can grow like gangbusters again, maybe not. But for now, people need to make fewer demands on nature and see if even our current numbers are sustainable over the long haul. After all, our situation is unprecedented – 6.9 billion people living or aspiring to live in modern, industrialized economies – and there is no guarantee that we aren't already in 'overshoot' mode."

Such convictions would only be strengthened by considering further evidence of global ecological degradation from the recent *Millennium Ecosystem Assessment* (MEA), including the depletion of important ocean fisheries, accelerating soil erosion, ongoing species extinctions throughout the world, the growth of immense "dead zones" at the mouths of many great rivers, and more. According to the MEA, humanity is currently degrading or utilizing unsustainably fifteen of twenty-four key ecosystem services.[5]

[2] Ibid.   [3] Ibid., p. 111.
[4] Donald Brown *et al.*, *White Paper on the Ethical Dimensions of Climate Change*, Rock Ethics Institute, Pennsylvania State University (2007).
[5] Walter Reid *et al.*, *The Millennium Ecosystem Assessment: Ecosystems and Human Well-Being: Synthesis* (Washington, DC: Island Press, 2005).

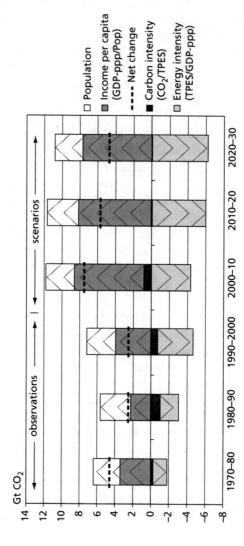

Figure 1 Decomposition of global energy-related CO_2 emission changes at the global scale for three past and three future decades.
Source: IPCC, *Climate Change 2007: Mitigation*, Technical Summary, figure 1.6. GDP-ppp/Pop = gross domestic product purchasing-power parity per person/total population; CO_2/TPES = total CO_2 emissions/total primary energy supply; TPES/GDP-ppp = total primary energy supply/gross domestic product purchasing-power parity per person.

However, neither GCC nor any of these other problems have led to a widespread re-evaluation of the goodness of growth.[6] Regarding GCC, we have seen a near-total focus on technological solutions by politicians, scientists, and even environmentalists. I contend that this is a serious mistake. Because business as usual with respect to growth probably cannot avoid catastrophic GCC or meet our other global ecological challenges, we need to consider a broader range of alternatives that include slowing or ending growth. Philosophers exploring this issue should ask what ethical role the notion of growth plays within climate change debates, and how widely held views of the sanctity of growth may close off viable or preferable courses of action. Continued neglect of this topic will undermine philosophers' attempts to specify a just and prudent course of action on climate change.

I THE WEDGE APPROACH

Of the many possible examples of mainstream approaches to climate change mitigation that we might consider, let us look at one of the most rigorous and influential, Stephen Pacala and Robert H. Socolow's "wedge" approach.[7] The wedge approach is a heuristic designed to help us compare alternative mitigation schemes. Pacala and Socolow's stated goal, in the face of numerous mitigation proposals and skepticism about whether any of them can succeed, is to provide a practical road map of choices that can facilitate successfully addressing the problem of GCC.

Each wedge in the "stabilization triangle" (Figures 2a and 2b) represents a technological change which, fully implemented, would keep 1 billion metric tons of carbon from being pumped into the air annually, 50 years from now. It would also prevent 25 billion metric tons of carbon from being released during the intervening 50 years.[8] The authors figure that eight such wedges must be implemented – not to reduce atmospheric CO_2, not to stabilize

---

[6] Brian Czech, *Shoveling Fuel for a Runaway Train: Errant Economists, Shameful Spenders, and a Plan to Stop them All* (Berkeley, CA: University of California Press, 2002); Gustave Speth, *The Bridge at the Edge of the World: Capitalism, the Environment, and Crossing from Crisis to Sustainability* (New Haven, CT: Yale University Press, 2009).

[7] This approach was first presented in Stephen Pacala and Robert Socolow, "Stabilization Wedges: Solving the Climate Problem for the Next 50 Years with Current Technologies," *Science*, 305 (2004), pp. 968–972. The latest iteration is Robert H. Socolow and Mary R. English, "Living Ethically in a Greenhouse," Chapter 8 in this volume. Recent research and even a downloadable version of the "Carbon Mitigation Wedge Game" can be found at the website for the Carbon Mitigation Initiative (www.princeton.edu/~cmi).

[8] In the most recent version of this approach, each wedge prevents 4 billion metric tons of CO_2 from being emitted annually, rather than 1 billion metric tons of carbon. Since 1 metric ton of carbon equals 3.67 metric tons of CO_2, the wedges are slightly larger in the new version.

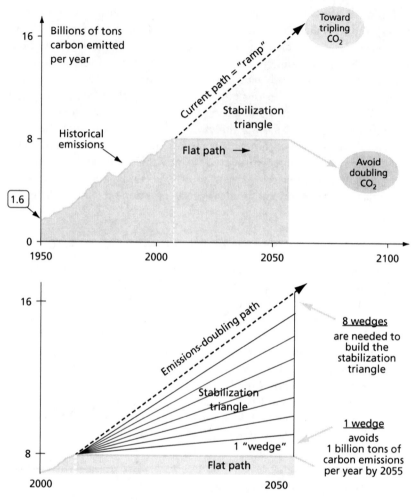

Figures 2a and 2b Stabilization triangles.
Source: Carbon Mitigation Initiative, Princeton University.

CO$_2$ levels – but simply to keep atmospheric carbon from doubling and pushing past potentially catastrophic levels during this period. In order to fully address GCC, in the following 50 years, humanity would have to take further steps and move to an economy where human carbon outputs do not exceed carbon uptakes in natural systems. The wedge approach buys us time and (allegedly) begins the transition toward such an economy. Table 1 shows the fifteen carbon reduction wedges proposed by Pacala and Socolow.

Table 1 *Potential conventional carbon mitigation wedges*

Option		Effort by 2055 for one wedge, relative to 14 gigatons of carbon per year (GtC/year) under business as usual
Energy efficiency and conservation	1. Efficient vehicles	Increase fuel economy for 2 billion cars from 30 to 60 mpg (12.75 to 25.5 km/l)
	2. Reduced vehicle use	Decrease car travel for 2 billion 30-mpg (12.75 km/l) cars from 10,000 to 5,000 miles (16,100 to 8,050 km) per year
	3. Efficient buildings and appliances	Cut carbon emissions by 1/4 in projected buildings and appliances
	4. Efficient baseload coal plants	Produce twice today's coal power output at 60% efficiency compared with 32% efficiency today
Fuel shift	5. Gas baseload power for coal baseload power	Replace 1,400 GW coal plants with natural gas plants (4 times current gas-based power production)
	6. Nuclear power for coal power	Add 700 GW nuclear power (tripling current capacity)
Carbon capture and storage	7. Capture CO_2 at power plants	Introduce CCS at 800 GW coal plants or 1,600 GW natural gas plants
	8. Capture CO_2 at hydrogen (H_2) plants	Introduce CCS at plants producing 250 million metric tons H_2/year from coal or 500 million metric tons H_2/year from natural gas
	9. Capture CO_2 at coal-to-synfuel plants	Introduce CCS at synfuel plants producing 30 million barrels per day from coal (200 times current capacity)
Renewable electricity and fuels	10. Wind power for coal power	Add 2 million 1-MW-peak windmills (50 times current capacity) occupying 30 million hectares, on land or offshore
	11. Photovoltaic (PV) power for coal power	Add 2,000 GW-peak PV (700 times current capacity) on 2 million hectares
	12. Wind-generated H_2 fuel-cell cars for gasoline-powered cars	Add 4 million 1-MW-peak windmills (100 times current capacity)
	13. Biomass fuel for fossil fuel	Add 100 times current Brazil or US ethanol production, with the use of 250 million hectares (1/6 of world cropland)
Forests and agricultural soils	14. Reduce deforestation, plus reforestation and new plantations	Halt tropical deforestation instead of 0.5 billion metric tons C/year loss, and establish 300 million hectares of new tree plantations (twice the current rate)
	15. Conservation tillage	Apply best practices to all world cropland (10 times the current usage)

Source: Carbon Mitigation Initiative, Princeton University (table modified).

Despite a stated desire to consider only alternatives that are technically feasible today, scaling up the carbon capture and storage options (wedges 7–9) appears to rely on future technological improvements that may not prove practicable.[9] While some of the wedges could pay for themselves over time, most, on balance, would involve significant economic costs. Most wedges also carry significant environmental costs, which in some cases may equal or outweigh the environmental benefits they would provide in helping mitigate GCC. This is arguably the case with the proposed nuclear wedge, given problems with waste disposal. Similarly, the coal wedges would encourage continued environmental degradation from coal mining. Even seemingly benign alternatives, such as wind or solar power expansion, will result in huge wildlife habitat losses if pursued on the scale demanded to achieve a full wedge.

One strength of the wedge approach is that it allows us to specify the costs and benefits of different courses of action and thus choose intelligently between them. So far this has mostly meant asking which wedges are cheapest economically. But the approach also allows us to compare alternatives based on environmental impacts, equitable sharing of costs and benefits, overall contribution to sustainability, or whatever criteria we deem relevant – if, that is, we are considering a reasonably complete set of alternatives.

Take another look at the fifteen proposed wedges. Fourteen focus on improvements in energy efficiency, or substitutions in energy and materials production; one or perhaps two wedges involve limiting consumption (cutting the miles driven by automobile drivers; maybe limiting deforestation); and none of them involve limiting human population growth. This is hardly a peculiarity of Pacala and Socolow. Most discussions of GCC neglect the possibility of limiting consumption or stabilizing populations. The goal, always, seems to be to accommodate *more* consumption by *more* people with *less* environmental impact.

Numerous illustrations can be cited from the IPCC's fourth assessment report itself. Its authors recognize agriculture as a major contributor to GCC, for example. Yet they simply accept projections for greatly increased demands for all categories of agricultural products (including a doubling in worldwide demand for meat over the next fifty years) and focus on changes in tillage, fertilizer use, and the like, as means to limit increased greenhouse

---

[9] Synapse Energy Economics, *Don't Get Burned: The Risks of Investing in New Coal-Fired Generating Facilities* (New York: Interfaith Center on Corporate Responsibility, 2008), pp. 29–30.

gas emissions.[10] Similarly, the assessment report notes that among significant greenhouse gas sources, aviation traffic is the fastest-growing sector worldwide. It considers numerous changes to aviation practices, including relatively trivial improvements in airplane technology and changes in worldwide flight patterns, while avoiding the obvious alternative of reducing the number of flights.[11] Many similar examples could be given.

2 CAUSES AND CONSEQUENCES OF THIS FAILURE

Now we need to be clear. The failure to consider policies designed to reduce consumption or limit population can't be chalked up to these factors' unimportance. The IPCC assures us that they are all-important in generating GCC. Nor is it because there aren't policies that might reduce consumption or slow population growth; there are many policy alternatives in these areas. Nor is it because such policies necessarily would be more expensive, harder to implement, more coercive, or in any other way less appealing than the technological approaches under consideration. Some may be, of course. But as I show below, there are almost certainly consumption and population wedges that could be developed and implemented at less economic, environmental, and social cost than most of the wedges proposed by Pacala and Socolow, and even with considerable overall benefit.

The real problem, I submit, is that the majority of policymakers and analysts are so ideologically committed to maximizing growth that it is impossible for them to consider the full range of alternatives. There are legitimate pragmatic worries that questioning growth will turn the public against climate change mitigation, and legitimate questions about how to reduce growth without reducing people's quality of life. Still, this failure could prove disastrous. Dire as the predictions in the fourth assessment report were, the scientific literature has grown even grimmer in the years since its publication, leading some climate scientists to argue that humanity must reduce greenhouse gas emissions more quickly and stabilize emissions at lower levels than previously thought.[12] We may not need eight wedges over the next fifty years, but ten or more, to avoid catastrophic GCC. Furthermore, like other approaches focused on midterm carbon reductions, the wedge framework

---

[10] IPCC, *Climate Change 2007: Mitigation*, ch. 8, "Agriculture."

[11] Ibid., ch. 5, "Transport and Its Infrastructure"; see also IPCC, *Aviation and the Global Atmosphere* (1999).

[12] James Hansen, "Scientific Reticence and Sea Level Rise," *Environmental Research Letters*, 2 (2) (2007), article 024002; James Hansen et al., "Target Atmospheric CO_2: Where Should Humanity Aim?," *Open Atmospheric Science Journal*, 2 (2008), pp. 217–231.

postulates a transition to *much* lower emissions in the long term, in order to meet the GCC challenge. But such a transition seems unlikely if we squeeze all the most easily achieved efficiency gains out of the system, while increasing the number of people and ratcheting up their per capita demands.

Pacala and Socolow try to meet this problem by advocating increased research and putting their faith in future technological breakthroughs to achieve a near-zero carbon society. I believe we cannot rest in such faith, given the stakes involved. I'm all for increased research into new energy technologies. But I think we'll probably need wisdom as well as cleverness, self-restraint as well as technological ingenuity, to avoid catastrophic GCC. And as many of the contributors to this volume argue, meeting this challenge is a moral imperative. Morality demands that this generation of wealthy human beings construct a plausible bridge to a sustainable future, and walk across it. It also demands that this bridge be one that the world's poor and other species can cross along with us.[13]

Proposals to trade in our old air-conditioned SUVs for new, air-conditioned *hybrid* SUVs and roar into the future with the hope of a Hail Mary techno-pass at the end do not meet our moral obligations.[14] This is so, because the potential harms are so great. The Fourth Assessment Report speaks of the likelihood that, during the coming decades, tens and perhaps hundreds of millions of poor people will have their lives threatened by GCC-induced droughts and famines, storms, and floods.[15] It estimates that 20 to 30 percent of the world's species will be extinguished or threatened with extinction by the end of the century, under business as usual emissions scenarios.[16] Such immense harms, should they occur, will have been caused (at least partly) by us, and they can be averted (at least partly) by us. The

---

[13] Brown *et al.*, *Ethical Dimensions of Climate Change*; Holmes Rolston III, "Duties to Endangered Species," in Rolston, *Philosophy Gone Wild: Environmental Ethics* (Buffalo, NY: Prometheus Books, 1989), pp. 206–219.

[14] Nor do radical geoengineering schemes to mitigate GCC. Stephen Gardiner details several such proposals and suggests that their proliferation exemplifies a "moral corruption" that seeks to avoid full responsibility for dealing with the problem ("Is 'Arming the Future' with Geoengineering Really the Lesser Evil? Some Doubts about the Ethics of Intentionally Manipulating the Climate System," in Gardiner, Simon Caney, Dale Jamieson and Henry Shue (eds.), *Climate Ethics: Essential Readings* (Oxford University Press, 2010). There are ample anthropocentric reasons for opposing climate geoengineering, based on its uncertain impacts on future human generations; I am also impressed by its breathtaking anthropocentrism. In effect, such proposals accept that the Earth, which over eons has been home to an amazing and ever-increasing diversity of hundreds of millions of species, will henceforth be managed for the benefit of one species: *Homo sapiens sapiens*. I call this interspecies genocide. But GCC policy should not be about securing *Lebensraum* for the master species.

[15] IPCC, *Climate Change 2007: Synthesis Report*, pp. 48–52.

[16] Ibid., p. 48. See also Chris Thomas *et al.*, "Extinction Risk from Climate Change," *Nature*, 427 (2004), pp. 145–148; *Millennium Ecosystem Assessment*, Biodiversity Synthesis, pp. 42–47.

scale of these potential harms and our responsibility for causing them demand a commensurate response, one with a *high likelihood* of averting catastrophic GCC, not just the *possibility* of doing so.

"Can Advances in Science and Technology Prevent Global Warming?" asks Michael Huesemann in a recent review article of the same name. After detailed analysis, he answers that an exclusive focus on efficiency improvements is unlikely to prevent catastrophic GCC. Indeed, "it is highly questionable that 12-fold to 26-fold increases in Gross World Product [over the twenty-first century, as predicted by the IPCC] are even remotely achievable because of biophysical constraints and the inability of technology to sufficiently uncouple energy and materials use from the economy."[17] Meeting the GCC challenge almost certainly depends on ending human population growth and either ending economic growth or radically transforming it, so that *some* economic growth in *some* sectors of the modern economy and in poorer countries that can actually benefit from it can be accommodated without radically destabilizing the Earth's climate. All the technological improvements we can muster will probably be necessary to enable this transition to a slow-growth or post-growth future – not as an alternative to it.[18]

Given the stakes involved, we should consider our full range of options. In the remainder of this chapter, I propose a number of alternative wedges focused on reductions in consumption, population growth, and economic growth. If techno-wedges are sufficient or superior, then they should be able to demonstrate that in a direct comparison with these alternative wedges. If not, then they should be supplemented or supplanted by the alternatives.

3 ALTERNATIVE WEDGES: CONSUMPTION

Consider four consumption wedges, the first two focused on food consumption and transportation, and the following two seeking to rein in consumption more generally. Once again, each of these wedges represents a technological or policy change that would prevent 1 billion metric tons of carbon from being emitted 50 years from now and 25 billion metric tons from being emitted over the next 50 years.[19]

---

[17] Michael Huesemann, "Can Advances in Science and Technology Prevent Global Warming? A Critical Review of Limitations and Challenges," *Mitigation and Adaptation Strategies for Global Change*, 11 (2006), p. 566.
[18] Bill McKibben, *Deep Economy: The Wealth of Communities and the Durable Future* (New York: Henry Holt, 2007).
[19] See the page "Alternative Climate Wedges" on my website (www.philipcafaro.com) for more detailed specifications and supplementary material on these and other wedges described in the following sections.

3.1 Meat wedge

According to a recent comprehensive study, agriculture currently contributes 18 percent of total world greenhouse gas emissions and livestock production accounts for nearly 80 percent of this.[20] Thus meat-eating contributes approximately 2.38 billion metric tons carbon equivalent to current greenhouse gas emissions. The UN Food and Agricultural Organization projects a worldwide doubling in animal production between 2000 and 2050, from 60 billion to 120 billion animals raised annually, which, under "business as usual," will double the greenhouse gas emissions from this sector to 4.76 billion metric tons. If, instead, we hold worldwide animal food production steady over the next 50 years, this would provide nearly two and a half carbon wedges (averting a 2.38 billion metric tons increase), while merely preventing half the projected doubling during that time would supply more than one full carbon wedge (1.19 billion metric tons averted).[21] Such wedges might be accomplished non-coercively by increasing the price of meat, removing subsidies for cattle production, banning confined animal feedlot operations (CAFOs) as the European Union is in the process of doing, and directly taxing meat to discourage consumption. These measures could accommodate a reasonable increase in meat-eating in poor countries where many people eat little meat, while providing environmental and health benefits in wealthy countries where people eat more meat than is good for them.[22] They could complement efforts to improve the conditions under which food animals are raised, changes that may be expensive, but which are arguably demanded by morality anyway.[23]

3.2 Aircraft wedge

According to the Fourth Assessment Report, civil aviation is one of the world's fastest-growing sectors of significant greenhouse gas emissions. Analysis shows that air traffic "is currently growing at 5.9% per year [and] forecasts predict a global average annual passenger traffic growth of around

---

[20] United Nations Food and Agriculture Organization, *Livestock's Long Shadow: Environmental Issues and Options* (Rome: 2006), p. 112.

[21] Gidon Eshel and Pamela Martin, "Diet, Energy, and Global Warming," *Earth Interactions*, 10 (2006), paper no. 9.

[22] Philip Cafaro, Richard Primack, and Robert Zimdahl, "The Fat of the Land: Linking American Food Overconsumption, Obesity, and Biodiversity Loss," *Journal of Agricultural and Environmental Ethics*, 19 (2006), pp. 541–561.

[23] Bernard Rollin, *Animal Rights and Human Morality*, 3rd edn. (Buffalo, NY: Prometheus Books, 2006). For a detailed discussion of the meat/heat connection, see the report *Global Warning: Climate Change and Farm Animal Welfare* from the group Compassion in World Farming (Godalming, UK: 2007).

5% – passenger traffic doubling in 15 years."[24] Under current projections, carbon emissions from aircraft might increase from 0.2 billion metric tons per year to 1.2 billion metric tons annually over the next 50 years.[25] In addition to emitting CO_2, airplanes increase "radiative forcing" through emissions of other greenhouse gases and by creating contrails and cirrus clouds, thus changing atmospheric conditions. Although the science remains uncertain, it appears that these contributions to global warming may be "2 to 4 times larger than the forcing by aircraft carbon dioxide alone."[26] Let's assume, conservatively, that the other effects of aviation add up to twice the impact of CO_2 emissions. Preventing half this increase would give us half a wedge from carbon alone (0.5 billion metric tons less carbon emissions annually, 50 years from now) and one and a half wedges overall, while holding total flights at current levels would supply a full wedge from carbon and three wedges overall. Once again, such reductions could be achieved by increasing the cost of air travel by taxing it. Such a proposal was recently put before the parliament of the European Union.[27] Alternatively, countries might decide that GCC is important enough to demand sacrifices from all their citizens, even rich ones, and strictly limit the number of allowable discretionary flights per person. The USA rationed gasoline use during World War II; perhaps GCC demands an equally strenuous and across-the-board response.

The Fourth Assessment Report does not consider such demand-reduction alternatives, nor do most governments or policy analysts. But they should. A recent study titled "International Aviation Emissions to 2025: Can Emissions Be Stabilized Without Restricting Demand?" answered with a resounding "no." With efficiency improving three times slower than the rate of increased demand and with no transformative technologies on the horizon, the air transport sector cannot make a sufficient contribution to mitigating GCC without limiting demand for air travel.[28] Meanwhile, total emissions may need to decrease by 60–80 percent in the next fifty years in order to avert catastrophic GCC. Clearly this cannot happen while major economic sectors *increase* their emissions.

---

[24] IPCC, *Climate Change 2007: Mitigation*, p. 334.   [25] IPCC, *Aviation and the Global Atmosphere*.
[26] Ibid., "Summary for Policymakers," section 4.8.
[27] Commission of the European Communities, "Proposal for a Directive of the European Parliament and of the Council amending Directive 2003/87/EC so as to include aviation activities in the scheme for greenhouse gas trading within the community" (Brussels), COM (2006) 818 final, 2006/0304 (COD).
[28] Andrew Macintosh and Lailey Wallace, "International Aviation Emissions to 2025: Can Emissions Be Stabilized Without Restricting Demand?" *Energy Policy*, 37 (2009), pp. 264–273.

3.3 Carbon tax wedges

A general carbon tax is by common consent one of the most economically efficient ways to cut carbon emissions. While such taxes are usually presented as means to force technological innovation and decrease pollution per unit of consumption, they also incentivize less consumption, and part of the success of a carbon tax in lowering emissions undoubtedly would come from getting people to consume less. According to the IPCC, a tax of $50 per metric ton of carbon dioxide equivalent could prevent from 3.5 to 7 billion metric tons of carbon equivalent from being emitted annually by 2030, while a tax of $100 per metric ton of carbon dioxide equivalent could prevent from 4.3 to 8.4 billion metric tons carbon equivalent from being emitted annually.[29] Projected out to 2060, such taxes could provide from 5.6 to 13.4 wedges of carbon reduction.[30] In other words, a right-sized carbon tax, by itself, could conceivably provide the 8 or more wedges needed to avoid catastrophic GCC.

A carbon tax is so effective because it affects consumption across the board, from airplane travel to new home construction to food purchases. It treats all these areas equally, from an emissions perspective, and does not distinguish between frivolous and important, useful or useless consumption. That is both its (economic) strength and its (ethical) weakness, and why it should probably be supplemented by measures that directly target unnecessary carbon emissions.

3.4 Luxury wedges

Like any general consumption tax, a carbon tax is regressive, hitting the poor harder than the rich. However, there is a case to be made that in a greenhouse world, everyone should do their part, including the wealthy, by limiting unnecessary consumption. Consider again airplane travel, where we saw that holding the number of flights steady over the next fifty years might provide three wedges. If we assume that the wealthiest 10 percent of the world's population (roughly those with an annual income of $10,000 or

---

[29] IPCC, *Climate Change 2007: Mitigation*, "Summary for Policymakers," pp. 9–10, tables SPM-1 and SPM-2.

[30] Note that these taxes would scale up in from two to two and a half decades, rather than the five decades in Pacala and Socolow's original wedges. I have (arbitrarily) assumed that the taxes would provide the same amount of annual carbon reductions in succeeding years. Strictly speaking, the resulting figure is not a triangular carbon "wedge" but a carbon trapezoid. A similar point applies to the economic growth reduction wedges discussed in a later section.

more)[31] account for 90 percent of flights and that much of this travel is discretionary, then we might construct our plane wedges in a way that transformed them, to some degree, into luxury wedges. For example, we might tax a person's first flight at a percentage n of the cost of a ticket, her second at $n \times 2$, her third at $n \times 3$, and not allow more than p personal flights annually (with medical or bereavement exceptions) – and ban the personal ownership and use of planes (again allowing for reasonable exceptions in the public interest). In a similar manner, Pacala and Socolow's "reduced vehicle use" and "efficient buildings" wedges might be partially transformed into luxury wedges, through some combination of progressive taxation and outright prohibitions. Automobiles priced above n that did not meet fuel efficiency standards could be taxed at a percentage q of the cost of the vehicle, at $2 \times q$ of the cost for even less efficient vehicles, etc. – while the least efficient vehicles could be banned (with exceptions for work-related vehicles and transportation for the handicapped). Houses with floor plans larger than n square meters could be taxed at a percentage r for the first extra hundred square meters, $2 \times r$ for the next hundred, and $3 \times r$ for the third extra hundred – while even larger houses could be prohibited (again, with reasonable exceptions).

While suggestions to prohibit unnecessary, high-emission consumption tend to be unpopular, I think they have considerable merit. It is possible to construct luxury wedges without prohibitions, solely through progressive taxation; and an "efficiency" case can be made that the extra tax revenues generated by luxury consumption justify allowing it to continue, particularly if those revenues are used to mitigate GCC or benefit the poor. On the other hand, setting strict limits to individual carbon pollution would acknowledge the seriousness of the problem and represent a society-wide commitment to avoiding catastrophic GCC. Supposedly efficient approaches which undermine such moral commitment – by encouraging the wealthy to burnish their status through luxury consumption, while the common folk ape or envy them – might wind up being ineffective, and hence not truly efficient.

It has become a commonplace in ethical discussions of GCC to say that we should not try to solve this problem on the backs of the poor. As Henry Shue puts it in his article "Subsistence Emissions and Luxury Emissions,"

---

[31] In 1993, the annual per capita income of people in the 90th percentile of income distribution worldwide was $9,110, according to Branko Milanovic, "True World Income Distribution, 1988 and 1993: First Calculations, Based on Household Surveys Alone," World Bank, Development Research Group, policy research working paper 2244 (November 1999), p. 30, table 19.

"the central point about equity is that it is not equitable to ask some people to surrender necessities so that other people can retain luxuries."[32] But it is probably also true that we won't be able to solve the problem solely on the backs of the world's striving middle classes, who are unlikely to forego desired consumption or pay much more for goods and services, without strong evidence that the wealthy are also doing their fair share – and not just buying their way out of doing so. Arguably, from a fairness perspective, from a wide buy-in perspective, and from a maximal emissions reductions perspective, it makes sense to consider the absolute prohibition of some high-energy, luxury consumption.[33]

Happily, we do have some possibilities for decreasing emissions by increasing people's consumption options. For example, building high-speed trains in Japan and Western Europe has apparently helped slow growth in domestic and regional plane travel, lowering overall carbon emissions. We need to push such win/win strategies (see the next section), while remaining realistic about how much they can achieve. Improved technologies and more options, by themselves, likely cannot achieve sufficient emissions reductions. We probably will need a mix of "goodies" (incentives and increased options) and "baddies" (general carbon taxes, luxury consumption taxes, and outright prohibitions) to find the most efficient and fair means to cut consumption and rein in GCC.

4 ALTERNATIVE WEDGES: POPULATION

When we turn to potential population wedges, we need to remember that population growth is one of the two main drivers of GCC.[34] Again according to the IPCC's Fourth Assessment Report, "The effect on global emissions of the decrease in global energy intensity (−33%) during 1970 to 2004 has been smaller than the combined effect of global per capita income growth (+77%) and global population growth (+69%); both drivers of increasing

[32] Henry Shue, "Subsistence Emissions and Luxury Emissions," *Law and Policy*, 15 (1993), p. 56.
[33] Surprisingly, I have not found any discussion in the scientific or philosophical literature of what role cutting back on high-end consumption might play in a comprehensive effort to mitigate GCC; hence my discussion of luxury wedges is underdeveloped. However, it seems to me such alternatives should be considered as a matter of basic fairness, and I would welcome suggestions on how to specify particular luxury wedges more rigorously.
[34] Brian O'Neill, Landis MacKellar, and Wolfgang Lutz, *Population and Climate Change* (Cambridge University Press, 2005); Brian O'Neill, "Climate Change and Population Growth," in Laurie Mazur (ed.), *A Pivotal Moment: Population, Justice and the Environmental Challenge* (Washington, DC: Island Press, 2009), pp. 81–94.

Table 2 *World population projections*

Projection	Annual growth rate	2050 population	2060 population
Low	0.40%	7.4 billion	8.0 billion
Medium	0.77%	8.9 billion	9.6 billion
High	1.12%	10.6 billion	11.8 billion

Source: UN Department of Economic and Social Affairs, Population Division, "World Population to 2300," p. 4.

energy-related CO_2 emissions."[35] When it comes to GCC and other environmental problems, "size (of the human population) matters."[36]

The current global population is approximately 6.9 billion people. Table 2 shows recent 50-year United Nations population projections at low, medium, and high rates of growth.[37]

The medium projection is presented as the "most likely" scenario, although all three projections are considered possible depending on a variety of factors, including public policy choices. Note that all three projections, even the highest, assume fertility decreases and lower annual growth rates than in recent decades, based in part on improving efforts to provide contraception and encourage family planning. If these efforts falter, birth rates may remain high and populations 50 years from now may balloon past 12 billion.

In 2000, world per capita greenhouse gas emissions were 1.84 metric tons carbon equivalent. Assuming this emissions rate, each 543 million people added to Earth's population adds another 1 billion metric tons of annual carbon emissions; conversely, preventing the existence of 543 million people 50 years from now provides a full carbon reduction wedge (543 million × 1.84 tons = 1 billion metric tons). If we follow the UN report and take 9.6 billion as our business as usual population scenario, then successfully holding world population growth to the lower figure of 8.0 billion would provide 2.95 global population wedges. Conversely, allowing the world's population to swell to the high projection of 11.8 billion (still within the realm of

---

[35] IPCC, *Climate Change 2007: Mitigation*, "Summary for Policymakers," p. 3.

[36] Lindsey Grant, *Too Many People: The Case for Reversing Growth* (Santa Ana, CA: Seven Locks Press, 2001).

[37] Original projections were to 2050; I projected out to 2060 using the annual growth rates provided. These figures might be somewhat rosy. More recently, the US Census Bureau projected that world population will grow from 6 billion in 1999 to 9 billion by 2040. See US Census Bureau, "World Population: 1950–2050," International Data Base (2009).

possibility) would create 4.05 population *de*-stabilization wedges and almost certainly doom efforts to mitigate catastrophic GCC. These figures show that reducing population growth could make a huge contribution to mitigating GCC.[38]

"Population control" tends to bring to mind coercive measures, such as forced abortions or sterilizations. In fact there are non-coercive policies that are almost as effective at reducing birthrates, and these are the ones we should pursue in constructing population wedges. First, providing free or low-cost birth control and accessible, appropriate information about how to use it has proven very effective in lowering birthrates in many poor countries.[39] Providing cheap birth control allows those who want to have fewer children to do so, increasing reproductive freedom while decreasing population growth. Second, policies which improve the lives of women have been shown to reduce fertility rates in many developing countries.[40] These include guaranteeing girls the same educational opportunities as boys, promoting female literacy, and improving women's economic opportunities (and thus their value and status in society). Third, making abortion safe, legal, and easily available has helped reduce birthrates in many countries. In fact, no modern nation has stabilized its population without legalizing abortion. All of these measures can directly improve people's lives at the same time that they help reduce population growth.

Given that these non-coercive methods have proven successful at reducing fertility rates in many places, and given the huge unmet need for contraception throughout the developing world, well-funded efforts to apply them globally seem capable of reducing population growth from the "most likely" scenario of 9.6 billion people to the lower projection of 8 billion people in 2060. Once again: 1.6 billion fewer people 50 years from now represents 2.95 carbon reduction wedges. That would make an immense contribution to mitigating GCC, equal to deploying all three of Pacala and Socolow's carbon capture and sequestration wedges. Unlike carbon capture, however, the proposed population reduction measures rely on proven technologies that are available right now. Population wedges would also provide numerous other environmental benefits, reducing

---

[38] Frederick Meyerson, "Population, Carbon Emissions, and Global Warming: The Forgotten Relationship at Kyoto," *Population and Development Review*, 24 (1998), pp. 115–130; O'Neill *et al.*, *Population and Climate Change*, ch. 6.

[39] Joseph Speidel *et al.*, *Making the Case for U.S. International Family Planning Assistance* (New York: Population Connection, 2009).

[40] Carmen Barroso, "Cairo: The Unfinished Revolution," in Laurie Mazur (ed.), *A Pivotal Moment*, pp. 245–259.

human impacts across the board, in contrast to the massive environmental harms that would be caused by continued coal and uranium mining under the carbon and nuclear power wedges, or the smaller but still substantial environmental harms caused by large-scale wind or solar power generation. Smaller populations would also make an important contribution longer-term, as humanity moves closer (we hope) to creating truly sustainable societies. *Ceteris paribus*, smaller human populations are more likely to be sustainable, while endlessly growing populations are unsustainable by definition.[41]

Securing women's rights and furthering their opportunities are the right things to do, independent of their demographic effects; they can also effectively help stabilize human numbers. Population wedges thus provide "win/win" scenarios with the potential to aid women and their families directly, increasing their happiness and freedom, while helping meet the grave danger of GCC.[42] Some of the very same aims written into the UN's Millennium Development Goals, such as improving maternal health and increasing the percentage of children receiving a full primary school education, turn out to be among the most effective means to reduce birthrates in poor countries.[43] In addition, a recent study from the London School of Economics, titled *Fewer Emitters, Lower Emissions, Less Cost*, argues that reducing population growth is much cheaper than many other mitigation alternatives under consideration.[44] Given all this, policies to stabilize or reduce populations should be an important part of national and international climate change efforts. These are some of the best wedges we've got.

Still, talk of limiting or reducing human numbers makes some people uncomfortable. Many of us have held a newborn baby and felt a sense of infinite possibility and value radiating out from that little form. How could the world possibly be better without him or her? Nevertheless, most of us do not have as many children as we are biologically capable of having. Resources are limited. There are high human costs when urban populations outgrow basic services, or large numbers of young people go unemployed; meanwhile, even confirmed anthropocentrists might well hesitate before accepting the total displacement of wild nature in order to maximize human

---

[41] Albert Bartlett, "Reflections on Sustainability, Population Growth and the Environment," in Marco Keiner (ed.), *The Future of Sustainability* (Dordrecht: Springer, 2006).

[42] Brian O'Neill, "Cairo and Climate Change: A Win/Win Opportunity," *Global Environmental Change*, 10 (2000), pp. 93–96.

[43] Colin Butler, "Globalisation, Population, Ecology and Conflict," *Health Promotion Journal of Australia*, 18 (2007), p. 87.

[44] Thomas Wire, *Fewer Emitters, Lower Emissions, Less Cost: Reducing Future Carbon Emissions by Investing in Family Planning: A Cost/Benefit Analysis* (London School of Economics, 2009).

numbers. People are wonderful, but it is possible to have too many people: in a family, a city, or a nation. GCC may be showing us that it is possible to have too many people on the Earth itself. Part of its message may be that with freedom to reproduce comes responsibility to limit reproduction, so as not to overwhelm global ecological services or create a world that is solely a reflection of ourselves.

5 ALTERNATIVE WEDGES: GROWTH

According to the US Department of Energy (DOE), "economic growth is the most significant factor underlying the projections for growth in energy-related carbon dioxide emissions in the mid-term, as the world continues to rely on fossil fuels for most of its energy use."[45] This suggests that one way to limit increased greenhouse gas emissions is to directly limit economic growth. In its report "International Energy Outlook 2009," the DOE quantifies the impact of various growth rates on carbon emissions out to 2030 as follows: at a high (4 percent) annual growth rate, energy use and CO_2 emissions both increase 1.8 percent annually; at a medium "reference" (3.5 percent) annual growth rate, energy use increases 1.5 percent annually and CO_2 emissions increase 1.4 percent annually; and at a low (3 percent) annual growth rate, energy use increases 1.2 percent annually and CO_2 emissions increase 1.0 percent annually.[46] The difference between a 3 percent or 4 percent annual growth rate adds up to 1.94 billion fewer metric tons of carbon emitted annually by 2030, or 19 percent less. By my calculations, limiting annual world economic growth to 3 percent rather than 4 percent over the next 50 years would lead to 73.64 metric tons less carbon emitted, or almost three carbon reduction wedges.

Both monetary policy and fiscal policy can be used to ratchet back growth. For example, central banks routinely tighten money supplies when rapid growth threatens to cause high inflation. This raises interest rates, reduces borrowing and spending, and slows overall growth. Conservative Chicago school economists argue bitterly with liberal Keynesians over whether monetary policy should primarily seek to control inflation, or whether it is also a proper tool for fighting unemployment. Happily, you can make a temporary peace between these warring factions by proposing that growth rates be manipulated to influence carbon

[45] United States Energy Information Administration, Department of Energy, "International Energy Outlook 2009" (Washington, DC: 2009), ch. 1.
[46] Ibid.

emissions: both sides will likely agree that you are crazy. More than once, economists have asked me "whether I want us all to live in caves?" when I've suggested that slowing growth might be part of slowing GCC. Still, once we have the idea that economic growth may be limited to further other important goals, the question becomes: which goals are important enough to trump growth? Preventing global ecological disaster would seem to be a good candidate.

Granted it sounds odd to say, "Let's embrace lower rates of wealth creation." For one thing, it makes hard choices about the fair allocation of resources even harder. For another, you are arguing for putting fewer overall resources at a society's command, and why would fewer resources be better? Well, here are two possible reasons. First, standard measures of economic growth lump all wealth together under a single monetary metric. But if by increasing overall resources, we lose some resources essential to our survival or well-being, we've failed, even if the financial balance sheet reads other-wise.[47] Second, we arguably also fail if in producing more total wealth, we harm poor people or future generations, or drive other species to extinction, through climate change. For what profits it a man if he gains the whole world and loses his soul? What profits it this current wealthy generation of men and women if we beggar our grandchildren, other defenseless people, or nature itself?

Directly reducing economic growth – by far the leading cause of GCC – is both possible and potentially very effective in reducing greenhouse gas emissions. Like similar efforts to fight inflation, reducing growth to fight GCC could be done in a limited and controlled way. It is telling that such an obvious wedge candidate is almost totally overlooked in analyses of the problem. To quote once again from the Fourth Assessment Report: "The challenge – an absolute reduction of global GHG emissions – is daunting. *It presupposes a reduction of energy and carbon intensities at a faster rate than income and population growth taken together.* Admittedly, there are many possible combinations of the four Kaya identity components, but with the scope and legitimacy of population control subject to ongoing debate, the remaining two technology-oriented factors, energy and carbon intensities, have to bear the main burden."[48] While the report's authors punt on population control, at least they mention it. They don't even seem aware

---

[47] The *Millennium Ecosystem Assessment* discusses this issue at length. For a recent attempt to specify better measurements of economic and social progress, see Joseph Stiglitz *et al.*, *Draft Report* (Commission on the Measurement of Economic Performance and Social Progress, 2009).

[48] IPCC, *Climate Change 2007: Mitigation*, Technical Summary, p. 109; emphasis added.

that we have a choice about whether or not to ease up on the pedal of economic growth. By defining "the challenge" of GCC as reducing greenhouse gas emissions while still accommodating endless growth, they avoid hard questions regarding the reigning economic orthodoxy – but only at the cost of rendering their problem unsolvable.

GCC should have awoken economists and the rest of us from our dogmatic slumbers regarding the goodness of maximal growth and the sustainability of endless growth.[49] As Tim Jackson writes in a recent report to the European Union Sustainable Development Commission, *Prosperity Without Growth? The Transition to a Sustainable Economy*, "The truth is that there is as yet no credible, socially just, ecologically-sustainable scenario of continually growing incomes for a world of nine billion people. In this context, simplistic assumptions that capitalism's propensity for efficiency will allow us to stabilize the climate or protect against resource scarcity are nothing short of delusional."[50] We can and should work to decouple economic growth from carbon emissions, but to assume we will achieve a complete decoupling anytime soon is irresponsible.

Clearly, we will need a new economic paradigm in order to create sustainable societies.[51] It is beyond the scope of this paper to wade into the debate about whether a truly sustainable economy must be post-growth, slow-growth, no-growth, or something else.[52] Here I make the more limited point that reducing (still relatively high) economic growth rates in the near- and midterm could make a significant contribution to reducing greenhouse gas emissions and avoiding catastrophic GCC. We don't "all have to live in caves" in order to do that.

---

[49] A few economists are awake to limits to growth, but the great majority is not. See Garrett Hardin, *Living Within Limits: Ecology, Economics, and the Population Taboos* (New York: Oxford University Press, 1993).

[50] Tim Jackson, *Prosperity Without Growth? The Transition to a Sustainable Economy* (European Union Sustainable Development Commission: 2009), p. 57.

[51] Valuable contributions to specifying the parameters of a sustainable economy include Herman Daly and John Cobb, *For the Common Good: Redirecting the Economy Toward Community, the Environment, and a Sustainable Future* (Boston: Beacon Press, 1989); and Samuel Alexander (ed.), *Voluntary Simplicity: The Poetic Alternative to Consumer Culture* (Whanganui, New Zealand: Stead & Daughters, 2009).

[52] There is a strong argument that economic growth can benefit poor societies, but it is funny how quickly economists turn the discussion to sub-Saharan Africa or Haiti, when anyone questions the need for continued growth in the overdeveloped world. See McKibben, *Deep Economy*, for an argument that economic growth no longer benefits people in wealthy contemporary societies such as the USA. McKibben admirably summarizes the scholarly evidence for this conclusion and clearly lays out the key issues. He also demonstrates that accepting limits to economic growth need not lead to personal or community stagnation. After all, increasing wealth, consumption, or numbers is neither the only nor the most important way people, or peoples, can grow.

6 COMPARING ALTERNATIVES

Table 3 shows the alternative carbon reduction wedges I have proposed in this paper; doubtless more are possible. All the wedges outlined above are achievable using current technologies. As with Pacala and Socolow's more conventional wedges, pushing reductions faster and further could result in additional wedges, while implementing some wedges limits the potential savings from others. My proposal is not to replace the original list with my alternative, but to combine the two, increasing our options.

Providing more wedges is important because we might have to consider relatively unpalatable choices, now that scientists are telling us we may have to ratchet down emissions faster than anticipated in order to avoid catastrophic GCC. However, some alternative wedges might actually be preferable to the usual proposals to mitigate GCC. For example, it is not clear

Table 3 *Potential alternative carbon mitigation wedges*

	Option	Effort by 2060 for one wedge, relative to 15 gigatons of carbon per year (GtC/year) under business as usual
Consumption reduction	1. Eating less meat	Prevent half of the projected world increase in meat eating
	2. Flying less often	Prevent one-third of the projected increase in commercial aviation
	3.–9. Taxing carbon	Tax greenhouse gas emissions at $50 per metric ton of CO_2 equivalent (a conservative estimate of the carbon reductions available using this method)
	10. Limiting luxury consumption	Tax or prohibit multiple personal plane flights, large homes, low-mileage vehicles, or other unnecessary consumption
Population stabilization	11.–13. Improving women's lives and reproductive freedom	Achieve UN's low rather than medium 2060 population projection (8.0 rather than 9.6 billion) by providing free, accessible birth control and improving women's economic and educational opportunities
Economic growth reduction	14.–16. Limiting the size of the world economy	Use monetary and fiscal policy to reduce growth rates from 4% to 3% annually (could be pushed further, providing more wedges)

that tripling the world's nuclear generating capacity is superior to simply cutting back on electricity use by reducing unnecessary consumption. Cutting consumption might be cheaper and less dangerous, and make a stronger contribution to creating sustainable societies. Again, some of us would prefer that our tax dollars go toward helping poor women in developing countries improve their lives, as in the population reduction wedges, rather than subsidizing energy companies' profits, as required by Pacala and Socolow's coal and nuclear wedges. Again, limiting consumption and population growth seems less selfish and more responsible than relying solely on efficiency improvements that pass significant environmental harms on to nonhuman beings and future generations, or on futuristic technologies that may not work.

Whether or not I'm right in these particular judgments, getting a full range of alternatives on the table would seem to be our best hope for finding the fairest and most efficient strategies to mitigate GCC. Pacala and Socolow's original framework allows us to consider whether to triple nuclear generating capacity or build 2 million new wind turbines; the expanded framework allows us to choose these options, or the option of paying more for electricity and using less of it. The original framework makes it hard to achieve eight wedges without committing to continued heavy use of coal; the expanded framework would allow humanity to phase out coal and uranium as fuel sources, provided we embrace population reduction. The new framework thus makes explicit some of the ecological and economic costs of continued consumption and population growth, and emphasizes that such growth is a choice.

As in the original framework, alternatives can be compared regarding monetary cost, total ecological impact (not just impacts on greenhouse gas emissions), and social equity. But now new alternatives are in play that would further sustainability, improve the lives of some of the world's poorest people, and demand greater contributions from wealthy global elites. My claim is that using this expanded list of wedges would allow us to come up with a better (because more just and more sustainable) climate change policy, and also that implementing more alternative wedges would put us in a better position to transition to truly sustainable societies in the second half of the twenty-first century.

7 CONCLUSION

A strength of the wedge approach is that it fosters intelligent choices among alternative courses of action. However, we need to expand the discussion to include alternatives that directly address the main forces driving GCC: increasing consumption, increasing populations, and rapid economic growth. Our continued failure to do so could lead to catastrophic GCC and delay the necessary transition to ecological sustainability.

Comparing alternative mitigation scenarios, done correctly, should make the ethical commitments behind our choices explicit. My own belief is that the best response to GCC is one that protects and enhances the flourishing of all life, human and nonhuman.[53] So in considering alternative courses of action, I believe we should choose wedges that:

- reduce the human appropriation of habitats and resources needed by other species, rather than increasing such appropriation
- avoid compromising the abilities of poor people to provide for their basic sustenance and well-being
- avoid relying on unproven technologies, or technologies that increase the risk of unintended or unexpected consequences.

Mindful of GCC's immense potential harms and the diminishing returns of increased wealth and consumption in wealthy nations, I would argue for more wedges rather than fewer. Looking to long-term sustainability, I would embrace some of Pacala and Socolow's technological efficiency wedges, but supplement them with wedges that build up our societies' capacities for demographic and economic restraint. The virtue of temperance is intrinsically valuable – commendable in itself, in a person or a nation – and its instrumental value will only increase in the more crowded and damaged world that we are busy making.[54] In order to create truly sustainable societies, we will need to become wiser, as well as more clever.[55]

---

[53] Philip Cafaro, "Thoreau, Leopold and Carson: Toward an Environmental Virtue Ethics," *Environmental Ethics*, 23 (2001), pp. 3–17; Eileen Crist, "Beyond the Climate Crisis: A Critique of Climate Change Discourse," *Telos*, 141 (2007), pp. 29–55.

[54] Joshua Gambrel and Philip Cafaro, "The Virtue of Simplicity," *Journal of Agricultural and Environmental Ethics*, 23 (2010), pp. 85–108.

[55] Thanks to Art Darbie and Steve Shulman for mathematical and economic advice, respectively; to Robert H. Socolow for generous critical comments; to Ron Sandler and his climate ethics class for suggestions on the proper wedge selection criteria; and to Denis G. Arnold and Kris Cafaro for timely and helpful editing that tightened and improved this chapter.

CHAPTER 10

Addressing competitiveness in US climate policy

Richard D. Morgenstern

I INTRODUCTION

Mandatory policies to reduce US greenhouse gas (GHG) emissions without full global participation – principally cap-and-trade systems, occasionally carbon taxes, and sometimes standards – are being seriously debated in the US Congress. However, even efficient market-based policies that effectively attach a price to GHG emissions will likely increase production costs for some domestic producers and give rise to competitiveness concerns where those producers compete against foreign suppliers operating in countries where emissions do not carry similar costs. These concerns are likely to be most acute in energy-intensive, trade-exposed manufacturing industries. While the impacts can be mitigated to some extent by the use of offsets or other flexibility mechanisms, it would be virtually impossible to eliminate the disproportionate burdens placed on certain sectors without undermining both the effectiveness and the cost-effectiveness of the policy. As Olson has eloquently argued, the more narrowly focused the adverse impacts of a policy, the more politically difficult it is to sustain.[1]

One of the key questions being asked is this: Why should US firms be disadvantaged relative to overseas competitors to address a *global* problem? The difficulty, moreover, is not just political: if, in response to a mandatory policy, US production simply shifts abroad to unregulated foreign firms, the resulting emissions "leakage" could vitiate some of the environmental benefits expected from domestic action. As is widely recognized, limiting emissions from the USA and other developed countries will not prevent dangerous interference with the climate system unless key developing countries also control their emissions.

The most obvious remedy is to encourage other countries to undertake to their own carbon reductions. Yet, the USA cannot compel such actions. In

---

[1] Mancur Olson, *The Logic of Collective Action* (Cambridge, MA: Harvard University Press, 1965), p. 47.

fact, there is ample precedent in international agreements for adopting differing obligations for developed and developing countries. Thus, a stated goal of a new US policy is

to craft legislation limiting U.S. carbon emissions that also *induces* [italics added] developing countries to limit their emissions growth 1) on a timetable that meets both environmental and competitiveness concerns; 2) in a manner that is reasonably certain to withstand challenge before the World Trade Organization (WTO); and 3) on terms that pose acceptable risks to U.S. interests in the event of a negative WTO determination.[2]

Not surprisingly, efforts to use the international trade system to induce policy changes in other nations raise concerns in both the free trade and the environmental communities. Free trade advocates fear that talk about climate change will be used as an excuse by some industries to gain protection for themselves against international competition, thereby undermining the global trade regime and provoking countervailing actions by other nations. The environmental community, in turn, fears that talk about free trade will be used as an excuse to give excessive emphasis to traditional economic considerations at the expense of climate protection goals.

Beyond the efforts to induce actions by other nations, proposals for modifications to the one-size-fits-all nature of an economy-wide, market-based carbon policy have also been advanced. These include adding particular design features to an economy-wide cap-and-trade system such as a generous allocation of allowances to particular industries, either based on historic emissions or via an updating approach. Alternatively, performance standards or other regulatory mechanisms might be substituted for price-based policies in some industries. While these options may not completely level the playing field with foreign competitors, they can ease the transition to a truly global system.

This paper analyzes alternative policy responses under consideration in the current debates, including both domestic and international oriented actions. Following this brief introduction, Section 2 summarizes some recent quantitative studies of potential competitiveness impacts of GHG mitigation policies in both the USA and the European Union (EU). While not an exhaustive review, the intent of this section is to provide rough estimates of the magnitudes involved in the competitiveness discussions.

---

[2] US House of Representatives, Committttee on Energy and Commerce, "Competitiveness Concerns/ Engaging Developing Countries," Climate Change Legislative Design White Paper, January 31, 2008, p. 2.

The core of the chapter, Section 3, examines potential policy responses, focusing on five alternative options: weaker overall program targets; partial or full exemptions from the carbon policy; performance standards not involving embodied carbon; free allowance allocation under a cap-and-trade system; and trade-related measures. The first of the options considered would involve the design of the policy as a whole; all of the remaining options attempt to target industries or sectors that would be particularly vulnerable to adverse impacts under a mandatory GHG reduction program. The last option is explicitly designed to engage developing nations. Although the policy landscape continues to evolve, reference is made to specific elements of a number of recent Congressional proposals. Section 4 offers some tentative conclusions and recommendations.

2 REVIEW OF RECENT QUANTITATIVE STUDIES ON COMPETITIVENESS

This section presents a brief review of recent quantitative studies of potential competitiveness impacts for both the USA and the EU. A more extensive analysis of this literature can be found in Fischer and Morgenstern.[3] Overall, the impact of a carbon price on the competitiveness of different industries is fundamentally tied to (a) the energy (and more specifically, the carbon) intensity of those industries, and (b) the degree to which firms can pass on costs to the consumers of their products, including to other firms. The latter issue hinges on the extent to which consumers can substitute other, lower-carbon products and/or turn to imports.

The scale of the potential impacts of a mandatory domestic climate policy is unprecedented in the history of environmental regulation, as is the range of industries that would be affected. Yet, quantifying potential impacts on an industry-specific basis is quite complex. By contrast, the debate leading up to the 1990 Clean Air Act Amendments was informed by extensive government- and industry-sponsored analyses of the likely effects of a cap-and-trade program for sulfur dioxide emissions on the electric power sector. These analyses were greatly simplified by the fact that the policy under consideration targeted a single, largely regulated industry that faced almost no international competition. A pricing policy for GHG emissions would not only have much more significant direct impacts on coal, and

---

[3] Carolyn Fischer and Richard D. Morgenstern, "Designing Provisions to Maintain Domestic Competitiveness and Mitigate Emissions Leakage," *Policy Brief 09–06, Energy Security Initiative* (Washington, DC: Brookings Institution, 2009), pp. 5–6.

other domestic energy industries, but it could also adversely affect the competitiveness of a number of large, energy-consuming sectors, particularly the energy-intensive, trade-exposed industries.

Industry-level studies of the competitiveness effects of carbon policies tend to focus on the energy-price impacts of a specific carbon policy. They typically do not consider what level of carbon price would be required to meet a particular emissions-reduction target or how overall program stringency is coupled with decisions about domestic or international offsets, price caps, or other mechanisms designed to reduce or contain costs.[4]

Ho, Morgenstern, and Shih estimate that expenditures on energy in most manufacturing industries (broadly defined at the two-digit level according to the North American Industry Classification System (NAICS)) are less than 2 percent of total costs.[5] However, energy costs are more than 3 percent of total costs in a number of energy-intensive manufacturing industries such as refining, nonmetal mineral products, primary metals, and paper and printing. For these industries, total production costs rise by roughly 1 percent to 2.5 percent for each $10 increment in the per metric ton price associated with CO_2 emissions (with less being known about the impacts of larger CO_2 prices). Cost impacts can be considerably greater within more narrowly defined industrial categories. For example, energy costs for the aluminum and chlorine industries are estimated to be more than 20 percent of total costs – more than ten times the average for all manufacturing.

Case studies in the EU by McKinsey & Co. and Ecofys (2006) find that a $10-per-metric-ton CO_2 price leads to a 4 percent increase in total costs for aluminum and a 6 percent increase in total costs for steel production using basic oxygen furnace (BOF) technology.[6] For cement, production costs would increase by 13 percent. Unlike the US analyses, the EU estimates include process emissions covered under the EU emissions-trading system (ETS). With free allowance allocation and some ability to increase prices, however, the researchers found that the adverse impacts on industry can be reduced substantially. For example, using simple demand models, a study by Reinaud found that output in most industries declined less than 1 percent – and by at most 2 percent in the most strongly affected

[4] Offsets are emission-reduction credits from unregulated sources. A price cap, also known as a safety valve, applies to a mechanism that directly limits costs under a cap-and-trade program by making an unlimited number of additional allowances available for sale at a fixed, predetermined price.

[5] Mun Ho, Richard D. Morgenstern, and Jhih-Shyang Shih, "Impact of Carbon Price Policies on US industry," *Discussion Paper 08–37* (Washington, DC: Resources for the Future, 2008), pp. 25–26.

[6] McKinsey & Company and Ecofys, *EU ETS Review: Report on International Competitiveness.* European Commission, Directorate General for Environment (2006), p. 4.

industries – for a $10-per-metric-ton CO_2 price with 95 percent free allocation.[7]

More generally, cost increases can be translated into impacts on production, profitability, and employment, using an explicit model of domestic demand and international trade behavior. Based on such a model of the USA, Ho, Morgenstern, and Shih generally find adverse effects of less than 1 percent when estimating the reduction in industrial production due to a $10-per-metric-ton CO_2 charge.[8] The exceptions are motor vehicle manufacturing (1.0 percent), chemicals and plastics (1.0 percent), and primary metals (1.5 percent). While not explicitly calculated, production losses in more narrowly defined industrial categories such as aluminum and chlorine are expected to be greater. Note that all these estimates represent near-term effects – that is, impacts over the first several years after a carbon price is introduced – before producers and users begin adjusting technology and operations to the new carbon policy regime. Longer-term effects in most industries are expected to be smaller, depending on technologies and costs. For example, pricing carbon may accelerate the substitution of plastics for steel in the manufacture of autos, which would soften cost impacts in the auto industry.

Naturally, impacts on domestic industries will generally be lower if key trading partners also implement comparable carbon reduction policies and/ or if particular design measures are included in domestic policies. Various options for lessening these competitiveness impacts are examined in the next section.

3 OPTIONS FOR ADDRESSING COMPETITIVENESS IMPACTS

Due to the great diversity of GHG sources, addressing climate change will – of necessity – involve many different types of actors, including industries, governments, and individuals. In general, pursuing a cost-effective approach that minimizes the overall cost to society of achieving a particular emissions-reduction target will minimize the burden imposed on businesses and consumers. Market-based strategies that effectively attach a price to GHG emissions, such as an emissions tax or cap-and-trade program, in particular offer significant cost and efficiency advantages. As part of a broad pricing

---

[7] J. Reinaud, *Industrial Competitiveness Under the European Union Emissions Trading Scheme* (Paris: International Energy Agency 2005), pp. 55–59.

[8] Ho, Morgenstern, and Shih, "Impact of Carbon Price Policies on US Industry," pp. 26–32.

policy, the use of additional flexibility mechanisms – such as recognizing offset credits from unregulated sectors or gases and/or from projects undertaken in other countries – can lower program costs for particular sectors as well as for the overall economy. A price cap can also lower overall costs. Close attention to cost and efficiency considerations in the design of an overall policy could thus be viewed as a first step to addressing competitiveness concerns.

Rather than focusing solely on the American Clean Energy and Security Act (ACESA), adopted by the US House of Representatives (HR2454) in June 2009 or other specific legislation, this section considers five potential options for addressing competitiveness issues. The first four have their primary impact on the design of domestic policy, while the last one involves trade-related approaches. While these trade-related approaches would be most responsive to the Congressional goal of inducing developing countries to limit their emissions, they are also likely to be most contentious in terms of WTO rules. The five options are:

- weaker overall program targets
- partial or full exemptions from the carbon policy
- performance standards not involving embodied carbon
- free allowance allocation under a cap-and-trade system
- trade-related options.

As we turn to a discussion of these options, an important caveat is in order. As compelling as the argument for protecting vulnerable firms or industries might be, few provisions or program modifications designed to accomplish this can be implemented without increasing emissions and/or raising costs in the sense that they will result in more expensive abatement options being used to achieve the same emission reduction goals. Trade-related actions are not costless either: they might raise legality concerns under WTO rules and/or risk provoking countervailing actions by other nations.

3.1 Weaker overall program targets

This option involves adjusting the stringency of the policy as a whole such that it results in a lower economy-wide emissions price without regard to the obligations of specific industries. In the case of a cap-and-trade system, a lower price can be achieved by allowing a greater quantity of emissions under the cap or by including a price cap. Other options for making the policy more flexible such as allowing a larger role for domestic or international offsets can also help to reduce domestic emissions costs. For example, the provision for sectoral offsets in HR2454 represents a potentially attractive mechanism for achieving low-cost emission reductions and for incentivizing developing country action. Compared to project-based activities such as the Clean Development Mechanism, sectoral offsets are more likely to meet relatively high standards for environmental integrity.

The lower emissions price associated with a less stringent policy will produce smaller economy-wide costs and should help ameliorate the competitiveness concerns of trade-exposed firms or industries. An important advantage of this option is that it does not require the government to identify particularly vulnerable industries, thereby avoiding the need to identify the truly disadvantaged. Further, this option does not require additional administrative mechanisms, nor does it diminish the cost-effectiveness of the underlying policy.

While weakening the policy may address the concerns of the most vulnerable industries, it does so at the expense of the overall environmental goal. Other options, considered below, attempt to deal more directly with vulnerable industries and would presumably be implemented as an alternative to weakening the overall policy.

3.2 Partial or full exemptions from the carbon policy

One option for addressing competitiveness concerns is simply to exempt certain industries from the broader GHG-reduction policy. The key challenge in implementing this approach – or indeed any of the targeted policies discussed in the remainder of this paper – is determining which sectors qualify for special treatment because of their vulnerability to competitiveness concerns. Applying a stringent threshold for exemption risks excluding vulnerable producers, while setting a loose threshold risks allowing too many sources to escape regulation, thereby increasing the difficulty of achieving the desired emissions reduction objective. Not surprisingly, such an approach opens the door to virtually unlimited lobbying for more favorable treatment.

The mechanics of actually providing exemptions, by contrast, are relatively easy. In a cap-and-trade system where downstream entities – primarily energy users – are regulated, exempt firms would face reduced requirements (or perhaps none at all) to submit allowances to cover their emissions. In the case of a carbon tax, eligible firms would face a reduced levy (or possibly none at all). In either approach, a procedure would need to be established to credit exempt downstream entities based on their emissions or fuel use. The credit could be payable in allowances (in the case of a cap-and-trade system) or via a tax credit or rebate (in the case of an emissions tax).

The principal advantage of partial or full exemptions is that they can be used to protect vulnerable industries in a targeted way. The principal disadvantage of the exemptions approach is that it would likely increase the total, economy-wide cost of achieving a given emissions target because exempting certain sectors would almost certainly leave at least some inexpensive mitigation options untapped. As a result, the program would be both less efficient and more costly overall. This approach may also raise equity concerns: If the same national target is pursued but some industries or firms are exempt from participating, this would place a greater burden on the remaining non-exempt industries.

Further, the difficulty of identifying truly vulnerable industries cannot be overemphasized. Technically, it would be quite challenging to adjudicate requests for exemptions on the basis of vulnerability to competitive harm. In part, the difficulties are due to limited data and modeling capabilities available at the industry-specific level.

Interestingly, some recent proposals call for significant exemptions but do not limit these exemptions to sectors that would seem most obviously at risk of suffering a business disadvantage under a mandatory domestic climate policy. For example, a bill introduced by Senators Feinstein and Carper (S.317, 110th) covers only the electricity sector – almost 40 percent of US emissions – and therefore exempts primary (non-electricity) energy use by households along with all transportation-related emissions. A bill introduced by Senators Lieberman and McCain (S.280, 110th), by contrast, covers all large facilities – defined as facilities in the electric power, industrial, and commercial sectors that emit at least 10,000 metric tons CO_2-e per year or more – plus transportation fuels at the refinery or importer (this program would cover an estimated 70–75 percent of the total US emissions). Only households, agriculture, and small non-transport emitters are exempt. In both these cases, however, the partial coverage appears to be motivated by strictly political considerations. For example, the premise underlying Feinstein–Carper is that it might be easier to focus on the electric power sector where competitiveness concerns are less of an issue. At the same time, focusing on the electric sector does not exempt large electricity users like the aluminum industry, which does face significant competition in international markets.

For a cautionary lesson concerning the political hazards of exemptions, it is useful to recall the energy (BTU) tax proposed by the Clinton administration in 1993. At that time, many firms and industries made claims of business hardship. As a result, the final House legislation included a long list of exemptions. Ultimately, the BTU tax was defeated in the Senate and the policy was never implemented – in part because its effectiveness was under-cut by the exemptions.

3.3 Performance standards not involving embodied carbon

Performance standards come in many varieties and may include minimum, average, and tradable standards for emissions or energy use per unit of output. Typically performance standards are tailored to specific products or groups of products. Unlike broad, market-based carbon policies, performance standards do not produce a direct increase in energy costs – therefore, they do not create as much pressure for firms to raise product prices. For this reason, performance standards may seem less likely than market-based policies to raise competitive-ness concerns and create incentives for shifting production abroad.

Well-crafted performance standards have the potential to encourage effi-ciency improvements without putting as much upward pressure on domestic production costs. In doing so, they may reduce the potential for domestic production to shift to countries without mandatory GHG-reduction poli-cies. Efficiency and cost considerations generally argue for some form of industry average or best performance standards rather than facility-specific regulations. Tradable and bankable performance standards – such as were used to effect the phase-down of lead in gasoline in the mid-1980s or as exemplified by current proposals for a national renewable portfolio standard (RPS) – provide flexibility and are generally quite cost-effective.

At the same time, even relatively simple performance standards can be more costly than broad market-based approaches because they do not encourage end users to reduce their consumption of GHG-intensive goods, and do not balance the cost of emissions reductions across different sectors. Relying on standards instead of market-based instruments to achieve emis-sions reductions will leave behind some low-cost abatement opportunities, thereby raising the overall cost incurred by society to achieve a particular emissions target. From an implementation standpoint, the process by which standards are set can be contentious and would impose potentially burden-some requirements on government to obtain detailed information on technologies, costs, and other metrics.

The academic literature provides abundant evidence that market-based mechanisms, especially broad-based ones, provide lower-cost emissions reductions than standards do.[9] Unlike performance standards, market-based

---

[9] Jody Freeman and Charles D. Kolstad (eds.), *Moving to Markets in Environmental Regulation: Lessons from Twenty Years of Experience* (New York: Oxford University Press, 2007), pp. 4–10.

instruments provide a continual incentive to reduce emissions – thus they promote technology innovations that, by their nature, take time to develop and deploy. Market-based instruments also offer maximum flexibility in terms of the means used to achieve reductions, including, for example, the shift to improved scrubbing technologies that occurred in the US sulfur dioxide program.

Notwithstanding these observations, it seems that many firms prefer standards to market-based policies. Arguably, they may fear that it will be more difficult to pass along increased energy costs under a market-based carbon policy; in addition, they may expect to be in a stronger position to negotiate the form and stringency of a regulatory program that is tailored to specific sectors rather than one designed for the economy as a whole.

3.4 Using free allowance allocation to address competitiveness concerns

Allocation refers to the approach used to distribute permits or allowances under an emissions-trading program. Here, two decisions are particularly important. The first concerns how many allowances (or what share of the overall allowance pool) will be given away for free. The second concerns the *methodology* to be used in apportioning free allowances to different industry sectors. In the current sulfur dioxide trading program, as well as most others operated at state or regional levels, allowances have been given away for free to directly regulated entities, primarily on the basis of historical emissions (an approach often called "grandfathering"). In the initial phases of the EU ETS, free allocation has been used to compensate firms believed to be vulnerable to trade impacts. Proposed plans for the post-2012 period call for a partial scaling back of free allocation.

Recent Congressional proposals, including HR2454, have sought to allocate free allowances in a way that recognizes changes in production over time, an approach known as updating allocation. Compared to an allocation based on grandfathering, updating can create incentives to maintain (or even expand) domestic production – and thus reduce competitiveness impacts. Although still quite limited in its application, a number of states, including Connecticut, Massachusetts, and New Jersey, are currently using annual updating approaches in their NO_x control programs.[10]

Under HR2454, emission allowances are initially distributed to eligible facilities on an output basis under an overall cap of 15 percent of total

[10] Dallas Burtraw *et al.*, "Economics of Pollution Trading for SO_2 and NO_x," *RFF Discussion Paper 05–05* (2005), pp. 28–35.

allowances. The allocation is updated biannually to reflect actual performance, with those facilities reducing output receiving relatively fewer allowances. The universe of covered industries will be based on a formal regulation to be issued by the administrator of the US Environmental Protection Agency (EPA), with the precise list to be derived from available data at the six-digit NAICS level. The formula for inclusion is based on trade-intensity of at least 15 percent and energy, or emissions, intensity of at least 5 percent. Some additional criteria apply, including an option to include subsectors based on individual petitions. Allowances are designed to cover both direct costs and those indirect carbon costs passed on by electricity suppliers, based on historical information. Special provisions apply to new entrants. Shutdown facilities (but not those with reduced output) would be required to surrender rebates.

In essence, an updating mechanism leaves the carbon price in place to signal efficiency improvements, while the rebates prevent operating costs from rising too high, which keeps the playing field level both at home and abroad. The principal advantage of using a free allocation of allowances to address competitiveness concerns is that it can compensate firms for losses suffered as a result of the new policy without changing the incentives to reduce overall emissions in response to the carbon pricing scheme. Thus free allocation avoids the efficiency losses and/or reduction in environmental benefit associated with other options, including weakening the goals, exempting some industries, or relying on traditional standards-based forms of regulation in some sectors.

At the same time, the use of rebates to avoid competitiveness impacts does come at the expense of opportunities to reduce consumption of emissions-intensive goods. In addition, the rebates themselves may raise WTO compliance issues under the Subsidies Code, which disciplines the use of subsidies and regulates the actions countries can take to counter the effects of subsidies.

In terms of the methodology used to distribute free allowances to individual firms, traditional grandfathering – which leaves the allocation fixed over time regardless of whether a business changes operations or even shuts down – can compensate firm *owners* for losses in value but does not necessarily discourage firms from moving their emissions-producing operations overseas to avoid costs associated with the regulatory program going forward.

The alternative of an updating output-based allocation, where allowance shares are periodically adjusted to reflect a firm's changing output, effectively subsidizes production. That is, firms stand to gain a larger allocation of free allowances if they expand their operations and a smaller allocation – or possibly none at all – if they move offshore, downsize, or shut down. While incentives of this type are generally regarded as distorting and hence inefficient – because they effectively reduce variable production costs and induce firms to produce above the level that would otherwise make economic sense – they may represent an attractive approach for dealing with competitiveness impacts precisely because they tend to encourage domestic production and discourage firms from moving operations (and emissions) overseas.

The benefits of an allowance-updating approach can accrue to domestic consumers as well as to firms that face competition from foreign suppliers, either in markets at home and/or in export markets abroad. That is, the cost increases that are passed on to consumers or absorbed by firms are smaller with output-based allocation than with allocation mechanisms not conditional on output; e.g., grandfathering. While the WTO legality of output-based allocation has not been tested, it is unlikely to be regarded differently than grandfathering, since they are both tied to an increased environmental obligation.

An important argument against free allocation, via either grandfathering or on an output basis, is that it misses the opportunity to auction allowances and use the revenue to provide benefits for the economy as a whole. From a strict economic efficiency perspective, it would make more sense to auction all allowances and use the proceeds to reduce taxes on income or investment. Compelling arguments can also be made for auctioning allowances and using the revenues to support certain other public policy objectives, such as energy R&D, offsetting the impact of higher energy prices on consumers (especially low-income households), and adapting to climate change.

Another concern is that overly generous free allocation risks conferring windfall gains on some firms, especially in situations where a firm is able to pass through most of the costs of regulation in the form of higher prices for its products. In that case, giving the firm free allowances would amount to a transfer of wealth from consumers – who will face higher prices for the firm's goods – to shareholders who do not really bear a substantial share of the cost burden associated with the policy. The early phase of the EU ETS has been subject to this criticism.

An updating-free allocation that subsidizes domestic production also gives rise to the same concerns noted in connection with other targeted responses that distort behavior relative to what would happen under a broad carbon-pricing policy. Namely, allocation decisions in practice may fail to target truly vulnerable industries. In that case, an updating allocation will create efficiency losses and increase the overall cost of the policy to society, while providing only limited benefits in terms of maintaining domestic production, preserving US jobs, and reducing the potential for emissions leakage.

Relative to targeted exemptions or reliance on performance standards instead of market-based approaches, using free allowances to compensate vulnerable industries as part of a broad, cap-and-trade or emissions tax program generally maintains efficiency. Among these three options, free allocation, via either grandfathering or updating, is the most efficient because it preserves the ability to trade off emission reductions throughout the economy – so that the environmental objective is achieved by exploiting the least-cost abatement opportunities.

3.5 Trade-related policies

The principal aim of trade-related policies, such as border adjustments, is to level the playing field between domestic and foreign suppliers. Such efforts would likely involve using financial or regulatory mechanisms to impose roughly equivalent burdens on imports as the domestic policy imposes on domestic production. Although not as widely discussed, a similar mechanism – presumably involving some type of export subsidy – could be used to level the playing field for US-produced goods that compete in foreign markets against goods produced in countries without comparable emissions control policies.

Even if they can be successfully defended under WTO rules, however, border adjustments have clear disadvantages as well. To the extent they act as barriers to trade (beyond correctly accounting for the cost of emissions), border adjustments are inherently inefficient and costly to US consumers and industries that depend on imported goods. For example, they would drive up costs for covered materials, e.g., steel, which, in turn, could adversely impact the auto industry or other domestic industrial steel consumers.

Moreover, there is a risk that the system could be abused by firms or industries – or even by other nations if they use it as grounds for instituting their own system of border adjustments, including for purely protectionist reasons unrelated to climate policy. These actions could work against long-sought-after free-trade objectives. They could also undermine the trust necessary to foster international cooperation and agreement on future international efforts to address climate-change risks.

Since any trade-related action risks a challenge by US trading partners before the WTO dispute settlement body, a central issue to consider is what kind of policy is most likely to be judged as WTO-legal. Even though WTO law is vague on this issue, the USA might be able to address the problem of offshore emissions associated with imported products by applying to imports a carbon tax or emissions-permit obligation that is *equivalent* to the requirements imposed on US-produced goods.

Further complexities arise in developing administrative procedures for assigning process carbon emissions to specific imported products. On the one hand, a border adjustment policy might be considered more acceptable if it were based on the processes and fuels used in the USA – the so-called US predominant method of production. Ironically, such an approach would likely be less onerous to foreign producers, since in many cases US production is less carbon intensive than the production of similar products abroad. At the same time, it might be necessary to establish procedures that would allow foreign producers to submit technical data to support their particular situation. Such determinations would be more defensible – and easier to calculate – if the focus were on basic products, such as steel, aluminum, and cement, rather than finished goods.

The size of any border adjustment would be diminished to the extent that domestic producers are effectively subsidized by a free allowance allocation. Thus, for example, if 50 percent of allowances are freely allocated to affected industries – either on the basis of grandfathering or updated allocation – an importer might have to surrender allowances equal to only half of estimated emissions associated with the imported product. Alternatively, if a broad-based carbon tax was imposed, importers would presumably face an equivalent adjustment at the border and there would be no need to account for offsetting benefits to US producers. A variety of other issues might also complicate the use of border adjustments, including the question of how to treat imports from a country with a limited domestic carbon policy versus imports from countries that lack such a policy altogether.

In the best case, border adjustments would effectively protect vulnerable domestic firms or industries against adverse competitiveness impacts from a domestic climate policy while simultaneously creating incentives for other nations to reduce their emissions. To improve the prospects for a successful WTO defense, a number of issues would need to be addressed, including the need to put major trade partners on notice and provide sufficient time for them to develop their own emission reduction policies.

HR2454 establishes a number of formal requirements that could possibly trigger a border adjustment mechanism. For example, the legislation directs the US President, beginning in 2017, to prepare a series of reports concerning the effectiveness of the rebate system for addressing carbon leakage and the extent to which other nations have adopted the same or similar measures. If a multinational climate accord is not in force by 2018, the legislation directs the administration to notify US trade partners that an International Reserve Allowance Program (IRAP) will be implemented. Under the program, imported products from eligible sectors would have to purchase emissions allowances according to their carbon intensity. No border adjustment for exports is provided. Import protection continues as long as less than 85 percent of imports for that sector are produced in countries that meet at least one of the following criteria:

- The country is a party to an international treaty to which the USA is a party and includes a nationally enforceable emissions reduction commitment that is at least as stringent as that found in the USA.
- The country is a party to an international agreement for that sector to which the USA is a party.
- The country has an annual energy or GHG intensity for that sector that is equal to or less than that of the USA.

4 CONCLUSIONS

Overall, there appear to be no simple answers to the competitiveness problem. While it is widely recognized that reliance on market-based domestic policies will generally minimize the costs of achieving any given emission target – and thus could be viewed as a first response to competitiveness concerns – such policies by themselves cannot mitigate the disproportionate impacts of the mandatory policy on energy-intensive, trade-exposed industries.

In terms of possible remedies, a weaker overall policy – less stringent emissions caps and/or lower emissions prices – represents the least focused approach available for addressing competitive impacts. While it has the advantage of not requiring policymakers to identify vulnerable sectors and thus avoids the potential for a "gold rush" of industries seeking relief, the less ambitious emission-reduction targets will produce smaller environmental benefits and weaker incentives for technology innovation.

Simply exempting certain sectors or types of firms provides a direct response to competitiveness concerns and the most relief to potentially affected industries. At the same time, such an approach is also the most costly in terms of reducing the economic efficiency of the policy.

More traditional (non-market-based) forms of regulation – such as emissions standards or intensity-based regulations for directly observable as opposed to embodied carbon emissions – can be used to avoid direct energy price increases and can deliver some emissions reductions. At the same time, regulated industries will still face compliance obligations, while the overall social cost of achieving a given environmental objective using these forms of regulation will tend to be higher than under a uniform, broad-based pricing policy.

Under a broad-based, cap-and-trade system, free allocation of allowances can be used to compensate adversely affected industries (even if those industries are not directly regulated under the policy) without necessarily losing the efficiency of the market-based approach. Different forms of free allocation – for example, an allocation based on historic emissions ("grandfathering") versus an updating allocation tied to current output (the type used in HR2454) – will have quite different incentive properties regarding domestic production and jobs.

Trade-related policies such as border adjustments can protect vulnerable domestic firms/industries and create incentives for nations without similar GHG policies to adopt their own emission-reduction efforts. However, such policies risk providing political cover for unwarranted and costly protectionism and may provoke trade disputes with other nations.

Finally, it is noteworthy that in practice it is not necessary to make stark choices among the alternative approaches. In fact, HR 2454 combines two of the major options discussed herein: a system of rebates (free allowance allocations subject to updating) is introduced initially, potentially followed by border adjustments in later years for imports in eligible sectors from certain countries. Such an approach represents a quite reasonable strategy to ease the domestic transition as our key trading partners move to adopt their own carbon reduction policies.

Reconciling justice and efficiency: integrating environmental justice into domestic cap-and-trade programs for controlling greenhouse gases

Alice Kaswan

As this volume demonstrates, the prospect of global climate change raises profound questions of international corrective and distributive justice.[1] At the same time, individual nations must grapple with the ramifications of domestic policies for reducing greenhouse gas (GHG) emissions. This chapter concentrates on the distributive and participatory justice challenges posed by one prominent US climate mitigation strategy: a GHG cap-and-trade program. While carbon dioxide emissions themselves are not harmful and do not create direct distributional concerns, they are invariably accompanied by hazardous co-pollutants. Policies that affect GHG emissions, therefore, indirectly impact co-pollutant emissions, raising distributive justice concerns for impacted communities. GHG mitigation policies also raise issues of participatory justice: Who participates in regulatory decisions about how and when industrial sectors should reduce emissions? At the facility-specific level, who controls and participates in decisions about facility emissions?

The cap-and-trade programs that are emerging as a core strategy for addressing climate change at the state and federal levels have long been considered antithetical to the environmental justice movement's distributional and participatory goals. Rather than concluding that the conflict is unbridgeable, however, I propose a reconciliation. I argue that a cap-and-trade program that is one component of a much larger climate change strategy, and that includes limitations to improve the distribution of co-pollutants, could balance efficiency and distributive justice.

---

[1] These profound questions about relative national obligations to address global climate change are addressed in Steve Vanderheiden, *Atmospheric Justice: A Political Theory of Climate Change* (New York: Oxford University Press, 2008) and in Eric A. Posner and Cass R. Sunstein, "Climate Change Justice," *Georgetown Law Journal* 96 (2008), 1565–1612.

The first section of this chapter describes the environmental justice movement's central claims for distributive and participatory justice and the relevance of those claims to climate policy. It goes on to describe market-based environmental policies designed to achieve efficiency, and the implications of such efficiency-driven mechanisms for climate policy. Next the fundamental conflicts between the environmental justice movement's equity and participatory goals, on the one hand, and market-based mechanisms' economic and administrative efficiency goals, on the other hand, are identified. The details of how these conflicts would be manifested under a GHG cap-and-trade program are then described.

The second section of this chapter focuses on reconciling justice and efficiency. This section provides a number of practical proposals to balance efficiency with distributive and participatory justice goals, addressing each mechanism's efficiency consequences. The chapter concludes that assumptions about the tension between justice and efficiency may be somewhat overstated. More broadly, it concludes that, rather than focusing solely on efficiency, cap-and-trade programs could be qualified to improve equity, providing benefits for disadvantaged communities and for society as a whole.

While the chapter focuses on the tension between equity and efficiency and its implications for domestic climate policy, I note that these issues are not the only ones of relevance to assessing the appropriate role of a cap-and-trade program or its design.[2] A third "E" – efficacy – is likewise essential.[3] Any set of climate change policies must achieve real emissions reductions, stimulate innovation and efficiency, and be easily and accurately enforced. A number of the proposals for reconciling equity and efficiency discussed below are driven by the goal of not only enhancing equity but also increasing effectiveness.

---

[2] Environmental justice advocates express a moral repugnance to cap-and-trade programs that could potentially allow companies to profit from selling pollution rights. Richard T. Drury *et al.*, "Pollution Trading and Environmental Injustice: Los Angeles' Failed Experiment in Air Quality Policy," *Duke Environmental Law and Policy Forum* 9 (1999): 270–71; "The California Environmental Justice Movement's Declaration on Use of Carbon Trading Schemes to Address Climate Change" (2008) (findings 10 and 11), EJ Matters for Climate Change, http://ejmatters. org/declaration.html; "Fact Sheet: The Cap and Trade Charade for Climate Change: 13 Reasons Why Trading and Offset Use Are NOT a Solution to Climate Change," EJ Matters for Climate Change (reason no. 4).

[3] Environmental justice advocates are deeply concerned that cap-and-trade programs will be less effective at reducing emissions and stimulating technological advancements than alternative policies ("Cap and Trade Charade").

I ENVIRONMENTAL JUSTICE, MARKET-BASED
MECHANISMS, AND THEIR INHERENT CONFLICTS

1.1 Environmental justice

The environmental justice movement emerged in the 1980s as communities of color began to challenge local and agency decisions to site industrial or disposal facilities near them.[4] These sporadic grass-roots initiatives prompted a wave of studies that confirmed what many had intuitively observed: that undesirable land uses, particularly environmentally hazardous land uses, are disproportionately located in poor communities of color.[5] The environmental justice movement raises central distributional[6] and participatory[7] justice claims. In this section, I consider these distinct claims in turn.

Existing environmental and civil rights laws have not been sufficient to address inequities in the distribution of pollution in the USA. The nation's pollution control laws focus primarily on improving general ambient conditions, not on particular facilities' consequences for adjacent communities. The Constitution's Equal Protection Clause requires proof of discriminatory intent, not simply evidence of disproportionate impacts.[8] Since existing inequities likely stem from a tangle of past zoning decisions, housing discrimination, and socioeconomic constraints rather than from current intentional discrimination, the Constitution provides little redress against siting or permitting decisions in communities of color.[9]

The distributional justice issues that have arisen in the context of traditional pollutants are salient in the climate context because of the integral

---

[4] For an excellent history of the environmental justice movement, see Luke W. Cole and Sheila R. Foster, *From the Ground Up: Environmental Racism and the Rise of the Environmental Justice Movement* (New York University Press, 2001).

[5] Michael Ash *et al.*, *Justice in the Air: Tracking Toxic Pollution from America's Industries and Companies to Our States, Cities, and Neighborhoods* (Los Angeles: University of Southern California Program for Environmental and Regional Equity, 2009); Robert Bullard, *et al.*, *Toxic Wastes and Race at Twenty: 1987–2007* (Cleveland, OH: the United Church of Christ, 2007); James P. Lester *et al.*, *Environmental Injustice in the United States: Myths and Realities* (Boulder, CO: Westview Press, 2001).

[6] Vicki Been, "What's Fairness Got to Do With It? Environmental Justice and the Siting of Locally Undesirable Land Uses," *Cornell Law Review* 78 (1993): 1001, 1029–55; Alice Kaswan, "Distributive Justice and the Environment," *North Carolina Law Review* 81 (2003): 1031.

[7] Sheila Foster, "Justice from the Ground Up: Distributive Inequities, Grassroots Resistance, and the Transformative Politics of the Environmental Justice Movement," *California Law Review* 86 (1998): 775.

[8] *Washington* v. *Davis*, 426 U.S. 229, 239 (1976).

[9] For a discussion of the Equal Protection Clause and its application to environmental justice claims, see Alice Kaswan, "Environmental Laws: Grist for the Equal Protection Mill," *Colorado Law Review* 70 (1999): 407–56.

link between GHGs and their associated co-pollutants. To the extent a GHG cap-and-trade program impacts industrial and electricity-generating facilities, it will impact the same combustion processes that generate co-pollutants. Since a GHG cap-and-trade program will inevitably impact the distribution of co-pollutants, its adoption involves distributive justice.

The environmental justice movement's distributive justice claims deserve serious consideration on both moral and social welfare grounds. As Steve Vanderheiden suggests, "*arbitrary* or undeserved inequality in outcomes is taken to be unjust."[10] From a moral perspective, there is no inherent justification for sharp inequalities in the distribution of environmental harms. Potential justifications for differences in the distribution of benefits – differing levels of need, or differing levels of entitlement based on desert[11] – do not apply where the distribution of environmental burdens is concerned. One would be hard-pressed to argue that a community "needs" or "deserves" more pollution.

Some may say that those who are *not* exposed to pollution "deserve" that benefit: they have earned and deserve the wealth that entitles them to live in pollution-free areas. There are two responses to such a view. First, an individual's economic standing cannot be deemed fully deserved because the underlying distribution of wealth in the USA is so heavily determined by forces outside individual control (e.g., by luck, as in the inheritance of wealth or privilege). Second, the wealthy's entitlement, such as it is, to be free of pollution does not mean that the poor "deserve" to be burdened.

Nor can existing inequalities in the distribution of goods be justified under a welfare-based theory of equality. Under a welfare-based theory of equality, which defines equality in terms of relative preference satisfaction rather than the tangible distribution of goods (or bads), an unequal distribution could be justified if the inequality reflected differing community preferences.[12] Differences in the distribution of polluting facilities would not be "unequal" if neighboring residents were equally satisfied. If communities

---

[10] Vanderheiden, *Atmospheric Justice*, 49.

[11] Nicholas Rescher, *Distributive Justice: A Constructive Critique of the Utilitarian Theory of Distribution* (Indianapolis, IN: Bobbs-Merrill, 1966), 73, 75–76 (discussing distribution on the basis of need); 53–55 (discussing distribution on the basis of "desert").

[12] Norman Daniels describes three potential egalitarian targets: of resources, of welfare, and of capabilities. Norman Daniels, "Equality of What: Welfare, Resources, or Capabilities?," *Philosophy and Phenomenological Research* 50 (Supplement 1990): 273–74. An approach that focuses on a numerical distribution of polluting facilities embodies an "equality of resources" approach, while one that focuses on relative degrees of preference satisfaction embodies an "equality of welfare" approach. The equality of capabilities approach is less relevant to the distribution of pollution. In this chapter, I do not evaluate the differing conceptions of equality, but simply indicate that unequal distributions of pollution are unlikely to be justified under any of the relevant theories.

were content with a disproportionate number of polluting facilities because they believed such facilities provided net benefits, due to employment or municipal tax consequences, then the community would be equally satisfied, and the difference in distribution would not be unjust.

As I have argued elsewhere, however, it is highly unlikely that existing inequalities in the distribution of environmental burdens correspond to differences in community preferences. Residents are rarely influential in siting processes, particularly if they have traditionally lacked political power. Moreover, residents of poor communities and communities of color dissatisfied with a siting decision rarely have the means or opportunity to find homes in less hazardous environments. Thus, whether one adopts a goods-based or a welfare-based theory of equality, the existing distribution of pollution is unjust.

It is important to recognize that the unequal distribution of environmental harms not only presents an abstract violation of principles of equality, but also imposes significant consequences for affected communities. Much of the air pollution that GHG policies are likely to impact is concentrated in communities of color.[13] While the existing Clean Air Act has improved overall air quality, it has failed to lead many areas into attainment of the nation's health-based standards.[14] The failure to attain health standards has led to a high incidence of asthma, heart disease, and cancer in impacted communities that is at least partly attributable to poor air quality.[15]

Addressing adverse pollution impacts on disadvantaged communities is justified not only by moral commitments to achieving equality and alleviating suffering, but also by a "social welfare" vision of efficiency that considers the overall costs and benefits of regulatory policies, rather than focusing solely on their cost-effectiveness for regulated entities. Concentrations of pollution impose social costs: the public health and welfare costs associated with increased levels of asthma, heart disease, and cancer.[16] A system that avoids concentrations of pollution and thereby reduces its public health and welfare costs could be more efficient from a social welfare perspective, even if it increases the cost of pollution control. Ironically, the "greatest good for all" may be best served by attending to distributive justice.

---

[13] "AP: Blacks Likely Breathe Most Unhealthy Air," *CNN.com*, December 14, 2005; "71 Percent of Hispanics Live in Areas Violating Federal Standards, Report Says," *Environment Reporter* 35 (2004): 1469.

[14] United States Environmental Protection Agency, *National Air Quality: Status and Trends Through 2007* (Washington, DC: Environmental Protection Agency, 2008), www.epa.gov/air/airtrends/2008/report/TrendsReportfull.pdf.

[15] J. Andrew Hoerner and Nia Robinson, *A Climate of Change: African Americans, Global Warming, and a Just Climate Policy for the U.S.* (Oakland, CA: Redefining Progress, 2008), 11, 13.

[16] Ibid., 11, 13.

Participatory justice – the ability to participate meaningfully in decisions that impact community well-being – is a second central goal of the environmental justice movement. Ultimately, participatory goals are linked to a deeper aspiration: empowering disadvantaged groups who have historically had little political power within local, state, and national decision-making structures.[17] The environmental justice movement supports regulatory approaches that involve the public in key environmental policy decisions. In traditional direct regulatory proceedings, the public has the opportunity for input into overarching industry standards at the national level. Most importantly, traditional regulatory approaches give communities the opportunity to participate in facility-specific permitting proceedings.[18] Even if communities do not have ultimate authority over permitting decisions, their ability to obtain information through public hearings and to voice their concerns creates a greater likelihood of industry accountability to impacted communities than is likely to exist without that input.

1.2 Market-based mechanisms

Market-based mechanisms present a very different approach to environmental policy. The most common market mechanism is a cap-and-trade program. After an initial cap on emissions is established, the implementing agency distributes allowances equal to the cap to facilities within the program. Each facility can choose whether to reduce emissions to meet its allowance allocation, reduce emissions by more than the allocation and sell the remainder, or maintain existing emissions and buy allowances to make up the difference between the number distributed and actual emissions. At the end of a specified compliance period, the facility must demonstrate that it had enough allowances to cover its emissions during the preceding compliance period.

Cap-and-trade programs have become a dominant trend in environmental policy. Many recent administrative initiatives for air and water pollution have incorporated pollutant trading.[19] Recent federal legislative proposals all included a cap-and-trade program as the primary mechanism for

---

[17] Luke Cole, "Empowerment as the Key to Environmental Protection: The Need for Environmental Poverty Law," *Ecology Law Quarterly* 19 (1992): 641.

[18] *A Citizen's Guide to Using Federal Environmental Laws to Secure Environmental Justice* (Washington, DC: Environmental Law Institute, 2002), 15–18.

[19] See, e.g., United States Environmental Protection Agency, "Rule to Reduce Interstate Transport of Fine Particulate Matter and Ozone," *Federal Register* 70 (May 2005): 25162; United States Environmental Protection Agency, Office of Water, "Water Quality Trading Policy" (January 2003),

controlling GHG emissions.[20] In the absence of a federal program, states in the Northeast, the Midwest, and the West have joined together to develop regional trading blocs.[21]

Market-based mechanisms are touted as more administratively and economically efficient than traditional regulatory approaches.[22] From an administrative standpoint, traditional regulations have required government agencies to determine the appropriate pollution control technology for an industry and translate it into an applicable performance standard, a time-consuming process that requires significant government resources and that requires government officials to develop technological expertise that is arguably already held by regulated industries. At least in theory, a cap-and-trade program is more administratively efficient because the government does not have to develop control standards for regulated facilities. The facilities themselves, with their internal knowledge of industry operations, make the critical decisions about whether and how to reduce emissions. The government's primary role is setting the cap, distributing allowances, and ensuring that facilities have sufficient allowances at the end of the compliance period. Greater administrative efficiency preserves limited government resources for other priorities or reduces the taxpayers' burden.

Cap-and-trade programs are also touted as more economically efficient than traditional technology-based regulation. Economic efficiency in this context means cost-effectiveness: reducing aggregate emissions at the lowest industry cost. Traditional regulatory approaches are deemed economically inefficient since they require all facilities to install control technology, even if some facilities could reduce at a lower cost than others. In contrast, cap-and-trade programs allow facilities with high reduction costs to buy allowances rather than make expensive reductions. Meanwhile, facilities that have lower control costs have an incentive to reduce emissions and sell the excess allowances. As a consequence, more of the reductions are accomplished by

www.epa.gov/owow/watershed/trading/finalpolicy2003.pdf. The particulate and ozone rule was, however, struck down as unauthorized by the Clean Air Act. *North Carolina* v. *EPA*, 531 F.3d 896 (D.C. Cir. 2008).

[20] For example, in 2009, the House of Representatives approved a bill that called for a cap-and-trade program to control stationary source and transportation emissions. The American Clean Energy and Security Act of 2009, HR2454, 111th Cong., 1st Sess., § 311 (June 26, 2009).

[21] Regional Greenhouse Gas Initiative, "Memorandum of Understanding," December 20, 2005, www.rggi.org/docs/mou_final_12_20_05.pdf; "Midwestern Greenhouse Gas Accord," November 15, 2007, www.midwesternaccord.org/midwesterngreenhousegasreductionaccord.pdf; "Western Regional Climate Action Initiative," February 26, 2007, www.westernclimateinitiative.org/.

[22] A classic article advocating market approaches is Bruce A. Ackerman and Richard B. Stewart, "Reforming Environmental Law," *Stanford Law Review* 37 (1985): 1333. For an overview of the history, principles, and assessment of trading systems, see T. H. Tietenberg, *Emissions Trading: Principles and Practice* (Washington, DC: Resources for the Future, 2nd edn., 2006).

low-cost reducers rather than high-cost reducers, lowering the overall cost of pollution control. For the participating firms, cap-and-trade programs reduce compliance costs. For society as a whole, fewer economic resources are devoted to achieving a given environmental goal.

Economic efficiency also indirectly affects wealth distribution. The energy and other consumer cost increases that are likely to result from climate change mitigation policies are likely to have a regressive effect: to have a disproportionately greater impact on the poor.[23] Lowering the costs of regulation would be one mechanism for softening the impact of climate change controls on the most vulnerable.[24]

Alternatively, the lower costs of an economically efficient reduction strategy could result in more stringent targets rather than cost savings. If policymakers tie the degree of acceptable stringency to the costs of control, then a more economically efficient control program could induce policymakers to accept more stringent emission reduction goals. The more stringent the emission reduction goals, the greater the benefit to the communities most vulnerable to the impacts of climate change. As the Intergovernmental Panel on Climate Change (IPCC) has noted, residents of developing countries are likely to be more adversely impacted by climate change's consequences than those in the industrialized world.[25] Within the USA, poor communities are also likely to be more vulnerable.[26] They live in areas more likely to be impacted by intense weather events, like hurricanes and heatwaves; live in more polluted areas that will be adversely impacted by heat-related increases in air pollution; will be more vulnerable to weather and disease consequences due to the lack of health insurance; and will, more generally, have fewer resources to adapt to the inevitable consequences of climate change.[27] Thus, if economically efficient policies lead to higher

---

[23] See Terry Dinan, *Trade-Offs in Allocating Allowances for CO_2 Emissions* (Washington, DC: Congressional Budget Office, 2005), 1, 3, 6–8.

[24] A cap-and-trade program will not, however, eliminate the regressive economic impact of climate change regulation on the poor. Moreover, the economic impact of climate change regulations could be directly addressed through government programs to increase energy efficiency for low-income residents or through compensation to low-income residents impacted by higher energy costs.

[25] "Summary for Policymakers," in *Climate Change 2007: Impacts, Adaptation and Vulnerability: Contribution of Working Group II to the Fourth Assessment Report of the Intergovernmental Panel on Climate Change*, ed. M. L. Parry *et al.* (Cambridge University Press, 2007).

[26] See Robert D. Bullard, "Climate Justice and People of Color" (2000), 3, www.ejrc.cau.edu/climatechg-poc.html (describing disproportionately adverse climate change impacts on the poor); Maxine Burkett, "Just Solutions to Climate Change: A Climate Justice Proposal for a Domestic Clean Development Mechanism," *Buffalo Law Review* 56 (2008): 169, 176–88; Rachel Morello-Frosch *et al.*, *The Climate Gap: Inequalities in How Climate Change Hurts Americans and How to Close the Gap* (Los Angeles: University of Southern California Program for Environmental and Regional Equity, 2009).

[27] Burkett, "Just Solutions to Climate Change," 176–88; Morello-Frosch *et al.*, *The Climate Gap*, 7–13.

reduction goals, they could mitigate the climate change impacts on the globe's most vulnerable regions and communities.

1.3 The conflict between environmental justice and market-based mechanisms

These two environmental policy trends, environmental justice and market mechanisms, do not co-exist harmoniously. Instead, they present fundamental conflicts, conflicts which have led members of the environmental justice community to oppose cap-and-trade programs for reducing GHG.[28] Although I ultimately argue that the two approaches' competing aims can be partially accommodated, it is important to recognize the inherent tensions. In this section, I focus on the potential conflicts and demonstrate how they are manifested in proposals for GHG cap-and-trade programs.[29]

The first fundamental tension is between distributive justice and the pursuit of economic efficiency. The environmental justice movement focuses on how fairly environmental burdens are distributed.[30] In contrast, market-based systems seek to achieve "the greatest good for the greatest number" by allowing society to achieve a level of pollution reduction at a lower cost.[31] To achieve such economic efficiency, however, market-based systems ignore distributional consequences.[32] If facilities with high costs of control are located in polluted areas and rely upon allowance purchases rather than reducing emissions, air quality will not improve. Meanwhile, if facilities with low costs of control are located in less polluted areas, then the emissions reductions will be concentrated in the areas where they are least necessary. Thus, pursuing economic efficiency could come at a cost to distributive justice.

---

[28] "The California Environmental Justice Movement's Declaration."

[29] A number of the arguments discussed below are detailed more fully in Alice Kaswan, "Environmental Justice and Domestic Climate Change," *Environmental Law Reporter* 38 (2008): 10291–312.

[30] Environmental programs raise an additional equity issue: the impact of pollution control obligations on regulated industry, an issue not addressed in this chapter. An article that focuses largely on the implications of considering fairness versus efficiency with respect to regulated industries is Shi-Ling Hsu, "Fairness Versus Efficiency in Environmental Law," *Ecology Law Quarterly* 31 (2004): 303.

[31] Daniel J. Dudek and John Palmisano, "Emissions Trading: Why Is This Thoroughbred Hobbled?," *Columbia Journal of Environmental Law* 13 (1988): 217, 223, 231–34 (describing efficiency goals for market-based programs).

[32] For further elaborations of this tension, see Lily N. Chinn, "Can the Market Be Fair and Efficient? An Environmental Justice Critique of Emissions Trading," *Ecology Law Quarterly* 26 (1999), 80; Drury et al., "Pollution Trading and Environmental Injustice," 231, 272; Stephen M. Johnson, "Economics v. Equity: Do Market-Based Environmental Reforms Exacerbate Environmental Injustice?," *Washington and Lee Law Review* 56 (1999): 111.

Some market advocates have suggested that cap-and-trade programs for GHG do not present issues of distributive justice because, unlike most other pollutants, they do not have local health consequences.[33] However, as discussed above, GHG emissions, and GHG trades, intrinsically implicate co-pollutant emissions. Every allowance a facility received for emitting GHG would be accompanied by co-pollutant emissions. The distribution of GHG allowances will impact the distribution of co-pollutants.

Overall, climate policies that reduce GHGs are likely to lead to co-pollutant reduction benefits, benefits that should accrue to the disadvantaged communities that are currently adversely impacted by pollution. Nonetheless, the distribution of that benefit involves distributive justice. If facilities in disadvantaged neighborhoods were to buy carbon dioxide allowances that maintain existing emissions, the surrounding neighborhood would not receive a co-pollutant reduction benefit. Conceivably, older facilities concentrated in urban poor areas could face higher costs of control and buy allowances rather than reducing emissions, while newer facilities in less polluted areas make reductions. The consequence of such trading patterns would be inequality in the distribution of the co-pollutant reduction co-benefit of climate change regulation and, potentially, an increasing disparity in the prevalence of air pollution. Thus, even if GHG reduction policies provide overall co-pollutant benefits, a trading program could fail to distribute those benefits equally or optimally.

It is also conceivable that trading in GHG allowances could lead to *increases* in associated co-pollutants at certain locations.[34] The risk of such hot spots depends, in part, upon how well the Clean Air Act (CAA) controls the associated co-pollutant emissions. A facility's ability to buy or sell GHG

---

[33] A. Denny Ellerman, Paul C. Joskow, and David Harrison, Jr., *Emissions Trading in the U.S.: Experience, Lessons, and Considerations for Greenhouse Gases* (Washington, DC: Pew Center on Global Climate Change, 2003), 40–41.

[34] It should be noted that the Acid Rain Program, a national trading program for controlling emissions of acid rain precursors, reportedly did not lead to hot spots. Byron Swift, "Emissions Trading and Hot Spots: A Review of the Major Programs," *Environment Reporter* (May 7, 2004). That that program did not lead to hot spots does not, however, mean that incidental co-pollutant increases would not occur in a GHG trading program. The Acid Rain Program likely avoided hot spots for several reasons. First, it achieved larger overall reductions than are slated for the early years of a GHG trading program, and the larger the overall reductions the less likely it is that emissions in any one place would increase. Second, many facilities were able to comply by obtaining low-sulfur coal, a widely available compliance option. If the ability to reduce GHG emissions varies more in the GHG context, then facility responses to emissions limitations are likely to vary more, increasing the possibility of hot spots. Third, as discussed below, federal GHG legislation will allow extensive use of offsets that could result in little actual emissions reductions from the covered sectors for the early years of a GHG control program. If actual emission reductions in regulated sectors are minimal, the risk of co-pollutant hot spots increases.

allowances would be constrained by existing CAA permits for the relevant co-pollutants. Since the existing CAA permit system does not fully constrain co-pollutant increases, GHG allowance trading could potentially lead to increases in co-pollutant emissions.

More specifically, co-pollutant permits generally dictate a facility's permissible *rate* of emissions, like a certain quantity of pollutant per amount of energy created.[35] Facilities can, therefore, increase their total absolute emissions, so long as they do not increase their emissions rate. For example, if the emissions rate were based upon the amount of energy generated, a facility could increase the amount of energy generated and the associated co-pollutant emissions so long as the ratio between the pollutants and the energy generated did not change.

Once absolute emissions increase beyond a certain threshold, however, additional co-pollutant controls could be required under the CAA's New Source Review (NSR) program and its stringent requirements for new and modified facilities. If a facility physically changes its plant and, as a consequence, increases its actual, absolute, emissions above a certain threshold amount, then it would be deemed to have been "modified," and would be subject to the additional controls associated with new pollution sources.[36]

Nonetheless, the NSR program does not impose limitations on all absolute increases in emissions. If a facility increases emissions due to increases in the hours of operation, not due to a physical change in the facility, then NSR would not apply even if emissions increased beyond the level deemed "significant."[37] In addition, the federal threshold for what constitutes a "significant" increase is relatively high, and could lead to a substantial impact on a local community even if it did not meet the statutory threshold.[38] Moreover, the integrity of the system relies upon proper enforcement of

---

[35] United States Environmental Protection Agency, *Tools of the Trade: A Guide to Designing and Operating a Cap-and-Trade Program for Pollution Control* (Washington, DC, 2003), 1–2, 2–5, and 2–9, www.epa.gov/airmarkets/resource/docs/tools.pdf.

[36] For example, see *U.S. Code* 42 (2006), § 7411(a)(4) (defining "modifications"), and *Code of Federal Regulations* 40 (2008), § 51.166(b)(3) (defining, under the Prevention of Significant Deterioration program, increase in emissions in terms of an increase in a facility's actual emissions).

[37] *Code of Federal Regulations* 40 (2008), § 51.166(b)(2)(iii)(f) (stating that emissions increases resulting solely from increases in hours of operation or increased production would not trigger new source requirements, even if the net actual increases were significant).

[38] See David Wooley and Elizabeth Morss, *Clean Air Act Handbook* (Eagan, MN: Thomas Reuters/West, 2007), § 1:113 (providing a table showing the threshold for determining a significant increase). The most common thresholds range from 15 to 40 metric tons per year of a given pollutant. New rules for determining the baseline from which to measure the emissions could increase the likelihood that significant increases would not be regulated. Facilities can now choose the emissions average from any 2-year period within the preceding 10 years. *Code of Federal Regulations* 40 (2008), § 52.21(b)(48)(ii)(c). Facilities increasing emissions could choose a baseline year in which emissions were higher than at present.

existing pollution control requirements, enforcement that has not always been effective.[39]

Thus, a GHG trading system in which a facility purchased allowances that allowed it to exceed its current emissions could lead to increases in co-pollutants, until and unless they are controlled by the CAA's NSR provisions. Given the existing disparity in the distribution of polluting facilities, GHG trading could lead to indirect increases in co-pollutant emissions in the nation's most vulnerable communities. While the co-pollutant increases are ultimately attributable to weaknesses in the CAA, they would nonetheless be an indirect consequence of the choice to control GHG through a trading program that is indifferent to distributional consequences.

Certain design features of cap-and-trade programs have additional distributive justice implications. For example, the use of offsets affects the distribution of co-pollutant reductions. Cap-and-trade programs generally cover particular sectors, like the electric utility sector, and allowance trading occurs among sources within the sector. Offsets are reductions or carbon sequestration achieved outside the regulated sector. If facilities are allowed to meet their emission reduction obligations by buying offsets, then the regulated facilities will not reduce either their GHGs or the associated co-pollutants.

Nor would most of the offset projects currently under consideration generate their own co-pollutant reduction benefits. Certain types of offsets, like biological sequestration, provide conservation benefits, but they do not generate co-pollutant reductions. For example, an electric utility might buy offsets from a timber company based on its commitment not to harvest carbon-sequestering trees. Assuming that forest conservation is a verifiable and legitimate mechanism for sequestering carbon (often a contested issue),[40] use of a timber-based offset would allow existing GHG emissions and their co-pollutants to continue while offsetting only the GHG emissions, not the co-pollutants. In contrast, if trading were limited to allowances within a sector, it would be more likely to generate co-pollutant reductions.[41]

Using the higher emissions level as a baseline would make it less likely for current increases to cross the significance threshold, and make it less likely for a facility's emissions increases to trigger the controls required under NSR.

[39] See, e.g., Tom Pelton, "State Gives Power Plants a Pass on Pollution," *Baltimore Sun*, May 28, 2006 (describing lax enforcement); Environmental Integrity Project, *Polluters Breathe Easier; EPA Environmental Court Actions Decline* (Washington, DC, 2004) (describing decrease in federal environmental enforcement actions).

[40] Kenneth R. Richards, R. Neil Sampson, and Sandra Brown, *Agricultural and Forestlands: U.S. Carbon Policy Strategies* (Washington, DC: Pew Center on Global Climate Change, 2006), 50–54.

[41] Requiring reductions to occur within the regulated sectors and limiting the use of offsets would also increase the incentive for facilities within the regulated sectors to adopt and develop sector-specific emission reduction techniques, rather than relying upon reductions in unregulated sectors.

Similarly, letting US facilities purchase international offsets, like offsets from reductions or emissions sequestration projects in developing countries, would result in fewer co-pollutant reductions in the USA. International trading clearly raises many profound questions of international justice, and allowing international offsets could stimulate desirable co-pollutant reductions and sustainable development in developing countries. Nonetheless, an important question is whether these goals should be furthered by mechanisms that complement, rather than replace, developed country GHG and co-pollutant reductions.

Recent federal legislative proposals would have permitted US facilities to make extensive use of both domestic and international offsets, with a potentially significant impact on the degree to which GHG policies led to actual reductions in facilities' GHG emissions and their associated co-pollutants. For example, a 2009 proposal in the House of Representatives would have allowed 30 percent of emissions to be covered by offsets at the beginning of the trading program, with even larger percentages permitted as the program became more stringent.[42] As long as the expected emissions reduction goal is less than the allowable use of offsets, facilities could comply with the law solely through offset purchases, without needing to make any actual GHG (or co-pollutant) emissions reductions.

A second central tension between environmental justice goals and market-based programs is the tension between participatory justice and administrative efficiency. Market-based mechanisms are designed to maximize private autonomy and administrative efficiency and to minimize the government's role. In a cap-and-trade program, government entities would set the emissions cap, but they would not design a system of industry-specific requirements through a public rule-making process. At the individual facility level, the public would continue to have a role in initial siting decisions. But since there is no opportunity for public participation in private allowance trading decisions, the public would not have any input into subsequent changes in GHG emissions unless those changes were substantial enough to trigger co-pollutant regulatory proceedings.

As discussed above, communities are accustomed, under traditional permitting programs, to having the opportunity to comment and respond to facility-permitting decisions. Such proceedings provide communities with key information about the emissions decisions of neighboring industries and

---

[42] Mark Holt and Gene Whitney, *Greenhouse Gas Legislation: Summary and Analysis of H.R. 2454 as Passed by the House of Representatives* (Washington, DC: Congressional Research Service, 2009), 84 (computing the percentage of allowable offsets as 30 percent in 2012 and larger in future years).

allow them to voice concerns that could help shape permitting conditions. While communities have never had veto power over facility emissions, the opportunity to participate in the proceedings is nonetheless of significant value. A cap-and-trade program's administrative efficiency and private autonomy come at the cost of public involvement.

2 RECONCILING JUSTICE AND EFFICIENCY

Battles over cap-and-trade policy present a fundamental tension between the utilitarian aspirations of market advocates, who seek economic and administrative efficiency to provide the greatest good for the greatest number, and the egalitarian aspirations of those in the environmental justice movement, whose focus is on the well-being of the disadvantaged. As the foregoing discussion makes clear, efficiency and equity are both important objectives. Rather than choosing one or the other, policymakers have the option of qualifying a cap-and-trade program's efficiency in order to better achieve equity.[43]

In the following discussion, I propose a number of practical mechanisms for alleviating the adverse distributional impacts of a cap-and-trade program and analyze their implications for administrative and economic efficiency. The first insight from the analysis is that policies to pursue equity do not always compromise efficiency to the extent assumed; the presumed tension is partially illusory. That said, the analysis acknowledges that the pursuit of equity could compromise efficiency in some instances and to a certain extent. Determining the appropriate trade-off between efficiency and equity is ultimately a question of values, values that policymakers will express through design choices that apportion the balance between efficiency and equity.

2.1 Combine a cap-and-trade program with direct regulation

The first proposal for improving equity is to combine a cap-and-trade program with more direct regulatory requirements. Fundamentally, a cap-and-trade program should be one component of a much larger and more

---

[43] As A. Dan Tarlock has stated, "A judgment that a certain allocation of resources is efficient but inequitable can constrain efficiency when the distributional impacts are deemed unacceptable." A. Dan Tarlock, "Environmental Protection: The Potential Misfit Between Equity and Efficiency," *University of Colorado Law Review* 63 (1992): 882.

comprehensive climate change agenda.[44] Some of the adverse distributional and participatory consequences of a cap-and-trade program could be mitigated if it were combined with direct regulations that require facilities, or at least some facilities, to adopt available cost-effective emission reduction mechanisms.[45] Existing cap-and-trade programs take a similar approach: virtually all of them supplement an existing regulatory system rather than replacing it. For example, in the Acid Rain Program, the trading program is designed to achieve reductions beyond those called for under existing CAA requirements.[46] In a GHG emission reduction program, facilities could be required to adopt available and cost-effective energy efficiency or other GHG emission reduction measures as a means of reducing energy consumption and thus co-pollutant emissions. Or regulations could be imposed on industries with the strongest links between GHG and co-pollutant emissions, or on facilities located in polluted areas, so as to maximize the co-pollutant benefits associated with GHG reductions.

Requiring facilities to adopt available mechanisms would distribute the co-pollutant reduction co-benefits of climate change regulation more equitably than a pure trading system since the most common type of regulation would require facilities to reduce their rate of GHG emissions. While regulated facilities would still be able to buy GHG allowances and increase GHG emissions and their associated co-pollutants, the regulation would have reduced the baseline level of emissions, lessening the impact of subsequent allowance purchases.

Another key benefit of a regulatory approach is the opportunity for public participation. The public could participate in the regulatory process for establishing initial regulatory standards. Most importantly, at the facility-specific level, communities could participate in individual permitting processes.

It is worth noting that "effectiveness" concerns provide an additional justification for supplementing a market approach with at least some degree of complementary regulation. Direct regulation is more effective at encouraging facilities to adopt known and cost-effective technologies than the

---

[44] I focus on those components that relate to emission reductions, since they directly involve the distribution of co-pollutants. Other key components of climate change policy include energy efficiency retrofits, land use planning to reduce sprawl, research and development of alternative technologies, adaptation planning and implementation, and other mechanisms.

[45] In the context of traditional pollutants, several scholars have advocated the maintenance of a regulatory safety net in conjunction with trading programs. Drury et al., "Pollution Trading and Environmental Injustice," 284–85; Johnson, "Economics v. Equity," 162, 165.

[46] EPA, Tools of the Trade, 3–21.

market's uncertain and sometimes ignored price signals.[47] Moreover, the extensive reductions necessary to address the threat of climate change will likely require all facilities to reduce, not just the lowest-cost reducers most likely to respond in a market-based program.

Combining a cap-and-trade program with direct regulation would have efficiency implications. Under ideal circumstances, cap-and-trade programs promise administrative efficiency because government regulators just set a cap and allocate the initial allowances; they leave the decision about how to comply to the regulated community. Government agencies therefore do not have to develop and defend initial industry- or facility-specific standards, and they do not have to engage in facility-specific permitting processes. Enforcement consists of ascertaining whether a facility has accumulated enough allowances to match its actual emissions. For example, the Acid Rain Program, which includes a limited number of sources, all of whom have been required to install continuous emissions monitoring that facilitates government enforcement, has been very administratively efficient.[48]

The relative administrative benefits of cap-and-trade should not, however, be overstated. Direct regulation is not always administratively complex and cap-and-trade programs are not always easy to administer. Industry standards for GHG reductions might focus on energy efficiency measures or process changes that are not as complex or contested as the pollution control technologies that have been developed for other pollutants.

And cap-and-trade programs could be more administratively complex than is generally assumed. If the program includes multiple diverse sources of varying sizes, including smaller players who lack information about the market or are unaware of available implementation choices, then government agencies may have to provide significant assistance.[49] Moreover, unless the participating facilities all have easily verifiable monitoring results, monitoring and enforcement can turn into a complex and time-consuming administrative process. In the RECLAIM program, a cap-and-trade program in Los Angeles, government officials concluded that the cap-and-trade program, which included a variety of types and sizes of sources and did not have an easily verifiable monitoring process for all sources, required more

---

[47] For example, in Los Angeles' RECLAIM trading program, regulators decided to directly impose pollution control requirements on large electricity-generating facilities when the market-based system failed to create sufficient incentives for technology adoption. Lesley K. McAllister, "Beyond Playing 'Banker': The Role of the Regulatory Agency in Emissions Trading," *Administrative Law Review* 59 (2007): 290.
[48] Ellerman, *Emissions Trading in the U.S.*, 16–17.   [49] McAllister, "Beyond Playing 'Banker.'"

administrative resources than a traditional regulatory approach.[50] Moreover, regulating the carbon market itself to ensure against market manipulation and fraud, and to prevent its collapse, is likely to present a new and challenging administrative burden. This is not to say that direct regulation is always more efficient than cap-and-trade programs, but that the relative administrative efficiency of a cap-and-trade program should not be presumed.

From the standpoint of economic efficiency, direct regulation could, theoretically, impact economic efficiency by requiring some sources to adopt emissions-reducing measures that could have been avoided if the sources had the option of purchasing allowances. The inefficiency of traditional regulation should not, however, be overstated. Traditional environmental regulations typically do take a given industry's overall pollution-control costs into consideration. Many standards are set on an industry-wide basis for an extensive list of industrial categories. The standards almost always require the regulatory agency to take costs of control into account. If costs vary *within* an industrial category, so that the required standards are more expensive for some firms within the same industry than for others, then traditional regulation could be inefficient. But, to the extent that regulators create cost-sensitive standards for industry groups composed of reasonably similar facilities, traditional regulation is not as inefficient as it is sometimes portrayed. Furthermore, some general standards are finalized on a facility-by-facility basis, providing regulators with some flexibility in determining the cost-effectiveness of control techniques as applied to that facility. While traditional regulation requires controls when a market program might let a facility forgo controls entirely, those controls are not blind to cost considerations and hence to overall efficiency.

In addition, regulations requiring cost-effective emission reductions through, for example, energy efficiency improvements, would likely require steps that would have been economically rational under a market-based approach, assuming a sufficiently stringent cap. A regulatory program that simply required such reductions, rather than relying upon the market to inspire potentially irrational actors, would provide a more certain and direct mechanism to accomplish currently achievable reductions that are already economically feasible. A regulatory program that achieves the same results as a market-based approach is not less economically efficient.

Finally, even if regulations require GHG reductions that are less cost-effective than the reductions that would have occurred in a pure market-based system, those reductions would not necessarily be "inefficient" from a

[50] Ibid., 294–304.

broader social welfare perspective that includes the costs associated with concentrated pollution. GHG regulatory programs that maximize co-pollutant reductions could achieve a socially optimal result even if they present marginally higher GHG reduction costs.

2.2 Impose limitations to maximize co-pollutant reductions

Even if complementary regulatory programs were adopted, they are unlikely to reduce GHG emissions sufficiently. It remains worthwhile to evaluate how improved co-pollutant distributions could be achieved within the confines of a trading program. One option is to evaluate communities' relative exposure to co-pollutants and impose trading restrictions in proportion to the risk.[51]

The trading restrictions could take a variety of forms. Facilities in polluted areas could be subject to individual facility emission caps that prohibit them from emitting more than they have in the past. Or fewer allowances could be distributed to facilities in polluted areas. For example, in a system in which allowances are distributed in proportion to existing emissions, facilities in polluted areas could receive proportionately fewer allowances than facilities in less polluted areas.[52] If allowances were auctioned, facilities in polluted areas could be allowed to purchase only a certain percentage of their baseline emissions.

Alternatively, more indirect reduction incentives could be developed. The regulatory agency could require facilities in polluted areas to submit more allowances per metric ton of emissions. For example, a facility might be required to obtain 1.2 allowances per metric ton of emissions in a polluted area, with the ratio depending upon the extent of the pollution in the area.[53] This approach would be similar to the differing offset ratios required for new sources in areas that have not attained the nation's air

---

[51] In the context of trading programs for traditional pollutants, scholars have noted the potential for trading limitations into overburdened areas while recognizing their potential adverse efficiency consequences. Drury *et al.,* "Pollution Trading and Environmental Injustice," 284; Johnson, "Economics v. Equity," 162; EPA, *Tools of the Trade,* 3–22. Jonathan Nash and Richard Revesz propose that all trades be prescreened to determine their impacts on a given region's ambient air quality, and be rejected if they would lead to a violation of air quality standards. Jonathan Remy Nash and Richard L. Revesz, "Markets and Geography: Designing Marketable Permit Schemes to Control Local and Regional Pollutants," *Ecology Law Quarterly* 28 (2001): 569. However, the very high transaction costs associated with this proposal render it impracticable.

[52] This approach has been proposed for traditional pollutants. Chinn, "Can the Market Be Fair and Efficient?," 119.

[53] If facilities in polluted areas use more than one allowance for each metric ton of emissions, that would lower national allowance supplies, effectively creating a more stringent overall cap. That result could be considered either an advantage or disadvantage, depending upon one's perspective.

pollution standards, known as nonattainment areas. In these areas, the number of offsets required depends upon the severity of the area's non-attainment status.[54] Alternatively, if allowances are auctioned, a higher rate could be charged for allowances to be used in disadvantaged areas.

Co-pollutant reduction benefits could also be enhanced by limiting the use of offsets so as to increase the actual GHG (and co-pollutant) reductions made in sectors subject to a cap-and-trade program. And offset policies could be tailored to promote co-benefits, like privileging offsets that provide desirable benefits (including co-pollutant reductions) and discounting off-sets that provide fewer collateral benefits.

Moreover, monitoring considerations could help define the appropriate scope of a cap-and-trade program. In a trading program, agencies must rely on the accuracy of a facility's contention that it has enough allowances to match its emissions. Unless emissions can be easily verified, emissions could exceed allowances, compromising the entire program's integrity. The risk of fraud or error would be minimized if a cap-and-trade program's scope were limited to those sectors and sources capable of providing accurate and verifiable emissions data.

These program parameters implicate administrative efficiency. The first administrative task would be assessing exposure risks to determine the areas to be subject to trading limitations. That burden is not particularly prob-lematic. The information is likely to be readily available in many jurisdic-tions and, where not, the value of the information for broader regulatory purposes is likely to justify the burden of obtaining it.

Once an agency determines the most-exposed areas, mechanisms that require a higher allowance ratio or charge more for allowances in polluted areas would be relatively straightforward to administer. However, mecha-nisms that turn upon a given facility's baseline emissions, like limiting a facility's allowance purchases to a certain percentage of its baseline, could be more administratively burdensome since they would require the agency to assess baseline emissions, an issue that is frequently contested. If the regulatory agency obtained baseline emissions data in any case, however, little additional administrative burden would be imposed.

Limiting offsets would not have adverse administrative efficiency con-sequences. Administering offsets is likely to require extensive government resources, so the smaller the role of offsets, the less the administrative burden. Certain sources, like biological sequestration, are inherently

---

[54] *U.S. Code* 42 (2006), § 7503(c) (establishing offset requirement).

difficult to quantify and verify, both initially and over time.[55] Government regulators also face jurisdictional challenges. Domestically, the agency responsible for verifying compliance, likely the EPA in a federal system, would not have historical jurisdiction over the entities providing the offsets, like agriculture or forestry entities. For example, if an electric utility were purchasing offsets from a farm's methane control program, the air pollution control agency is unlikely to have had traditional jurisdiction over the farm's unregulated methane emissions, and the government will have to develop new administrative systems for monitoring and verifying the farm's reductions.

To the extent that a trading program permits international offsets or links to other international trading programs, additional administrative challenges are likely. The agency accepting international offsets or allowances will have to develop systems for determining their legitimacy. While some progress has been made on this front through the Kyoto Protocol's Clean Development Mechanism, through which the European Union has been purchasing offsets from developing countries, that process has been viewed as insufficient.[56]

Limiting the scope of a trading program to facilities that can be easily and accurately monitored would also be more administratively efficient. Distributing allowances will be more straightforward if regulators have confidence in information about existing baseline emissions. In addition, enforcement will be far easier if regulators have easily verifiable emissions data. If sectors that cannot be easily monitored are included in the trading system, then regulators will face a much greater administrative burden in determining baseline emissions and verifying emissions associated with trades.

These mechanisms for increasing the equity of co-pollutant consequences, like limiting trading into polluted areas, limiting offsets, and limiting the program to easily-monitored sources, also raise economic efficiency considerations. If facilities in polluted areas have to spend more to cover their emissions than facilities in other areas, then they are more likely to invest in emissions reductions – even if other facilities could have reduced more cheaply. Limitations on offsets would also preclude the use of inexpensive reduction alternatives. Limiting the scope of the program to easily monitored sources would deny the economic benefits of a cap-and-trade program to the omitted sectors. These restrictions thus pointedly raise the

---

[55] Jonathan L. Ramseur, *The Role of Offsets in a Greenhouse Gas Cap-and-Trade Program: Potential Benefits and Concerns* (Washington, DC: Congressional Research Service, 2009), 12–16.
[56] Michael Wara, "Measuring the Clean Development Mechanism's Performance and Potential," *University of California, Los Angeles Law Review* 55 (2008): 759.

trade-off between economic efficiency and distributive justice. The extent of the conflict would depend upon the stringency of each restriction. While there is no objectively correct answer to the question of "how stringent?," some impact on efficiency is justified by the distributive justice benefits of alleviating concentrations of co-pollutants in disadvantaged areas and by the overall social welfare benefits that derive from addressing co-pollutants along with GHG emissions.

2.3 Use auction revenue to further distributive justice

If allowances are auctioned, rather than given away for free, then facilities must pay for the right to pollute. The government receives revenue that it can use for a variety of purposes. Auction revenue could be directed to impacted communities for alternative co-pollutant reduction activities.[57] Auction revenues could be used to subsidize industrial co-pollutant reductions, finance less-polluting public transportation, finance less-polluting private transportation, or finance other pollution-reducing activities. Subject to overarching specifications, local communities could play a role in deciding how the funds should be used, creating a public participation opportunity.

This approach to the potential adverse distributional impacts of trading would have the least impact on the trading system's internal administrative and economic efficiency. Other co-pollutant-reducing activities are not always easily available in all communities, however, so it may not suffice to address a trading system's co-pollutant impacts. And while it might make a trading system more efficient, it would create a separate administrative process that might not reduce overall governmental costs.

More generally, auction revenues could be used to redress the economic impacts of climate mitigation policies on poor communities by, for example, financing energy efficiency improvements in poor communities. Funds could also be used to provide green job training for disadvantaged communities, to create other green economic opportunities in historically disadvantaged communities, to finance necessary adaptation measures for poor communities, or for other purposes that assist those most vulnerable to climate change and its regulation.[58]

---

[57] Market Advisory Committee to the California Air Resources Board, *Recommendations for Designing a Greenhouse Cap and Trade System for California* (2007), 57.

[58] See Burkett, "Just Solutions to Climate Change"; and Van Jones, *The Green Collar Economy: How One Solution Can Fix Our Two Biggest Problems* (New York: HarperCollins, 2008).

2.4 Public accountability

Participatory justice is the Achilles' heel of cap-and-trade systems. Notwith-standing the critical importance of participatory justice, it is ˙virtually impossible to balance with the essential characteristics of a trading system. Requiring public participation in connection with facility allowance pur-chases and trades would significantly impact administrative efficiency. That impact would, in all likelihood, be substantial enough to lead facilities to forego economically efficient trades. As a consequence, public participation could also compromise a trading program's economic efficiency.

The proposals for reconciliation articulated above therefore do not include any robust mechanisms for public involvement within the confines of a trading program. The absence of direct public participation renders public accountability, and public access to trading information, critical. If the public loses the ability to have input in facility decisions before they are made, then it is imperative for surrounding communities to be able to keep track of the consequences of trading for local GHG and co-pollutant facility emissions. Allowance purchases and their emissions consequences should be frequently and publicly recorded.[59] Although public accountability is not the same as public participation, it could provide an important check on potential abuse, and provide information that regulators could use to address adverse pollution concentrations if and when they emerge.

3 CONCLUSION

Achieving GHG emissions reductions efficiently is an important and laud-able goal that benefits society at large. In theory, pursuing efficiency could even serve distributive justice goals if it accelerates reductions and lowers the economic costs of regulation, results that benefit disadvantaged com-munities. But pursuing a narrowly defined version of economic efficiency, with efficiency defined solely in terms of lowering costs for industry and lowering the total cost of a given level of reduction, could sacrifice equity in the distribution of co-pollutants and fail the most vulnerable.

This chapter indicates that climate policies could be designed to balance the sometimes conflicting aims of equity and efficiency. Careful scrutiny of a number of measures reveals that, if carefully crafted and administered, they could have a less adverse impact on administrative and economic

---

[59] Similar proposals have been made in connection with trading programs for traditional pollutants. Johnson, "Economics v. Equity," 150.

efficiency than is commonly supposed. Nor should any impact on efficiency doom measures to achieve equity. Efficiency is only one of several relevant considerations.[60] Climate change legislation is likely to be a vehicle for fundamental industrial transformation. Holistically integrating principles of distributive and participatory justice into the design of regulatory policies will better serve an ethical commitment to equality and enhance overall social welfare.

[60] Dietz, Hepburn, and Stern observe that "by its very nature climate change demands that a number of ethical perspectives be considered, of which standard welfare economics is just one." Simon Dietz, Cameron Hepburn, and Nicholas Stern, "Economics, Ethics, and Climate Change" (2008), http://ssrn.com/abstract=1090572. Their essay addresses the role of economic and ethical principles in determining the appropriate degree of emissions reduction, rather than the design of specific climate change policies. Their observation about the importance of moving beyond economic metrics is nonetheless applicable.

CHAPTER 12

Ethical dimensions of adapting to climate change-imposed risks

W. Neil Adger and Sophie Nicholson-Cole

Historical, present, and future emissions of greenhouse gases, principally through energy use, are altering the climate. Hence they are imposing harm and will impose harm in the future. This volume concentrates, rightly, on the ethical issues of stewardship, responsibility, and precaution that are relevant to questions of reducing emissions and avoiding climate change. But the remit of ethical questions relevant to the climate change challenge is wider: the ethics of science, evidence, and foresight; the ethics of public discourse and media; and the ethics of representation of the future and nonhumans, to name but a few. Here we examine issues around how society will adapt to a changing climate and its related impacts and challenges.

The impacts of human-induced climate change are with us, and in anticipation of future risks, adaptation to those changes is under way. While individuals and societies have always adapted to risks associated with changing climates and environments, adaptation to human-induced climate change is categorically different in that it involves the avoidance of harm imposed by others. In this chapter we explore the ethical issues raised by imposed harm and the need for sustainable adaptation. In particular, we explore the landscape of issues around the governance process and procedures by which climate justice for adaptation could potentially be implemented. We focus on the three issues of (1) identifying climate change adaptation; (2) identifying vulnerability; and (3) managing the responsibility for action. We examine a case of coastal planning in the east of England where the details of resettlement and compensation are in flux. The analysis of the three dimensions limiting fair adaptation is based on a review of the scientific evidence for multiple stresses in coastal zones; a documentary analysis of UK policy responses and practice from central and local government and statutory agencies; and observations from action research engaging with coastal communities in East Anglia as part of ongoing consultations for coastal futures. The analysis highlights the difficulties in distinguishing climate change impacts from other natural and human stresses on the environment and hence the identification of the vulnerable, the polluters, and the ultimate responsibility for action.

I ADAPTING TO A CHANGING WORLD

There is now incontrovertible evidence, from the reports of the Intergovernmental Panel on Climate Change (IPCC) and scientific consensus, of observed warming of the world over the past century. Furthermore, there are observed trends in changing climate variability and the impacts of these trends on plants, animals, water, glaciers, and ice.[1] These observed impacts may well be amplified in the future through further changes in means of climate parameters, though the possibility of significant and even catastrophic impacts outside the range of many scenarios is also real.[2]

The changing geography of the world associated with climatic changes clearly brings opportunities as well as risks, and there are likely to be winners as well as losers in material outcomes when one looks at fine scales.[3] No one asked for the opportunity to adapt to climate change – adaptations to a changing climate are in some senses involuntary actions forced upon society, caused by past and present human-induced change. Hence adaptation to climate change is a manifestation of past inequitable use of the Earth's resources.[4] But adaptation is becoming a central part of the changing landscape of planning, economic development, and resource management in a changing climate.

In the face of the observed impacts of climate change today (as documented by the IPCC), it would be inconceivable that humans would not adapt, both in reaction to change, but also in anticipation of change.[5] And indeed, the most recent IPCC reports also demonstrate that adaptation is already under way, though to nowhere near the degree necessary to avoid significant impacts and with considerable pain and cost to vulnerable

---

[1] E.g., Cynthia Rosenzweig et al., "Assessment of observed changes and responses in natural and managed systems," in Climate Change 2007: Impacts, Adaptation and Vulnerability. Contribution of Working Group II to the Fourth Assessment Report of the Intergovernmental Panel on Climate Change, ed. Martin Parry et al. (Cambridge University Press, 2007): 81–131; Peter Stott et al., "Detection and attribution of climate change: a regional perspective," Wiley Interdisciplinary Reviews: Climate Change (2010): 192–211.

[2] Tim Lenton et al., "Tipping elements in the Earth's climate system," Proceedings of the National Academy of Sciences 105 (2008): 1786–1793.

[3] Karen O'Brien and Robin Leichenko, "Winners and losers in the context of global change," Annals of the Association of American Geographers 93 (2003): 89–103.

[4] W. Neil Adger, Irene Lorenzoni, and Karen O'Brien, eds., Fairness in Adaptation to Climate Change (Cambridge University Press, 2006).

[5] Rosenzweig et al., "Assessment of observed changes and responses," 81–131.

communities.[6] Decisions to adapt to climate change to maintain present settlements are being taken in every part of the world and the equity implications of these decisions are becoming manifest and critical.

Given pervasive exposure of vulnerable populations, issues of responsibility, fairness, and governance come to the fore. As Roger Pielke and colleagues argue, there is an implicit assumption that only governments have the foresight and the long-term vision to be able to provide strategic leadership and resources for adaptation.[7] But evidence suggests that adaptation to climate change, as experienced through a series of unforeseen stresses and experiences, relies on government, market, and civil society networks, and social capital in equal measure. In the UK, for example, Emma Tompkins and colleagues documented more than 300 separate adaptations to known and projected risks.[8] The motivation for these actions was variously to increase the capacity to handle future risks, build resilient infrastructure, and change behaviour. The majority of observed adaptations in that analysis were initiated by the public sector. This result suggests both that adaptation is within the competence of government and within its responsibility and that an environment safe from risks of climate impacts exhibits characteristics of public goods.

But other evidence suggests that it will be individuals and informal networks that are the key to adaptive actions focused on individual insecurity. So-called social capital is an important determinant of how vulnerable individuals and communities actually adapt.[9] Experience of impacts also affects how individuals perceive climate change and even whether they are likely to act on it.[10] The breakdown of social capital leads to impacts of extreme weather events, such as the high mortality rate experienced in Chicago where more than 700 excess deaths occurred in a heatwave in 1995,

---

[6] W. Neil Adger *et al.*, "Assessment of adaptation practices, options, constraints and capacity," in *Climate Change 2007: Impacts, Adaptation and Vulnerability. Contribution of Working Group II to the Fourth Assessment Report of the Intergovernmental Panel on Climate Change*, ed. Martin Parry *et al.* (Cambridge University Press, 2007): 719–743.

[7] Roger Pielke *et al.*, "Climate change 2007: lifting the taboo on adaptation," *Nature* 445 (2007): 597–598.

[8] Emma Tompkins *et al.*, "Observed adaptation to climate change: UK evidence of transition to a well-adapting society," *Global Environmental Change* 20 (2010): 627–635.

[9] W. Neil Adger, "Social capital, collective action and adaptation to climate change," *Economic Geography* 79 (2003): 387–404.

[10] Lorraine Whitmarsh, "Are flood victims more concerned about climate change than other people? The role of direct experience in risk perception and behavioural response," *Journal of Risk Research* 11 (2008): 351–374; Johanna Wolf, "Climate change and citizenship: a case study of responses in Canadian coastal communities," PhD thesis, School of Development Studies, University of East Anglia, Norwich, 2007.

and has certainly been suggested to have contributed to the high excess mortality rate in France during the 2003 heatwave.[11]

Clearly there are limits to adaptation among the most vulnerable. Hence adaptation to the risks posed by climate change has profound implications for social justice and for responsibility and governance. In one sense the world is in uncharted territory in adapting to climate change in that the risks are unknown and there is no historical precedent for a changing climate due to human action. But in another sense, the decisions that need to be taken to adapt to climate change (and the dilemmas for equity, resource efficiency, and sustainability) are well known and understood.[12] As Mike Hulme has pointed out, most societies in the world are not well adapted to their present climates, yet manage the risks of weather by multiple means and through technologies and institutions that are deeply embedded in cultures.[13] Hence separating climate change as a new phenomenon outside standard risk management may be unhelpful. We explore these issues below.

2 JUSTICE IN THE CONTEXT OF IMPOSED HARM

Responsibility for climate change extends well beyond the reduction of emissions of greenhouse gases. Equity in sharing the burden of climate change impacts and the costs of adapting to these impacts are more urgent and less well understood challenges to international action than that of achieving a fair distribution of the burden of mitigating greenhouse gas emissions.[14] Can the principles of equity and justice, so contested in sharing the burden of mitigation, be applied to sharing the burden of climate change impacts? There are both ethical and instrumental dimensions of the problem. From an ethical perspective, justice in the context of imposed impacts requires consideration of irreversibility, precaution, and the protection of the most vulnerable in the context of uncertain climate and climate science.[15]

---

[11] Eric Klinenberg, *Heat Wave: A Social Autopsy of Disaster in Chicago* (University of Chicago Press, 2002); Marc Poumadère et al., "The 2003 heat wave in France: dangerous climate change here and now," *Risk Analysis* 25 (2005): 1483–1494.

[12] W. Neil Adger, Nigel Arnell, and Emma Tompkins, "Successful adaptation to climate change across scales," *Global Environmental Change* 15 (2005): 77–86.

[13] Mike Hulme, *Why We Disagree About Climate Change* (Cambridge University Press, 2009).

[14] Richard Starkey, *Allocating Emissions Rights: Are Equal Shares, Fair Shares?*, Working Paper 118, Tyndall Centre for Climate Change Research (Norwich: University of East Anglia, 2008).

[15] Kirstin Dow, Roger Kasperson, and Maria Bohn, "Exploring the social justice implications of adaptation and vulnerability," in *Fairness in Adaptation to Climate Change*, ed. W. Neil Adger et al. (Cambridge, MA: MIT Press, 2006): 79–96; Simon Caney, "Human rights, climate change and discounting," *Environmental Politics* 17 (2008): 536–555; Marco Grasso, *Justice in Funding Adaptation Under the International Climate Change Regime* (Dordrecht: Springer, 2010).

From an instrumental perspective, a globally equitable outcome on the recognition and financing of adaptation in the least developed and most vulnerable countries is central to any political solution that requires legitimacy among diverse nations and interests.[16] The challenge of climate change is clearly one of governance, science, and politics, but it is also one of finding fair processes and achieving just outcomes.

In reviewing the elements of equity in the context of adaptation, Paavola and Adger have previously suggested that adaptation presents a number of justice dilemmas, including the responsibility of polluters for climate change impact; the principles by which developed countries (the major polluters) should give assistance to developing countries (suffering greatest impacts) for adapting to climate change and the distribution of burden among the developed countries; how assistance should be distributed between recipient countries and adaptation measures; and what procedures are fair in planning and making decisions on adaptation.[17] Each of these issues involves significant challenges. Here we address the specific issues of identifying danger and vulnerability; the framing of adaptation responses and rights; and the governance of adaptation actions and collective response to risk.

Arguments from human rights principles for responsibility for adaptation action suggest that adaptation to human-induced climate change is categorically different from previous adaptation to risk in that it involves the avoidance of harm imposed by others.[18] Simon Caney makes the case that this set of equity issues around climate change can be framed as the right "not to suffer from dangerous climate change" or as the right to avoid dangerous climate change.[19] Caney makes the case that climate change impacts jeopardize fundamental interests of individuals in their lives and livelihoods (such as impacts on disease burden, malnutrition, and food security): rights to life, health, and subsistence as a minimum set.[20] This literature, in effect, suggests that fundamental interests are significant enough, and universal enough to warrant obligations on others, even without recourse to the

---

[16] Richard Klein and Annett Mohner, "Governance limits to effective global financial support for adaptation," in *Adapting to Climate Change: Thresholds, Values and Governance*, ed. W. Neil Adger, Irene Lorenzoni, and Karen O' Brien (Cambridge University Press 2009): 465–475.

[17] Jounni Paavola and Neil Adger, "Fair adaptation to climate change," *Ecological Economics* 56 (2006): 594–609.

[18] Simon Caney, "Climate change, human rights and moral thresholds," in *Human Rights and Climate Change*, ed. Stephen Humphreys (Cambridge University Press, 2010): 69–90, and Adger *et al.*, eds., *Adapting to Climate Change*.

[19] Simon Caney, "Human rights, climate change and discounting," *Environmental Politics* 17 (2008): 537; see also W. Neil Adger, "The right to keep cold," *Environment and Planning A* 36 (2004): 1711–1715.

[20] Caney, "Human rights, climate change and discounting," 536–555.

polluter pays principle. As Caney points out, there is a strong case for rights to basic human needs and risks to them. The rights become more contested if consideration is given to persons in the future who are not represented now, or to the interests of the natural world.

The framing of the ethical dimensions of climate change in terms of rights raises a number of difficult issues in practice in implementing them and in seeking to balance and distinguish between competing fundamental rights and how governments enact the rights of their citizens through so-called social contracts.[21] The first of these is in deciding what constitutes dangerous climate change. The physical and biological sciences of climate change have their greatest traction and explanatory power on the global scale. This is one reason why much of the discussion around climate change has focused on single numbers that may constitute dangerous climate change ($2°C$ of warming or 350 ppmv CO_2).[22] Clearly, who defines danger to all and danger to some is at the crux of defining rights in this area, and cannot readily be resolved by climate and allied sciences.[23]

Further dimensions of fairness relate to what constitutes vulnerability to climate change, and to responsibility to avoid such vulnerability. The methods for measuring differential risk have evolved in the past decade and have developed new ways of portraying mobility and distance from thresholds of risk.[24] These methods and vulnerability frameworks have been applied in all areas of climate-related and environmental risks, from hazardness of place to vulnerability of resources and livelihoods.[25] The forefront of this research is to identify replicable measures and indicators of vulnerability in terms of hazard, individual livelihood, and the appropriateness of institutional response. But such external objective measures do not do justice to how vulnerability is experienced in terms of personal stress, dislocation, and perceptions of marginalization.[26] Here the process of defining and attributing vulnerability is the key issue.

[21] Karen O'Brien, Bronwyn Hayward, and Fikret Berkes, "Rethinking social contracts: building resilience in a changing climate," *Ecology and Society* 14 (2009): 12.

[22] Johan Rockström *et al.*, "A safe operating space for humanity," *Nature* 461 (2009): 472–475.

[23] Suraje Dessai *et al.*, "Defining and experiencing dangerous climate change," *Climatic Change* 64 (2004): 11–25.

[24] Hallie Eakin and Amy Luers, "Assessing the vulnerability of social-environmental systems," *Annual Review of Environment and Resources* 31 (2006): 365–394.

[25] Roger Few, "Health and climatic hazards: framing social research on vulnerability, response and adaptation," *Global Environmental Change* 17 (2007): 281–295; Susan Cutter *et al.*, "A place-based model for understanding community resilience to natural disasters," *Global Environmental Change* 18 (2008): 598–606.

[26] Greg Bankoff, Georg Frerks, and Dorothea Hilhorst, eds., *Mapping Vulnerability: Disasters Development and People* (London: Earthscan, 2004).

Finally, there remains the issue of how interests are safeguarded within public and private actions and interventions. The most vulnerable countries will have difficulty in adapting to the impacts of climate change, including extreme events, without significant international assistance.[27] Within the public policy literature, there has been a direct examination of the principles by which governments should intervene to assist both their citizens, and those outside their national jurisdictions, to adapt to impacts. Nicholas Stern, for example, makes the case that adaptation will not happen autonomously because of missing and misaligned markets and that international transfers are justified on the basis of the principles of interdependence of the world economy and the public-good nature of resources required for adaptation.[28] Hence Stern suggests that governments have justified roles to provide high-quality climate information; to provide planning and other regulations; to protect climate-sensitive public goods; and to protect the most vulnerable in society.[29] But is adaptation indeed a public good? Some resources, such as water supply, change from being public to private depending on national regulations and circumstances. The strength of the public-good aspects of adaptation are clearly mutable.[30] Nevertheless, the ideas of interdependence on public goods suggested by Stern and others are widely adopted and shared within the proposals on global shared responsibility.[31]

Potential principles for allocation of the responsibility and cost of adaptation, according to Dan Farber, include: (1) the beneficiaries of adaptation, whether local governments or individual property owners; (2) governments through an international taxation on the basis of ability to pay; (3) polluters, in this case those who emit greenhouse gases and hence ultimately cause the harm; or (4) "climate change winners," i.e., those who benefit rather than lose out as a result of climate change (such as northern municipalities or landowners).[32] Farber concludes that having emitters pay for adaptation (in effect the polluter pays principle) has the greatest normative appeal, but

---

[27] Shardul Agrawala and Maarten van Aalst, "Adapting development cooperation to adapt to climate change," *Climate Policy* 8 (2008): 183–193; Joyeeta Gupta *et al.*, "Mainstreaming climate change in development co-operation policy: conditions for success," in *Making Climate Change Work for Us*, ed. Mike Hulme and Henry Neufeldt (Cambridge University Press, 2010): 319–339.

[28] Nicholas Stern, *The Economics of Climate Change: The Stern Review* (Cambridge University Press, 2007): 467.

[29] Ibid., 471.

[30] Adger *et al.*, "Successful adaptation to climate change across scales"; Frans Berkhout, "Rationales for adaptation in EU climate change policies," *Climate Policy* 5 (2005): 377–391.

[31] Joyeeta Gupta *et al.*, "Mainstreaming climate change in development co-operation policy: conditions for success," in *Making Climate Change Work for Us*, ed. Hulme and Neufeldt.

[32] Daniel Farber, "Adapting to climate change: who should pay?," *Public Law Research Paper No. 980361*, School of Law, University of California, Berkeley, 2007.

also suggests that it has significant benefits and the feasibility of global compensation and risk spreading.[33]

In summary, the literature on environmental justice points to the emerging issues of adaptation as a critical, but overlooked, element of climate justice. As we have discussed, the issue of vulnerability is complex. Vulnerability is a relative term, and the judgement of what constitutes vulnerability (and danger) and who is vulnerable, is in itself subject to issues of fair process. Acting to reduce vulnerability will have winners and losers; hence issues of fair allocation of scarce resources for adaptation raise distributional issues. The principle of protecting the vulnerable as a central goal of public policy may seem obvious, yet it is not easy to implement in practice. We demonstrate the ambiguities and the political ramifications of these ambiguities below with reference to the case of managing risks in coastal areas.

3 MULTIPLE STRESSES, VULNERABILITY PROCESSES, AND RESPONSIBILITY IN COASTAL REGIONS

Coastal regions are on the front line of climate change impacts. Up to 40 percent of the world's population live in coastal areas, including many of the world's major cities and ecosystems, alongside estuaries and coastal seas which are at risk from multiple stresses. These areas will be directly affected by changes in sea level, rises of which are already being observed and measured around the world. These will continue over the coming centuries, even with strong mitigation of greenhouse gas emissions, due to time lags in the thermal expansion of the oceans.[34] Coastal storms are likely to increase in most regions of the world under all scenarios of climate change, and impacts will be closely associated with sea-level rises of 3–4 mm per year. The direct impacts of sea-level rise on coastal regions will exacerbate the already increasing intensity of coastal storms in tropical and subtropical regions and will interact with other biological and physical processes in coastal regions. Since the publication of the most recent IPCC reports in 2007, there has been some evidence from global assessments suggesting that observed and projected sea-level rises may in fact exceed those reported by the IPCC, and such rises would exacerbate the vulnerability of coastal communities.[35]

[33] Ibid.
[34] Robert Nicholls and Jason Lowe, "Benefits of mitigation of climate change for coastal areas," *Global Environmental Change* 14 (2004): 229–244; Stefan Rahmstorf et al., "Recent climate observations compared to projections," *Science* 316 (2007): 709.
[35] Rahmstorf et al., "Recent climate observations"; James Hansen, "Scientific reticence and sea level rise," *Environmental Research Letters* 2, no. 2 (2007).

Governments, other public and private organizations, and individuals around the world are already aware of these risks, and plans for the protection of infrastructure are under way. But they also face difficulty in isolating climate change impacts from other challenges facing the ecosystems, economies, and societies of coastal zones; in identifying what and who is vulnerable; and in negotiating responsibility for action in the implicit social contract between state and citizen in protecting the vulnerable.

3.1 Dimension 1: identifying and distinguishing climate change as a distinct driver of change

The first issue in implementing fair adaptation is that of being able to identify climate risks among trends and pressures in coastal zones. This is evident for multiple challenges for the coastal zone of East Anglia in England.[36] The region includes some protected cliffs but is predominantly low-lying, and is also sinking very slowly due to isostatic readjustment.[37] The coastal zone is particularly susceptible to coastal erosion and inundation as sea levels rise, and storm surge events become more regular.[38] The extent to which climate change is driving coastal change in the region (and further afield) is, however, difficult to quantify and project into the future in light of a historically altered natural system and changes in the national policy for coastal protection.

Dating back to the nineteenth century, the east coast of England demonstrates a complex legacy of engineered coastal defence.[39] These protection works have allowed urban development and other land uses to become established in protected areas. They have also, however, interrupted the natural balance of sediment transport along the coastline, which no longer functions naturally. Natural processes of erosion in some places would normally allow for beach replenishment in others and greater natural protection of low-lying areas at risk of inundation. However, efforts to protect erosive sections of the coast mean that sediment is not being supplied to the flood-risk coast, which consequently also has to be defended at great cost. Existing defence works are starting to reach the end of their serviceable life,

---

[36] Sustainable Development Round Table for the East of England and UK Climate Impacts Programme, *Living with Climate Change in the East of England* (Oxford: UKCIP, 2004).

[37] Ian Shennan, "Holocene crustal movements and sea-level changes in Great Britain," *Journal of Quaternary Science* 4 (1989): 77–89.

[38] Mustafa Mokrech *et al.*, "Regional impact assessment of flooding under future climate and socio-economic scenarios for East Anglia and North West England," *Climatic Change* 90 (2008): 31–55.

[39] Ruth Brennan, "The North Norfolk Coastline: A Complex Legacy," *Coastal Management* 35 (2007): 587–599.

and are becoming more and more expensive to maintain or replace, and "coastal squeeze" – as coastal habitats are squeezed against hard defences – is becoming a concerning factor.[40] The increasing standards of protection needed in order to protect against the threats presented by sea-level change, higher tidal surges, and saline intrusion are adding to this cost.

The UK national strategic approach to flood and coastal erosion risk management, encapsulated in a policy called "Making Space for Water," signals a move toward designing a more adaptable and naturally functioning coastline, promoting a more consistent risk-based framework, and new forms of cost-benefit analysis. This strategy has significant implications for sections of England's coast, particularly in the eastern region, where many places face the prospect of defence withdrawal, retreat, and realignment of the coastline.

In sum, the present and future challenges to the sustainability of England's east coast incorporate natural coastal evolution, natural processes of climate change, human-induced climate change, policy change, and a historically managed coastline. Attributing particular aspects of change to particular drivers over long timescales is difficult, even with coastal simulation science, which takes all these factors into account in an attempt to understand more deeply the interplay between them.[41] This complexity presents a significant difficulty when considering questions related to adaptation and the liability for adaptation. The proportion of human-induced climate change driving adaptation will become greater over time as the attribution of physical and biological processes to climate change outside the range of "normal" climate increases.[42] But the fact remains that adaptation takes place in a complex landscape of competing knowledge and competing values. Both the knowledge and the values and preferences for conservation and change themselves evolve over time.

3.2 Dimension 2: the challenge of identifying who is vulnerable, to what, and when

The east of England's coastal zone is home to some of the region's most important wildlife habitats, landscapes, and archaeological sites. Tourism is

---

[40] Edward Evans et al., Foresight. Future Flooding. Scientific Summary (London: Office of Science and Technology, 2004).

[41] Mike E. Dickson, Mike Walkden, and Jim Hall. "Systematic impacts of climate change on an eroding coastal region over the twenty-first century," Climatic Change 84 (2007): 141–166.

[42] See, for example, the discussions of attribution and liability in Myles Allen and Richard Lord, "The blame game," Nature 432 (2004): 551–552.

an important industry, as are agriculture, forestry, offshore oil and gas, and some manufacturing. This part of England's coastline houses some large towns and many remote rural and coastal communities, with some wealthy areas but many others in need of regeneration. These tend not to be wealthy communities and have a high demographic of retired people.[43] The changing physical risks and uncertainty driven by natural coastal processes, climate change, and policy change are, therefore, only one part of a more complex set of contributors to the vulnerability of coastal communities. A parliamentary inquiry on Coastal Towns (House of Commons Community and Local Government Committee, 2007), for example, highlighted the vulnerability of many coastal communities. These include physical isolation, high deprivation levels and lack of affordable housing, inward migration of older people, high levels of transience, outward migration of younger people, poor-quality housing, and fragile, isolated coastal economies.

The complexity underlying the drivers of present and future coastal change, and the myriad factors which contribute to the underlying vulnerability of the region, mean that the task of identifying who or what may be vulnerable (and in what ways) to future coastal change, what adaptation should entail, and when adaptation actions should be taken is an exceptionally difficult one to tackle. The present governance arrangements for coastal management in England are not in themselves adaptive, and consequently do not easily lend themselves to responding to the challenge of climate change, among the noise of so many other factors.[44] Further, the threat of change, even 100 years into the future and in the absence of government policy yet being reflected in reality (i.e., actual erosion or permanent coastal inundation), is already triggering perceptions of the increased and extreme vulnerability of individuals and even whole communities, where blight on property prices is being widely experienced.[45]

3.3 Dimension 3: who is responsible for developing and implementing adaptation, and how should it be governed?

The interpretation of the recent change in England's national strategy for flood and coastal erosion risk management at local levels happens via the Shoreline Management Planning process. These plans are the official

---

[43] Sustainable Development Round Table, *Living with Climate Change.*
[44] Sophie Nicholson-Cole and Tim O'Riordan, "Adaptive governance for a changing coastline: science, policy and the public in search of a sustainable future," in *Adapting to Climate Change*, ed. Adger et al., 368–383.
[45] Patrick Barkham, "Waves of destruction," *Guardian*, April 17, 2008.

mechanism for guiding decision-making about coastal protection for sub-sections of the coastline. They provide assessments of the risks associated with coastal flooding and erosion, and present a strategic and long-term policy framework to reduce these risks in a sustainable manner. They are nominally reviewed at five-yearly intervals to incorporate new scientific research results and revised regional and national policy guidance in line with national strategy.[46] In some places, as the implications of the national strategy have filtered down to local levels, this has meant making some locally contentious policy recommendations including suggestions for no active intervention and managed realignment along parts of the frontage within the next 25, 50, or 100 years where previously the line was held. In these locations, this means that existing defence works will no longer be maintained, or will be removed and the coast will be allowed to retreat.

The national shift of priority and consequences of the Shoreline Management Plan policy recommendations have not been matched by any compensation package or initiatives to promote the development and delivery of adaptation mechanisms to manage the change in the longer term, for locations facing significant coastal change.[47] Consequently, many coastal communities of varying size in eastern England and elsewhere are now facing a situation of great unease and anxiety about their future. In the county of Norfolk, for example, the change in policy for some areas has led to much difficulty in coming to a consensual acceptance of the proposals contained within the recently reviewed Shoreline Management Plan for the area.[48] The development planning system in England and Wales has, even recently, allowed for development in potentially risky locations, particularly since coastal defences have been in place. In some places, such as the village of Happisburgh where defences have been allowed to fail, property and community infrastructure have already been lost.

An up-welling of concern and discontent is very evident in eastern England and is partly visible through the mobilization of various coastal concern action groups in the area, as well as research into the impacts that the Shoreline Management Planning process and outcomes have had on

[46] Nicholas Cooper et al., "Shoreline management plans: a national review and engineering perspective," Proceedings of the Institution of Civil Engineers: Water Maritime Engineering 154 (2002): 221–228; Richard Leafe, John Pethick, and Ian Townend, "Realizing the benefits of shoreline management," Geographical Journal 164 (1998): 282–290.

[47] Jessica Milligan et al., "Nature conservation for sustainable shorelines: lessons from seeking to involve the public," Land Use Policy 26 (2009): 203–213; Tim O'Riordan, Sophie Nicholson-Cole, and Jessica Milligan, "Designing sustainable coastal futures," Twenty-First Century Society 3 (2008): 145–157.

[48] Anglian Coastal Authorities Group, "Kelling to Lowestoft Ness Shoreline Management Plan. First Review. Final Report, November 2006."

coastal communities.[49] This work has revealed that as people's knowledge and awareness of the situation have increased, their perception of their vulnerability has escalated. This is being reflected in terms of both property value blight and other individual factors such as anxiety, tension, help-lessness, depression, and loss of personal and local business confidence – which has knock-on effects into the wider community:

I don't feel secure here anymore, I worry constantly. We have sunk our life savings into this house and we still have a mortgage. We are desperate. If this house goes, what will happen to us? We are too old to just get up and go.[50]

A widespread sense of not having been adequately involved, consulted, or valued in the Shoreline Management Planning process has added to this sense of discontent and distrust of government and government agencies tasked with the strategic management of the coastline. The feeling that consultations have been inconsequential, tokenistic, and dismissive of local knowledge and interests pervades the debate around the process of Shoreline Management Planning in eastern England. In some senses, then, meaningful recognition and consultation need to go beyond "the illusion of inclusion" in order to ensure meaningful public participation processes in coastal manage-ment in the context of climate change adaptation.[51]

Alongside the change in policy in many places from holding the line to no active intervention, there has been no right to government compensation for the loss of land or property, and no longer-term support mechanisms in place to enable what could be a dramatic transition for many communi-ties. The government line on compensation is that they will not contem-plate it as an option: flood and coastal defence legislation in England and Wales does not confer a right to protection and there is no provision for compensation to offset the disadvantage suffered by any landowner. There is no statutory obligation on coastal defence operating authorities to provide coastal protection. The Shoreline Management Plan for Norfolk provides itself a clause that suggests that protection is not guaranteed in perpetuity,

[49] Jessica Milligan *et al.*, "Implications of the draft Shoreline Management Plan 3b on North Norfolk Coastal Communities. Final Report to North Norfolk District Council," Tyndall Centre for Climate Change Research, 2007; Sophie Nicholson-Cole, Jessica Milligan, and Tim O'Riordan. "Investigating the scope for community adaptation to a changing shoreline in the area between Caister and Hemsby," *Tyndall Centre for Climate Change Research Report for Great Yarmouth Borough Council* (Norwich: University of East Anglia, 2007).
[50] Resident of coastal Norfolk, quoted in Sophie Nicholson-Cole *et al.*, "Investigating the scope for community adaptation".
[51] Roger Few, Katrina Brown, and Emma Tompkins, "Public participation and climate change adaptation: avoiding the illusion of inclusion," *Climate Policy* 7 (2007): 46–59.

going on to add that individuals must recognize that there is no obligation for the rest of society to protect that place of residence if it is located in an area of risk:

Those who have made property purchases/development assuming that future protection was guaranteed are unfortunately misinformed. Whilst current policy at the time may have been for continued protection, there can never be a guarantee that funding will be available indefinitely or that the information upon which any decision is made will not be superseded in the future.[52]

Any government-funded compensation package or funding for adaptation measures would have to be evaluated against other national demands upon taxpayers' money such as education and the health service. Cooper and McKenna highlight this tension, asserting that local demands for defence works in particular locations on the basis of social justice arguments particularly are flawed, because the broader society stands to incur losses other than purely financial if there is public intervention for the benefit of coastal property owners.[53] While coastal residents may face a direct financial loss, arguments for public intervention weaken at geographically larger and longer timescales where the costs to society increase. A critical problem, in the context of the fairness of flood and coastal erosion risk management in England, is that different social groups differentially benefit, and are differentially burdened by system interventions across space and time. Hence there is a growing demand for policies that enable the fair distribution of the benefits and burdens of flood and coastal erosion risks.

There has been little direct consultation in relation to national strategy with regard to the balance between individual or private interests and the broader public good, perhaps because there are some significant ethical hurdles ahead, as well as logistical ones. Consequently, some individuals affected by the threat of losing property to imminent coastal erosion, in the absence of any compensation or other support, have begun to pursue these issues via legal channels on the basis of a breach of their human rights. Increasingly, questions are being asked about the fairness and justice of a change in policy in the absence of any financial assistance or other kinds of support to adapt and there are currently some policy changes under way.[54]

---

[52] Anglian Coastal Authorities Group, "Kelling to Lowestoft Ness Shoreline Management Plan, p. 18.
[53] Andrew Cooper and John McKenna, "Social justice in coastal erosion management: the temporal and spatial dimensions," *Geoforum* 39 (2008): 294–306.
[54] Sophie Nicholson-Cole and Tim O'Riordan, "Adaptive governance for a changing coastline, 268–283.

Mark Stallworthy explores some of the justice arguments surrounding sustainability, coastal erosion, and climate change.[55] He argues that "on the face of it, protecting an occupier threatened with loss by process of erosion on the basis of property entitlement has little justification" since, in English law, the foreshore rights are vested in the Crown, and in effect that foreshore is moving and has always moved.[56] Among other things, he raises questions of trade-offs, the consequences of making choices, burden distribution, knock-on effects of protection at other coastal locations, and disparities in advantage, such as we see in this case, where relatively more investment in coastal protection is being directed toward larger urban and economic centres, rather than smaller, rural ones. Indeed, the economic justification of some of the policy options recommended through the Shoreline Management Plan is subject to much debate.

Much of the national spending on flood and coastal erosion risk management is characterized by high benefit-cost scores and not on the basis of other, more intangible factors such as landscape value, community identity, and sense of place. The impacts of climate change and perhaps the greatest challenges in fair and legitimate adaptation may therefore come through social and natural phenomena that people care deeply about, but which are lost in the calculus of material costs and benefits.[57]

For local people in eastern England who deeply value the place in which they live, coming to terms with an explanation that coastal defence for that area cannot be afforded because it has not been deemed to be economically worthy has been a bitter pill to take. Such challenges are likely to be replicated worldwide.

4 CONCLUSIONS

Climate change adaptation plays, perhaps rightly, a relatively minor role in the international regime on climate change compared to the imperative to reduce emissions to avoid dangerous rates of change. Yet adapting to change will be a reality of the twenty-first century as the impacts of changing the world's atmosphere and climate system become apparent. The issues

---

[55] Mark Stallworthy, "Sustainability, coastal erosion and climate change: an environmental justice analysis," *Journal of Environmental Law* 18 (2006): 357–373.

[56] Ibid., 365.

[57] Neil Adger, Jon Barnett, and Heidi Ellemor, "Unique and Valued Places," in *Climate Change Science and Policy*, ed. Stephen Schneider *et al.* (Island Press: Washington, DC, 2010): 131–138; Julian Agyeman, Patrick Devine-Wright, and Julia Prange, "Close to the edge, down by the river? Joining up managed retreat and place attachment in a climate changed world," *Environment and Planning A* 41 (2009): 509–513.

raised here, of identifying who and what is vulnerable, and who has responsibility for action, will likely therefore have increasing prominence in the international discourse on climate change, but also in the everyday local politics of planning and development. These considerations are important not just in a coastal context, but in many others where attribution to climate change is difficult but where transformation is required. Agriculture, settlement planning and urbanization, shared water resources, and direct health risks all require significant planning and implementation.[58]

It is clear in the case we examine here that coastal governance arrangements at national to local scales have not yet adequately responded to the new strategic outlook and its implications. Consequently, they are experiencing a steep learning curve where adapting to a changing coastline is concerned. There is a plethora of plans, strategies, organizational responsibilities, agendas, and interests, not to mention different organizations at different institutional levels involved in the management of England's coast. These are, as yet, uncoordinated and lack any clear and coherent vision for the future. Encouraging efforts are emerging at local to national scales, however, which are pushing the adaptation agenda forward and searching for some real answers to the tricky questions that must be tackled.[59] There is also a realization that present coastal governance arrangements must evolve to become more adaptive themselves, before it is clear what adaptation will involve and how it can be funded and delivered.

So what is to be done to implement fair adaptation? First, there needs to be a recognition of the scale and the nature of the risks and, hence, the liability associated with climate change. As Jon Barnett has summarized the problem, responsibility for climate change is not equally distributed and some groups will continue to emit greenhouse gases. While the responsibility is changing over time, it remains and will remain highly skewed toward the wealthy of the world.[60] Second, in the same way, climate change will not affect everyone equally – vulnerability is unevenly distributed. And this vulnerability is at least partly driven by the same causes that drive economic inequality in society more generally. Finally, actions to reduce specific vulnerabilities to climate change will have winners and losers and raise issues of both distributive and procedural justice. In addition, experience demonstrates that there is little guarantee that adaptation actions

[58] Stephen Dovers, "Normalizing adaptation," *Global Environmental Change* 19 (2009): 4–6.
[59] O'Riordan, *et al.*, "Designing sustainable coastal futures."
[60] Jon Barnett, "Climate change, insecurity and injustice," in *Fairness in Adaptation to Climate Change*, ed. W. N. Adger, J. Paavola, S. Huq, and M. J. Mace (Cambridge, MA: MIT Press, 2006): 115–129.

will be effective, and may even create new sets of vulnerabilities.[61] The solutions to these problems at the global scale, or even at the local level, are not obvious.

For fair adaptation to be implemented for the UK, as much as for all vulnerable places throughout the world, we believe certain changes need to be made in our approaches to adaptation. These include a far better understanding of the risks of climate change relative to other environmental stresses; wider consideration of the social justice dimensions of policy; financial support and compensation as well as direct assistance for those most vulnerable; and processes that support creativity, vision, and inclusive and participatory decision-making. This is a long and difficult list that in and of itself highlights the challenges of adaptation.

---

[61] Jon Barnett and Saffron O'Neill, "Maladaptation," *Global Environmental Change* 20 (2010): 211–213.

CHAPTER 13

Does nature matter? The place of the nonhuman in the ethics of climate change

Clare Palmer

I INTRODUCTION

Ethical discussion about climate change has focused on two highly significant sets of questions: questions about justice between existing peoples and nations, and questions concerning the moral responsibilities of existing people to future people. However, given the likely planetary effects of climate change, one might also expect to find a third area of ethical debate: questions about the impact of climate change on the nonhuman world directly. But on this subject, very little has so far been said. Of course, ecosystems and species have been important in existing political and ethical debate about climate, because climate change may affect them in ways that have serious implications for human beings. Floods and droughts may cause widespread human hunger; invasive species can spread human disease; loss of biodiversity may threaten ecosystem services. But in all these cases, the nonhuman world is understood to be of *indirect* moral concern; to use a distinction made by Jan Narveson, even though the *object* of the concern is ecosystems or species, the *ground* of the concern is human beings.[1] Here, in contrast, I'm concerned with the ethical implications of climate change for species, ecosystems, organisms, and sentient animals *directly*, independently of the possible harms that such impacts might cause humans.

Certainly, concerns about the impact of climate change on the nonhuman world directly have been expressed, both by environmentalists and by ethicists. The Nature Conservancy, for instance, aims to "safeguard Nature from irreversible harm from a changing climate."[2] The Humane Society of the United States (HSUS), in a recent declaration on climate change, worries that "climate change is already adversely affecting animals around the globe ... increasing numbers of extreme weather events are displacing or

---

[1] Jan Narveson, "On a Case for Animal Rights." *Monist* 70, no. 1 (1987): 35.
[2] Jonathan Hoekstra for the Nature Conservancy at: www.nature.org/initiatives/climatechange/features/art26193.html.

killing unprecedented numbers of farm animals, companion animals and wildlife."[3] Some ethicists writing about climate change have similar concerns. Stephen Gardiner notes that "deciding what trajectory to aim for [in terms of long-term global carbon emissions] raises issues about our responsibilities with respect to animals and nature."[4] John Broome emphasizes that, independently of effects on humans, "damage to nature may well be one of the most harmful consequences of climate change."[5] The White Paper on the Ethical Dimensions of Climate Change, an influential, collaboratively written document intended to influence climate policy, maintains repeatedly that climate change raises ethical questions about "harm to" and "duties to protect" plants, animals, and ecosystems directly, and that their interests should be taken into account alongside humans' in climate policymaking.[6]

But although many claims of this kind have been made, there have been few attempts to work through what such claims might actually mean – and how far they are right.[7] Will climate change really "harm nature"? If so, which parts of nature, and in what ways? Are wild animals, and the world's natural systems, really facing a "crisis of immense proportions," as the HSUS claims? In this paper I will take some first steps in investigating these claims. I will argue, in particular, that we cannot always move straightforwardly from anthropogenic emissions of greenhouse gases to morally significant harms to the nonhuman world. Though climate change does raise ethical problems in a nonhuman context, these problems are more restricted in scope and more complex in form than we might intuitively think.

2 THREE ASSUMPTIONS

Given the limited space available here, I will begin by bracketing three issues about which I will make assumptions rather than present arguments. These

---

[3] Humane Society of the United States at: www.humanesociety.org/about/policy_statements/statement_on_climate_change.html.
[4] Stephen Gardiner, "Ethics and Climate Change: An Introduction." In *Wiley Interdisciplinary Reviews: Climate Change*, ed. Michael Hulme. Draft version online at http://faculty.washington.edu/smgard/index2.shtml.
[5] John Broome, *Valuing Policies in Response to Climate Change: Some Ethical Issues*, 2009, www.hm-treasury.gov.uk/d/stern_review_supporting_technical_material_john_broome_261006.pdf, p. 12.
[6] Donald Brown et al., *White Paper on the Ethical Dimensions of Climate Change*, 10, 16, 18. http://rockethics.psu.edu/climate/whitepaper/whitepaper-intro.shtml.
[7] The two fullest accounts I can find are Robin Attfield, "Mediated Responsibilities, Global Warming and the Scope of Ethics." *Journal of Social Philosophy* 40, no. 2 (2009): 233–235, and Dale Jamieson, "Climate Change, Responsibility and Justice." *Science and Engineering Ethics* (forthcoming). Jamieson mentions the loss of wild nature as a problem for climate change in a number of places, including prominently in Dale Jamieson, "Ethics, Public Policy and Global Warming." *Science, Technology and Human Values* 17, no. 2 (1992): 139–153.

issues are controversial (although to different degrees) but have been widely discussed elsewhere. Setting them on one side here will allow me to focus on the less-explored ethical issues in which I'm principally interested.

(1) I'll assume that climate change really is happening and that human beings are largely causally responsible for it. So, I'll reject general skepticism about climate science. I'll also assume that, despite recent controversy, the Intergovernmental Panel on Climate Change (IPCC) 2007 reports provide our best source of reliable information on the likely impacts of climate change. See the introduction to this book for detailed discussion.

(2) When we think about climate ethics, problems are raised not only with respect to the *effects* of climate change, but also in terms of the complicated way in which human beings are *actors*. Unlike many ethical problems, in the case of climate change causal responsibility is distributed over many actors, over long time spans, and over space. However, I don't have space to discuss this "causal end" of the problem here. I'll just assume here that human beings *do* have moral responsibility for climate change.

(3) Thirdly, and most controversially, I'll assume that the nonhuman world, or at least parts of it, can be of direct moral relevance, independently of its usefulness to humans. Specifically, I'll assume that there are four possible objects of direct moral concern: species, ecosystems, nonconscious living organisms, and conscious, sentient animals. I'll restrict the way in which I understand moral concern by focusing only on claims about *moral considerability*. I take moral considerability in Harley Cahen's sense to mean "the moral status x has if and only if (a) x has interests, (b) it would be *prima facie* wrong to frustrate x's interests (to harm x) and (c) the wrongness of frustrating x's interests is direct."[8] I'm not, therefore, directly concerned with accounts of nature's "existence value," nor with some interpretations of "intrinsic value" in nature, where these are distinct from claims about morally important interests (though what I say may be relevant for these kinds of accounts as well).

Of course, I recognize that the claim that species, ecosystems, and nonconscious living organisms (less so sentient animals) have morally relevant interests is highly controversial. I'm not persuaded myself of such claims, except in the case of sentient animals. But what's interesting here is that *even if we grant* that these parts of the nonhuman world are morally considerable,

---

[8] See Harley Cahen, "Against the Moral Considerability of Ecosystems." In *Environmental Ethics: An Anthology*, ed. Holmes Rolston and Andrew Light (Oxford: Blackwell Publishers, 2003 [1988]), 114.

this doesn't always get us directly to the position that climate change causes harm. And it's this that I want to explore further. As even my very superficial study here will indicate, the ways in which human-originating climate change are likely to affect the nonhuman world generate puzzling and difficult ethical problems. Much more work will be required in order to draw satisfactory conclusions with respect to these problems.

3 CLIMATE CHANGE AND THE NONHUMAN WORLD: FIVE KEY FACTORS

I'll begin here by highlighting a number of important and related factors we need to consider when thinking about the impact of climate change on the nonhuman world:

(a) *Harm and change*: On this account, for climate change to be directly morally problematic, it must set back interests, or *harm*. But with respect to at least some of the entities at issue here it's not clear what might be "in" their interests, and whether climate change is the sort of phenomenon that actually might harm them – as opposed, for instance, to bringing about a particular form of *change*.

(b) *Climate change as productive*: Climate change is usually viewed as a *destructive* force. But it is also *productive*, bringing new things and beings into existence that would not otherwise have existed. Inasmuch as these things and beings have the same capacities or properties that we think makes existing things and beings morally relevant – in this case, being interest-bearing – some of what climate change *produces* will be morally relevant as well as what it eliminates.

(c) *Numbers questions*: Climate change may have an impact on *numbers* of particular things and beings in the world. Climate change might alter total numbers of organisms, total numbers of particular kinds of organism, and numbers of things and beings of different complexity and psychological sophistication.

(d) *Non-identity questions*: Climate change will have substantial global impacts. One result of this is that *different* things and beings will exist in the future than would have existed had climate change not occurred. But this raises questions about whether things and beings can be harmed by phenomena that were actually responsible for, or necessary conditions of, their existence.

(e) *Uncertainty questions*: The future effects of climate change on the nonhuman world are highly uncertain. For instance, we know very little about the temperature tolerance of many species. Some of the concerns

that might be of interest to ethicists (such as: Will climate change result
in the existence of more or fewer psychologically complex beings in the
world?) are not the kinds of questions that current scientific researchers
are addressing, are likely to address, or would know how to go about
addressing. So, there's little hope of any progress on reducing uncertainty
in some areas that are especially ethically relevant.

Having introduced these factors here, I'll now consider the ethical questions
climate change raises with respect to each of the four possible "objects of
moral concern" – species, ecosystems, nonconscious living organisms,
and experiencing animals – in turn. To reiterate: I'm not attempting here
to defend arguments that these beings and entities *actually are* of moral
relevance, in the sense that they have interests that can be set back, and can
thus be harmed. I'm asking what climate change might mean ethically *if
we accept that they do* have morally relevant interests.

4 CLIMATE ETHICS AND THE NONHUMAN: SPECIES

I'll begin by considering species, because they are one of the (somewhat)
more straightforward cases here. If we accept the idea that species are in
some sense interest-bearing, and that the possession of interests is sufficient
for moral status, then it does seem as though climate change *could* harm
species in morally significant ways.[9]

First, let's sketch the current and projected effects of climate change on
species. Already, climate change has probably affected species; some have
become more abundant and widely distributed; others have become less
abundant; a handful may have already become extinct. Future climate
change, involving alterations in the timing of seasons and in rainfall patterns, and increases in temperature, ocean acidification, and extreme
weather events, is likely to have significant effects on the range, habitat,
and survival of many more species.[10] It will also affect the bodily forms of
species members; for instance, recent research indicates that many US bird
species are smaller in size and weight, and have shorter wings, than 50 years

[9] David Hull, among others, argues that a species is a kind of individual. See David Hull, "A Matter of
Individuality." *Philosophy of Science* 45 no. 3 (1978): 335–360. Some environmental ethicists have built on
this argument to maintain that a species is morally considerable. See Lawrence E. Johnson, "Future
Generations and Contemporary Ethics." *Environmental Values* 12 (2003): 471–487. However, this
argument is problematic. Ron Sandler and Judith Crane, "On the Moral Considerability of *Homo
sapiens* and Other Species." *Environmental Values* 15 (2006): 69–84, put forward a persuasive critique.
[10] See the discussion in Jean Christophe Vie, Craig Hilton-Taylor, and Simon N. Stuart, "Species
Susceptibility to Climate Change Impacts." *Wildlife in a Changing World: An Analysis of the 2008
IUCN Red List of Endangered Species* (Gland: IUCN, 2009), 77–89.

ago (though their populations remain constant), an alteration attributed to changes in climate.[11] Climate change may also result in the loss of some species; the IPCC Fourth Assessment Report notes that "There is medium confidence that approximately 20–30% of species assessed so far are likely to be at increased risk of extinction if increases in global average warming exceed 1.5–2.5°C (relative to 1980–1999)."[12]

However, the actual proportion of the world's existing species that may become extinct on account of climate change is widely contested; major uncertainties relate to how slowly and how modestly climate will actually change, and how climatic changes will interact with other human-originating factors that already threaten species, such as habitat fragmentation and loss.[13] For my purposes here, however, firm numbers are not critical. It seems safe to assume that climate change will commit some species to extinction. Further, recent research suggests that rates of species loss are already outstripping rates of species evolution (though it's not clear how much of this is currently a consequence of climate change).[14] So, fewer species overall are likely to exist, at least in the medium-term future. Inasmuch as climate change will cause at least some species extinctions, and threaten others with extinction, this looks to be a relatively straightforward way in which climate change is of direct moral relevance in the nonhuman context.

We should not move too quickly here, though, because while in this "broad-brush" sense climate change does look as though it will harm some species in morally significant ways, further questions are raised. First, what actually is "in the interests" of a species? The most plausible primary interest of a species must surely be in not becoming extinct. But even this is not always obvious. Suppose that for some species to continue to exist, all the individual organisms that would compose it, present and future, would have such extremely painful, distressing lives that, as individuals, they would be better off dead, since their lives are not worth living. Would it still be in the interests of the *species* to persist, even though it would not be in the

---

[11] See Josh Van Buskirk, Robert S. Mulvihill, and Robert C. Leberman, "Declining Body Sizes in North American Birds Associated with Climate Change." *Oikos* 2010 (on early view).

[12] IPCC, *Climate Change 2007: Synthesis Report: Summary for Policymakers* at: www.ipcc.ch/publications_and_data/publications_ipcc_fourth_assessment_report_synthesis_report.htm.

[13] Chris D. Thomas *et al.*, "Extinction Risk from Climate Change." *Nature* 427 (2004): 145–148 predicted that climate change would doom 15–34 percent of species to extinction. However, these figures are contested. See O. T. Lewis, "Climate Changes, Species-Area Curves and the Extinction Crisis." *Philosophical Transactions of the Royal Society B* 361 (2006): 163–171.

[14] Simon Stuart, Chair of the International Union for the Conservation of Nature's Species Survival Commission. Quoted at: www.telegraph.co.uk/earth/earthnews/7397420/Worlds-nature-becoming-extinct-at-fastest-rate-on-record-conservationists-warn.html.

interests of *any* of the members of the species that would ever live? Or take a case suggested by Bryan Norton, and perhaps a more plausible concern with respect to climate change: Would it be in the interests of a species to come under a steady adaptational pressure that both fuels its decline in population and, simultaneously, increases the likelihood that it will speciate before it becomes extinct?[15] Does a species have an interest in speciation? Because of the kind of thing a species is, even if one thinks it does have morally relevant interests, there are cases relevant to climate policy where it is difficult to work out what those interests actually are.

Second, there's a further question about what recognizing species' interests would commit us to. Is the primary claim just that species should not be harmed? Or can species' interests be weighed against one another to produce best outcomes? And other things being equal, would it be better if there were more species in the world rather than fewer? These questions are difficult, and not presently addressed in the ethical literature on species, but they seem to be important questions in policy terms.

So, for instance: It might be that our primary goal with respect to species – roughly speaking, a deontological one – is not to *harm* them. On this view, harms to one species could not be weighed against benefits to another. On an alternative view – a broadly consequentialist one – the aim might be to bring about the maximization of the flourishing of species' interests – either by promoting or protecting the flourishing of existing species, and/or by creating new species, and/or by allowing/promoting speciation, as appropriate. Climate change, of course, could be problematic on both deontological and consequentialist views. But it's more obviously a difficulty on a broadly deontological view than on a consequentialist one. Of course, on the consequentialist view, climate change still looks problematic, since it seems implausible that, overall, climate change would promote the flourishing of species. Even though some will benefit, more will lose out, as speciation won't keep pace with extinction, at least not on any humanly comprehensible time scale. (Over millennia, though, speciations may well overtake extinctions, so on a long time frame we might think the losses now will be made up in the future.)[16] But even in the shorter term, if species' interests can be weighed against one another, wider policy options present themselves than if avoiding species harms is the central focus.

---

[15] Bryan Norton, *Why Preserve Natural Variety?* (Princeton University Press, 1992), 171.

[16] Over *geological* time, human-induced extinctions may shape evolutionary processes. See David Jablonski, "Lessons from the Past: Evolutionary Impacts of Mass Extinctions." *Proceedings of the National Academy of Sciences of the USA* 98 no. 10 (2001): 5393–5398.

Climate change will be "good for" some species and "bad for" others. "A species' individual susceptibility to climate change depends on a variety of biological traits including its life history, ecology, behavior, physiology, and genetic make-up."[17] Species that are habitat generalists, tolerant of wide temperature ranges, not dependent on very specific environmental triggers or cues for processes such as breeding, and resilient in the face of extreme weather events, and that can disperse or colonize new habitats relatively easily, are likely to flourish in a changed climate, even as other species are threatened.[18] Climate change is not, therefore, unequivocally "bad for" species – even if some of the species for which it might be thought to be "good" are not ones that humans generally welcome, such as the American dog tick, a primary vector for Rocky Mountain spotted fever. Some species are likely to benefit from the very same phenomena that harm other species. And if harms and benefits can be balanced to produce best overall outcomes, it might be better (in resource terms, for instance) to reintroduce a wild species that is now extinct, for instance, from seeds in a seed bank – such as the recent reintroduction of the Malheur wild lettuce into Oregon – rather than to expend substantial resources to prevent a different, currently endangered, species from being harmed by becoming extinct.

In general, though, if one accepts that species can have morally relevant interests, then climate change does threaten at least some of those interests, even though it promotes others. The difficulties here concern the theoretical framework within which one understands those interests, and thus what this might mean in terms of policy responses. But in terms of harms, at least, species raise less troubling ethical questions than some of the other possible objects of moral concern that I'll now go on to consider.

5 CLIMATE ETHICS AND THE NONHUMAN: ECOSYSTEMS

It is very likely that climate change is already affecting many of the world's ecosystems. The IPCC Fourth Synthesis Report notes with "high confidence" that "recent regional changes in temperature have had discernible impacts on physical and biological systems."[19] In particular, there have been changes in Arctic and Antarctic ecosystems, especially those based around sea ice; spring peak discharge in snow and glacier-fed rivers is earlier and higher; events such as leafing, migration, and egg-laying in birds are also happening earlier. These changes, and many others to come, including the

---

[17] Vie, Hilton-Taylor, and Stuart, "Species Susceptibility," 77–89.   [18] Ibid., 79.
[19] IPCC, *Climate Change 2007*, 31.

effects of more extreme weather events, floods, and fires, will have a significant impact on ecosystems. Some existing ecosystems may change their nature altogether (so, for instance, some coastal ecosystems are likely to become part of marine ecosystems). Many ecosystems will be rather different on account of human-originating climate change in the future than they are now, and than they otherwise would have been.

But are these changes of direct moral relevance? On the terms with which I've been working, to answer this question positively we would need to think *both* that ecosystems are, in some sense, directly morally considerable (that they have interests, and can be benefited and harmed in morally significant ways) *and* that, if they are morally considerable, then the effects of climate change would actually *be* of ethical significance – primarily, that climate change would *harm* at least some of them. Since we're *assuming* the moral considerability of ecosystems, let's focus on the second issue here. What's at stake here is the kind of alteration that anthropogenic climate change is likely to cause.

If we accept that ecosystems are morally considerable, then there must be some way in which ecosystemic interests can be set back, or in which ecosystems can be harmed. A clear-cut case of harm, on this view, would be the completion of a dam project that quickly and permanently inundated an entire grassland ecosystem, completely destroying it. The question is, though, whether climate change is likely to have this kind of harmful effect. At first sight, we might think not. After all, climate change is not nearly such a swift and dramatic process as inundation from a dam project. It seems more likely that the shifts precipitated by climate just push ecosystems to develop in one way, rather than another. Some species will do well, others will do badly, creating new ecosystem compositions. But why see a "push" in one way rather than another as "setting back their interests," as being "worse" for them than other disturbances and climate patterns that would have occurred in a different, non-anthropogenic, climate regime?

One response to this is to maintain that the ways in which climate change affects at least some ecosystems differs from the changes encountered in "normal" disturbance regimes. Many of those who argue that ecosystems have some kind of moral status draw on Aldo Leopold's account of the land ethic, published in 1949. A key part of Leopold's land ethic is the claim that "A thing is right when it tends to preserve the integrity, stability and beauty of the land community. It is wrong when it tends otherwise."[20] The concern raised by climate change, then, might be that it will upset the

[20] Aldo Leopold, *A Sand County Almanac* (Oxford University Press, 1989 [1949]).

integrity and stability of ecosystems, in "abnormal" ways. But we need to qualify this concern. Changing emphases in ecological theory since Leopold's time have led to questions about the appropriateness of "stability" or "integrity" seen as measures of ecological "goodness" (or, for our purposes here, of ecological harm). It is unclear how much cohesion, regularity, and organization ecosystems have; they seem to exhibit "varying degrees of historical particularity, stochasticity, regularity, cohesion and hierarchical organization."[21] Pickett and Ostfeld argue, influentially, that we should understand ecosystems to be open, unpredictably changing and variable, internally heterogeneous and patchy, and containing species redundancy. In light of these shifting ideas, J. Baird Callicott suggests that we should adapt Leopold's land ethic to read: "A thing is right when it tends to disturb the biotic community at normal spatial and temporal scales. It is wrong when it tends otherwise."[22] So, one thought here is that anthropogenic climate change could disturb biotic communities at *abnormal* spatial and temporal scales, more quickly, more dramatically, more comprehensively, and more permanently, than the fluctuations of regular disturbance regimes. These major *scale* changes to ecosystems, it might be argued, are what is directly morally problematic, and that we should think of as the "wrong thing" that results from climate change in the nonhuman world.

But in the context of climate change, this reworked land ethic is also problematic (as Callicott himself now maintains).[23] Much turns on the word "normal." Certainly, in the past few centuries, global climate has been relatively stable. But analysis of pollen deposits and Greenland ice cores strongly indicates that abrupt climate change – on a human-relevant scale of a decade or two – has happened in the past. So, for instance, during the Younger Dryas 12,000 years ago, there was very significant and abrupt cooling, followed by fairly abrupt warming, over the space of about 50 years in northern Europe and North America, with some climate changes also recorded globally.[24] The National Academy of Sciences notes: "Recent

---

[21] Kevin De Laplante and Jay Odenbaugh, "What Isn't Wrong with Ecosystem Ecology." www.public. iastate.edu/~kdelapla/research/research/pubs_assets/wiwwee.pdf.

[22] J. Baird Callicott, "From the Land Ethic to the Earth Ethic." In *Gaia in Turmoil: Climate Change, Biodepletion, and Earth Ethics in an Age of Crisis*, ed. E. Crist and H. B. Rinker (Cambridge, MA: MIT Press, 2009), 184.

[23] Ibid., 178. Callicott's argument differs from mine, however; he suggests that given the planetary scale and long time span over which climate change is likely to occur, we need a "Gaian ethics" rather than a land ethic, and that the land ethic is in this context irrelevant.

[24] There's a large literature on this. See, for instance, Jeffrey P. Severinghaus *et al.*, "Timing of Abrupt Climate Change at the End of the Younger Dryas Interval from Thermally Fractionated Gases in Polar Ice." *Nature* 391 (January 8, 1998): 141–146.

scientific evidence shows that major and widespread climate changes have occurred with startling speed. For example, roughly half the north Atlantic warming since the last ice age was achieved in only a decade, and it was accompanied by significant climatic changes across most of the globe."[25] Such abrupt past climate changes are not, of course, thought to be anthropogenic. But – on the long time scale on which it seems appropriate to think about climate change – these climate changes are not *abnormal* either. Anthropogenic climate change, though different in origin, is not otherwise unlike at least some other "natural" episodes of climate change. So, if it's the *normality* of spatial and temporal scales that matter, it's hard to see how climate change would be "a wrong thing" in an ecosystemic context.

But perhaps it's not "normality" on which we should focus here. What's important is surely whether we think that abrupt climate changes – whether "normal" or not, and whether they are of the natural or the human-originating kind – can set back the morally significant interests of ecosystems. If we think that they cannot, because, however dramatic these kinds of changes might be, they all constitute ecosystemic processes of change that cannot "harm" the system, then they are not of direct ethical concern whatever their origin (though they might be of critical *indirect* concern to *us*). But if we think that abrupt climate change *can* set back the interests of ecosystems, then for humans to create such climate change would be of direct moral concern. That such climate changes can also occur naturally would be, in this case, irrelevant. If a human arsonist were to ignite a deadly wildfire that predictably set back human interests by destroying property and lives, we wouldn't think that the arsonist had no moral responsibility because lightning starts similarly destructive fires naturally. Likewise with human-originating climate change; humans are moral agents, so their actions can be morally culpable in ways that natural climate changes cannot be.

This, then, takes us back to the question: what kind of effect could climate change have that would harm ecological systems? This question is difficult to answer, partly because of our uncertainty about the effects of climate change, and, but more deeply, because of the problem, in the context of ecosystems, of identifying any sustainable distinction between changes and harms, especially over extended time scales. Of course, some cases – where changes are relatively quick and clearly destructive – are less difficult to assess. For instance, the sudden bleaching of a coral reef, leading to the loss of reef-dependent organisms and ecosystem function, does look

---

[25] See *Abrupt Climate Change: Inevitable Surprises*. Committee on Abrupt Climate Change, National Academy of Sciences, 2002. www.nap.edu/openbook.php?record_id=10136&page=1.

like a harm. But the slow extension of a forest ecosystem into what was previously a tundra ecosystem is much less obviously harmful. And ecosystemic processes can be maintained in systems that look very different from the ones we currently have, as Nott and Pimm argue: "We can imagine a future pantropical forest system composed of a few score of tree species, the same introduced insect, bird and mammal species. Such a world might exhibit acceptable ecosystemic processes, yet contain a tiny fraction of the current tropical biodiversity."[26]

So, there are some cases where, *if* we think that ecosystems have morally relevant interests it looks as though climate change could cause harm. But it would be problematic to think that *all* such changes are harms, given our general lack of knowledge about ecosystemic change over time, and the difficulties of identifying both the boundaries and the identity of any particular ecosystem. It will be difficult, and controversial, to come up with anything like a set of conditions to separate ecosystem *harm* from ecosystem *change*.

An alternative way to think about the effects of climate change on the nonhuman world is to focus not on ecosystems, but rather on the organisms of which they are composed. We might think, more plausibly, that it's individual organisms that will be harmed, rather than ecosystems. There is much wider general acceptance of the idea that living individual organisms – in particular sentient animals – are the kinds of beings that *can* be harmed, and it's much easier in the case of individual organisms (rather than species or ecosystems) to identify what a harm actually *is*. So, I'll now turn to consider the effects of climate change on individual organisms. Here questions about numbers, creativity, identity, and uncertainty in the context of climate change arise with particular force.

6 CLIMATE CHANGE AND THE NONHUMAN:
INDIVIDUAL LIVING ORGANISMS

Before moving on to consider the effects of climate change on individual living organisms, I first need to comment on how the moral relevance of individual organisms is usually understood. On most accounts, what matters are the particular inherent *capacities* individual organisms possess (rather than, for instance, their relational properties, or some property

---

[26] M. P. Nott and S. L. Pimm, "The Evaluation of Biodiversity as a Target for Conservation." In *The Ecological Basis of Conservation*, ed. Steward T. Pickett, Richard S. Ostfeld, Moshe Shachak, and Gene Likens (New York: Chapman and Hall, 1997), 127.

that is not a capacity, such as species membership). Those who argue, for instance, that nonconscious living organisms are morally considerable, and have interests that can be set back, usually claim that living organisms are goal-directed or that they have their own goods. Those who argue that only certain animals are morally considerable maintain that capacities such as being able to experience conscious pain are what is morally important. Almost everyone agrees that the focus should be on capacities, even though they disagree about what the relevant capacities are. A further source of disagreement concerns whether some capacities are more valuable than others. On some views, the possession of sophisticated mental capacities gives an organism added moral significance. So, for instance, while a plant might have bare moral status because it is goal-directed, a chimpanzee's sophisticated psychological capacities give it a much higher degree of moral significance.

In what ways, then, might climate change have morally relevant impacts on individual organisms? Some possibilities are:

(a) *Numbers*: Climate change may affect the *total* number of organisms that ever live.

(b) *Identity*: Climate change means that different individual organisms come into being than would otherwise have existed. These organisms may have different capacities than organisms that would otherwise have lived (they might be more, or less, psychologically sophisticated, for example). In any case, climate change is a necessary condition of the organisms' existence.

(c) *Harm*: Climate change may have harmful effects on those organisms that *actually do* live.

These three impacts are closely related to the difficulty of *uncertainty*. That is, we have no real idea what effect climate change will have in terms of (a) and (b) – at least.

Let's begin, then, by thinking about *numbers*. One worry – on which I don't think we need linger – is that climate change might bring about a world in which fewer organisms ever live. For some consequentialist positions this could matter morally: a world in which fewer organisms ever lived would be, other things being equal, worse than a world in which more organisms ever lived. But this outcome is empirically very implausible. Certainly, *some* ecosystems may contain fewer individual living organisms (for instance, marine ecosystems affected by ocean acidification, or desertified land areas). But others will likely contain *more* living organisms. Recent research published in the UK, for instance, indicated that warmer, wetter weather between 1997 and 2007 virtually doubled the number of invertebrate soil organisms in existence (although their diversity reduced).[27] We can't foresee (in the way that we might were we planning on causing a nuclear winter) that on account of anthropogenic climate change, fewer organisms will ever live. It seems unlikely that a *total numbers* argument should be seen as an ethical concern generated by climate change.

So, let's assume a roughly same-numbers future world. What ethical concerns might there be here? Two possibilities:

(i) Although this is a same-numbers world, simpler nonconscious organisms will come into being instead of more complex nonconscious or conscious ones, and this is of ethical significance. Let's call this the *simplification worry.*

(ii) Organisms will be harmed, and this is of ethical significance.

For the simplification worry to be of moral concern, we would need to think that some organisms have more moral significance than others, *and* that climate change was likely to result in organismic simplification. The idea that different living organisms have different value is not universally accepted. Paul Taylor, for example, argues that all (wild) living things have equal inherent worth and "all are held to be deserving of equal moral consideration."[28] On this view, that less complex beings come to replace more complex beings would not, in itself, be problematic. But there are many other ethicists who would be worried about organismic simplification. Robin Attfield argues that the flourishing of all living organisms has some value, the more complex the organism, the more valuable the flourishing. And, as a consequentialist, Attfield also maintains that good can be summed.[29] So one reading of his view would be that – other things being equal – a world in which more complex organisms flourished would be a more valuable world than one with the same number of organisms, but where at least some organisms were simpler.

Is it, though, empirically likely that climate change would actually bring into being simpler organisms than would otherwise be the case? In one sense, this is more plausible than the total numbers worry. Climate change is likely to endanger or make extinct some complex, psychologically sophisticated species whose members would be particularly important here. But on the other hand, climate change will also provide the necessary conditions

[27] Research from the Center for Ecology and Hydrology, UK National Environment Research Council as reported in "Weight of Bugs in Britain's Soil Has Nearly Doubled in Just Ten Years," *The Observer*, February 28, 2010. www.guardian.co.uk/science/2010/feb/28/soil-biodiversity-invertebrates-countryside-survey.

[28] Paul Taylor, *Respect for Nature* (Princeton University Press, 1986), 79.

[29] See Robin Attfield, *A Theory of Value and Obligation* (London: Croom Helm, 1987).

for the expansion of the range of other complex, psychologically sophisticated, and potentially invasive species. It's very likely that in the plant world more "weedy" species will dominate. But again, these may be no more or less complex in terms of their morally relevant capacities than the native plants they displace. So although some species may become more abundant and others less abundant – or extinct – this doesn't in itself imply a loss of organismic complexity.

If the simplification worry did turn out, though, to have an empirical basis, it's surely a worry that humans could relatively easily fix – though not in ways that would necessarily please those morally concerned about the environment. After all, it's easy enough for humans to breed relatively complex and psychologically sophisticated organisms to make up for wild ones that are going extinct. Indeed, we could even breed more of ourselves. However simplification is read, it has no necessary relationship with "wildness"; the concern here is with lost individual capacities, not with lost relations or contexts. In fact, in expanding animal farming, we're likely already increasing the numbers of psychologically complex organisms in the world (though most versions of this view would also require that the organisms flourished, questionable in the case of intensive animal agriculture). The simplification worry, then, like the total numbers worry, need not detain us long.

Let's turn, then, to questions about harm, and here divide organisms into two groups: the nonconscious ones, such as higher plants and algae, and the sentient, experiencing ones, particularly mammals and birds. First: Will climate change harm nonconscious organisms? The answer to this question, again, isn't entirely straightforward. Undoubtedly, some organisms will be harmed or killed by the impacts of climate change. This is most obvious in the case of longer-living organisms such as trees; some existing trees will struggle to stay alive, and will (probably) die sooner because of changes in temperature, water availability, advancing pests, or extreme weather events.[30] It's much more difficult to make such claims about organisms with short lives of months or days, where the time scales on which gradually shifting climate impacts operate are just too long for effects to be felt (though even short-lived organisms could be negatively affected by extreme weather events such as storms, floods, or fires).

But climate change will also bring organisms into existence that wouldn't otherwise have existed, and will promote the interests of other organisms

---

[30] There are problems with the counterfactual (who knows what would have happened without anthropogenic climate change?), but I won't address this here.

that do actually exist. And climate mitigation policies (for instance, the use of some new energy-generating technologies), if implemented, are also likely to take some individual organisms' lives while preserving the lives of others.[31] One question here, then, is whether it is possible or appropriate to add together these deaths and new lives, these harms and benefits, to create some kind of overall sum of the organismic effects of climate change.

Forms of consequentialism that value the lives of nonconscious organisms would accept that such calculations are appropriate; what matters from such consequentialist perspectives is whether climate change *actually will* promote or frustrate organismic interests overall. Here the critical problem is one of deep uncertainty, where we can't even be clear about what outcomes are *likely*. But other non-consequentialist ethical approaches maintain that what's of moral importance is not the increase or decline in overall organismic well-being, but rather the human duty not to harm organisms that have a well-being; for Paul Taylor, for example, this is the central principle of non-maleficence.[32] On *this* view, if climate change harms organisms, and humans are responsible for climate change, then climate change is a direct wrong to some organisms in the nonhuman world.[33] Since it seems very likely that climate change will directly harm some living organisms, in views like Taylor's climate change is ethically problematic. However, this ethical problem may not persist over time. That is, it may only apply in the case of organisms *currently in existence* to which climate change may cause harm. It may not apply to organisms if the effects of climate change are a necessary condition of their existence. To avoid repetition, I'll return to this shortly when discussing sentient animals, for at least some of what I say about this problem in the animal case may also apply to nonconscious organisms.

To move on to conscious, experiencing animals then, it's widely accepted that mammals and birds have interests that matter to them. In this sense, arguments about the effects of climate change on animals can be distinguished from those relating to species, ecosystems, and nonconscious living organisms, where there's much deeper skepticism about moral considerability. But there's still work to do, even accepting that sentient animals are morally considerable and that they can be harmed in morally significant

---

[31] The Natural Resources Defense Council produced a set of maps in 2009 showing areas of the western USA where the development of alternative energy would threaten endangered wildlife such as the western sage grouse. www.nrdc.org/media/2009/090401a.asp.

[32] Taylor, *Respect for Nature*, 172.

[33] Although, as noted, mitigation may also harm individual organisms.

ways, in reaching the conclusion that climate change – in any general sense, at least – produces morally significant harms to them.

Let's think first about the effects climate change is likely to have on sentient animals. Inevitably, it will cause some animals to suffer (from heat, lack of water, melting ice, lack of food, reduction in or disappearance of usable habitat, flooding, fires, climate-related disease, or other extreme weather events). Climate change will kill some animals. The majority of these animals will be undomesticated, and will live in relatively wild places. Climate change will also benefit some sentient animals. It will reduce their suffering, and result in more food, warmth, and more, or more suitable, habitat. Some of these benefited animals will be wild, others feral; changing climate and ecosystems will permit some animals to move into and colonize new territories. On yet other animals, climate change is likely to have little impact: experimental animals and others that live mainly inside climate-controlled buildings will be largely unaffected (even if their lives go experientially very badly or very well for other reasons).

Some similar issues arise here as with nonconscious organisms. One way of thinking about the effect of climate change on sentient animals is, essentially, consequentialist – a focus on producing best outcomes.[34] Even if some animals suffer and die, climate change might result in an overall net gain in pleasure, or preference satisfaction (for instance) in the context of sentient animals. This may be unlikely, but it's not impossible. Certainly, climate change may result in fewer American pika and more brown rats; but while many human beings would dislike this outcome, from the perspective of how sentient animals experience their own lives – how their lives go "from the inside," there's no good reason that we know of to privilege "pika" experience over "brown rat" experience. What matters, on this view, is whether climate change creates a worse or better world, overall, over time, in terms of whatever is thought to be directly valuable about the experience of sentient animals.

An alternative view – say, a rights view that accepts animals as rights-holders – would reject this "summing" approach. Here the worry is about infringing animals' basic rights, such as their right to life.[35] If climate change kills animals, or harms them by, for instance, depriving them of vital habitat, then the fact that some *other* animals benefit does not make good the harm. If climate change seriously harms animals – as it plausibly does –

---

[34] Although not all forms of consequentialism are totalizing in this way.
[35] Tom Regan is the best-known exponent of such a view. See Tom Regan, *The Case for Animal Rights* (Berkeley, CA: University of California Press, 1983).

and humans are responsible for it, then those chiefly responsible are wronging animals (in similar ways, it might be argued, to the ways in which fellow humans are also being wronged by climate change: few of them benefit, they carry a disproportionate burden, some are extremely vulnerable, and so on).

But we need to be careful when considering exactly how this claim about harm is framed, since it raises what's called – in the human case – the "non-identity problem."[36] Climate change will not only harm some existing animals and benefit others. It will also cause different particular individual animals to come into existence than would otherwise have been the case. Just one example: birds that migrate earlier will likely meet different mates than they would otherwise have done, and will produce fertilized eggs from different sperm. So, different individual birds will end up existing. And the more wide-ranging and intense the effects of climate change are, the more effects of this kind will come about, at least in the context of animals whose breeding is not tightly controlled by human beings, and is strongly influenced by temperature, the seasons, and climate-affected changes in ecosystems. Now, if what's thought to be morally important is *total* positive (or negative) animal experience over time, understood "impersonally" (in the sense of not belonging to any *particular* individual), this productive effect of climate change may not be ethically important. But if we're concerned about climate change harming particular individuals, or infringing rights, then this productive effect does seem very important.

Let's compare individual animals in two different situations. First, let's take a mature adult Alaskan polar bear. One effect of climate change is to shift patterns of sea ice; ice thaws earlier and forms later. These changes in sea ice make life more difficult for the polar bear, so that its survival becomes precarious. Its life is going less well than it would otherwise have done; ultimately its life might be shorter.[37] We can claim, therefore, that this polar bear is being harmed by climate change in ways that matter morally. Without climate change, this bear's life would have gone better; now it is going comparatively badly, and human activities are causally responsible.

But now, let's imagine a different situation. A decade from now, changes in climate as a consequence of human activities have led to the expansion of nine-banded armadillo populations northwards in the USA. (I'll just assume that armadillos are sentient, and that climate changes would have this effect.) Two armadillos move into new territory, meet, mate, and

---

[36] As discussed by Derek Parfit, *Reasons and Persons* (Oxford: Clarendon Press, 1984).

[37] Some of what I say here is influenced by John Broome, *Counting the Cost of Global Warming* (Cambridge: White Horse Press, 1992), though Broome only discusses human cases.

produce offspring. But while the warming climate has opened up new habitat possibilities for the young armadillos, it is also more liable to produce extreme weather events, such as droughts and floods. Armadillos are particularly vulnerable to drought; and a drought begins to develop. The young armadillos struggle to survive; they have to travel long distances to find sufficient water; they have more difficulty locating sources of food.

Although different factors are at work here – loss of sea ice versus drought – we might think that our choices have made these young armadillos, like the polar bear, worse off than they would have been had we adopted different policies. But there's a telling difference. The polar bear was already in existence before the sea ice began to change and to make its life go badly. But climate change is a necessary condition of the *very existence* of these particular armadillos. Without the change in climate, some other armadillos, or indeed some other animals altogether, would have existed. It's not that anthropogenic climate change has *worsened* their lives (as we can say about the polar bear), but rather that, without climate change, these particular armadillos would never have existed.

This argument is significant inasmuch as it suggests that we should be wary about making claims that climate change will harm animals – and perhaps nonconscious organisms – that have not yet come into being. For at least some of these animals (and for an increasing number of animals over time) climate change will be a necessary factor in their existing at all. So, while we might reasonably be concerned that existing animals (such as individual polar bears) are *currently* being harmed by climate change, if we look into the further future, it's going to be increasingly difficult to think of climate change as a factor that's harming, or infringing the rights of, particular individuals.

However, this doesn't mean that we should not be (as it were) *impersonally* concerned at the prospect of a future world with more animal suffering in it, rather than less, on account of the changes in the world that climate change might bring.[38] Consequentialists could still see this as a troubling prospect. But there's deep uncertainty here. We can't tell whether climate change will cause more suffering to nonhumans than it will relieve. We don't know whether more animals capable of suffering will ever live. Nor do we know whether, on account of climate change, more or fewer cognitively sophisticated animals will ever live. If climate change were both to cause a proliferation in the numbers of cognitively simpler armadillos while

---

[38] There are a number of other possible responses to non-identity problems that it would be interesting to develop in this case; unfortunately, I don't have space to pursue these here.

simultaneously bringing about a decline in the numbers of cognitively sophisticated elephants or chimpanzees, this would be particularly pernicious in some ethical views. But we lack any evidence about the plausible effects of climate change that could help us to make judgments of this kind; nor is it at all obvious how such evidence could be obtained. Certainly, no one is conducting research directly into such questions, nor are they likely to do so.

7 IN CONCLUSION

I began by noting that very little has been published so far concerning the effects of climate change on the nonhuman world. While this silence is not unexpected in political and philosophical circles where only present and future humans have moral status, it's very surprising in the context of environmental ethics. However, looked at more closely, this silence seems less surprising. For the ethical implications of climate change in the context of the nonhuman world are extremely unclear. One reason for this lack of clarity is the high degree of uncertainty about how the climate will actually change, and what effect that will have on species, ecosystems, and living organisms. Second, while climate change will have significant negative impacts, it will also drive speciation, change some ecosystems without destroying them, produce organisms that would not otherwise have existed, and promote the flourishing of at least some species, ecosystems, and individual organisms. In part, ethical judgments here will depend on whether one takes the view that one can weigh the benefits of climate change against the harms it causes, or not. A consequentialist must here struggle with the questions of deep uncertainty climate change raises, where we don't have a good sense of what outcomes are even plausible, let alone predictable or to be expected. A non-consequentialist, though, will have to think through the non-identity problems that climate change raises, at least in the context of living organisms. All I've been able to outline here are some very preliminary steps in thinking about such problems. What's needed is a much more careful and detailed account both of the likely effects of climate change on the nonhuman world, and whether – and why – these effects might be of ethical importance.

CHAPTER 14

Human rights, climate change, and the trillionth ton

Henry Shue

The desultory, almost leisurely approach of most of the world's national states to climate change reflects no detectable sense of urgency. My question is what, if anything, is wrong with this persistent lack of urgency. My answer is that everything is wrong with it and, in particular, that it constitutes a violation of basic rights as well as a failure to seize a golden opportunity to protect rights. I criticized the outcome of the initial climate conference in Rio de Janeiro in 1992, the Framework Convention on Climate Change, for establishing "no dates and no dollars: no dates are specified by which emissions are to be reduced by the wealthy states and no dollars are specified with which the wealthy states will assist the poor states to avoid an environmentally dirty development like our own. The convention is toothless."[1] The general response to such criticisms was that the convention outcome was a good start.

Nearly two decades later, the outcome of the 2009 Copenhagen Conference of the Parties to the Convention is equally toothless, once again containing no dates and no dollars. The *New York Times* described the twelve-paragraph Outcome Document as "a statement of intention, not a binding pledge to begin taking action."[2] It contains a vague commitment to the end of keeping the global temperature rise to no greater than 2°C beyond pre-industrial levels but no specification of, much less commitment to, the means necessary to that end.[3] And the USA has with great fanfare vaguely committed itself to contributing to a fund to assist poorer nations, while

---

"Ton" here, as throughout this book, denotes "metric ton."

[1] H. Shue, "Subsistence Emissions and Luxury Emissions," *Law and Policy*, 15 (1993), 39; reprinted in S. M. Gardiner, S. Caney, D. Jamieson, and H. Shue (eds.), *Climate Ethics: Essential Readings* (New York: Oxford University Press, 2010), 200.

[2] A. C. Revkin and J. M. Broder, "A Grudging Accord in Climate Talks," *New York Times*, December 20, 2009.

[3] Para. 1 "recognizes" "the scientific view that the increase in global temperature should be below 2 degrees Celsius." Para. 12 requires the assessment of the implementation of the accord to "include consideration of strengthening the long-term goal . . . including in relation to temperature rises of 1.5

providing no specification of how much it might put in the fund, and pitifully pledged laughable emissions cuts equivalent to a reduction of 4 percent below 1990 levels.[4] Yet, once again, the defense of the outcome has been that it is a good start.

But after nearly two decades of delay in confronting the fact that we are ourselves creating an on-rushing problem for those who come after us, good intentions, a first step, and a good start are not nearly good enough. After nearly two decades of denial, the time has passed for perpetual first steps. Indeed, we wrong the people of tomorrow by doggedly persisting in contributing to conditions in which they will be unable to fulfill their basic rights. If we all continue simply to consume fossil fuels like there is no tomorrow, there may indeed be no tomorrow for those who will have to try to live in it. And if we persist in business-as-usual, we miss a spectacular opportunity to provide protection for people who come after us that they will be in no position to provide for themselves. Here is one general argument.

Rapid climate change places current and future generations in precisely the kind of general circumstances that call for the construction of rights-protecting institutions.[5] While the identities of future individuals are not yet determined and are thus not knowable,[6] we know that as humans they will all be entitled to human rights, including rights that depend on a functioning economic system, which itself depends in turn on a planetary environment within which those future humans can adapt and support themselves. They need, therefore, to inherit from past generations an environment that is neither radically inhospitable nor radically unpredictable. Now, however, we are producing through our carbon-fueled economic activities

---

degrees Celsius," whatever this means. See United Nations, Framework Convention on Climate Change, Conference of the Parties, Fifteenth Session, Agenda item 9, 'Draft Decision -/CP.15', FCCC/CP/2009/L.7 (December 18, 2009).

[4] Bill McKibben, "Heavy Weather in Copenhagen," *New York Review of Books*, 57:4 (March 11–24, 2010), 32–4.

[5] For a reasonably accessible broad general introduction to the science of climate change, see J. Houghton, *Global Warming: The Complete Briefing*, 4th edn. (Cambridge University Press, 2009). Also see Save the Children UK, *Legacy of Disasters: The Impact of Climate Change on Children* (London: Save the Children UK, 2007).

[6] Many philosophers have been much taken with what is known as the non-identity problem, which was formulated by Thomas Schwartz in two contemporaneous pieces, "Obligations to Posterity," in R. I. Sikora and B. Barry (eds.), *Obligations to Future Generations* (Philadelphia: Temple University Press, 1978), 3–13, and "Welfare Judgments and Future Generations," *Theory and Decision*, 11 (1979), 181–94; and later by D. Parfit, *Reasons and Persons* (New York: Oxford University Press, 1984). As far as I can see, our current lack of knowledge of particular identities in later centuries has no implications at all for what we ought to do now about future climate change. At most, it has some implications for how we explain our more basic moral judgments. For a different view, see Edward A. Page, *Climate Change, Justice and Future Generations* (Northampton, MA: Edward Elgar, 2006), 132–60.

increasingly rapid environmental change that currently has no fixed or otherwise predictable limit. The members of future generations of humans are completely vulnerable to the choices made by earlier generations, including ours, which is the very first generation in human history to have acquired the knowledge necessary to understand either the problem of climate change or possible solutions to the danger of excessively rapid change.

A situation in which some humans are utterly at the mercy of other humans, but those others have the capacity to create institutions to protect the vulnerable against the forces against which they cannot protect themselves, is the paradigmatic situation calling for the recognition and institutionalization of rights. Moreover, in the case of climate, the forces against which future humans will otherwise be defenseless are being unleashed by us ourselves, so we bear especially great responsibility for ceasing to make matters worse, compensating those whom we will have wronged by what we have done and will still do, and constructing social institutions, including most especially a regime for dealing with human-induced climate change, to protect those who could not otherwise protect themselves because the causal lead times in matters affecting the climate extend far beyond single generations. The opportunity to replace short-sighted, rights-undermining practices reliant on primitive fossil-fuel energy technologies with far-sighted, rights-protecting institutions resting on alternative energy technologies is historic, unprecedented. We can take steps now to protect rights for those who come after us, steps that it may be too late for anyone to take later.

The purpose of a right is to provide protection for human beings against a threat to which they are vulnerable and against which they may be powerless without such protective action. For this protection to endure, rather than needing constantly to be improvised and reimprovised, the protection needs to take the form of an intergenerational institution. Human rights are an expression of human solidarity; we commit ourselves to each other to try to guarantee to all the urgent interests that some cannot provide for themselves. It is especially sad, as a matter of Western intellectual history, that the criticism that Marx derived from Hegel that individual rights are a kind of juridified or legalized form of the war of all against all, authorizing each of us to battle to be left alone by the others, should have taken as deep a hold among academic theorists as it has.[7] Degenerate systems of rights in which rights are reduced to purely negative rights of non-interference can have

---

[7] K. Marx, "Zur Judenfrage," in K. Marx/F. Engels, *Werke*, Bd. 1 (Berlin: Dietz Verlag, 1972 [1844]), 347–77; translated and abridged but with commentary as "On the Jewish Question," in J. Waldron (ed.),

such an 'atomistic' form in which each merely tries to protect himself *from* the others, especially if rights are narrowed to rights to negative liberty. But developed systems of human rights, while they contain some protections against those others who are in fact predatory (even then in the form of solidarity among potential victims against the potential predators), are much more about cooperating with others in solidarity to create social institutions and practices that provide protection *to* others when they face dangers they cannot handle on their own.[8]

One question naturally is: Which rights of the people to come are threatened by climate change, and in which particular ways? Fortunately, a strong contribution to answering this question in detail has been made by Simon Caney, who has carefully shown how climate change will specifically threaten at least three rights, the right to life, the right to health, and the right to subsistence.[9] Here I shall simply rely on Caney's arguments about which rights so that I can focus on two other questions as they arise in the context of climate change: Which features do rights-protecting institutions need to have and what specifically are the tasks that need to be performed to protect rights against the threat of rapid climate change?

I NECESSARY FEATURES OF RIGHTS-PROTECTING INSTITUTIONS: I, INTERNATIONAL

We cannot of course protect each other against all threats to all interests, and I have referred to the dangers for which protection ought to be provided as "standard threats."[10] In an important recent book Charles Beitz has stated

Nonsense upon Stilts: Bentham, Burke and Marx on the Rights of Man (London: Methuen, 1987), 137–50. Prominent scholars adopting the Marxian criticism include C. Taylor, "Atomism," in C. Taylor, *Philosophy and the Human Sciences, Philosophical Papers*, vol. 2 (New York: Cambridge University Press, 1985 [1979]), 187–210; and M. A. Glendon, *Rights Talk: The Impoverishment of Political Discourse* (New York: The Free Press, 1991), 47. For elaboration of this point, see H. Shue, "Thickening Convergence," in D. K. Chatterjee (ed.), *The Ethics of Assistance* (Cambridge University Press, 2004), 217–41.

[8] For a fuller discussion see H. Shue, "Solidarity Among Strangers and the Right to Food," in W. Aiken and H. LaFollette (eds.), *World Hunger and Morality*, 2nd edn. (Upper Saddle River, NJ: Prentice-Hall, 1996), 113–32.

[9] S. Caney, "Climate Change, Human Rights and Moral Thresholds," in S. Humphreys (ed.), *Human Rights and Climate Change* (Cambridge University Press, 2009), 69–90. Also see Oxfam International, *Climate Wrongs and Human Rights: Putting People at the Heart of Climate-Change Policy*, Oxfam Briefing Paper 117 (Oxford: Oxfam International, 2008); P. Baer, T. Athanasiou, and S. Kartha, *The Right to Development in a Climate Constrained World: The Greenhouse Development Rights Framework* (Berlin: Heinrich Böll Foundation, 2007); and International Bank for Reconstruction and Development/The World Bank, *Development and Climate Change: World Development Report 2010* (Washington, DC: The World Bank, 2010), Box 1.4, 53 [mentioning "the right to food, the right to water, and the right to shelter"].

[10] H. Shue, *Basic Rights*, 2nd edn. (Princeton University Press, 1996), 29.

the purpose of human rights as "to protect urgent individual interests against certain predictable dangers ('standard threats') to which they are vulnerable under typical circumstances of life in a modern world order composed of states."[11] This rich characterization contains more elements than can be commented upon here, but I want to begin from Beitz's primary emphasis on human rights as "matters of international concern."[12] I shall argue, first, that any institutions to protect the rights threatened by climate change must be international; second, that they must also be intergenerational; and, third, that we must begin to build imaginative international intergenerational institutions immediately.

For better or for worse, the "modern world order" assigns individual human beings to the control, possible protection, and possible predation of national governments. At the level of theory of the ideal, one can ask whether the best global order would be such an international system of national states; I seriously doubt it. On the other hand, be that as it may, at the level of what is practical for the foreseeable future, one can instead ask how, given the global order we actually have, the human rights of individuals can best be protected. The arrangement that we in fact have is what Beitz calls "a two-level model" of protection for rights: "The two levels express a division of labor between states as the bearers of the primary responsibilities to respect and protect human rights and the international community and those acting as its agents as the guarantors of these responsibilities."[13] I have sometimes referred to these second-level, back-up responsibilities as "default duties."[14] When a national government fails to carry out its primary responsibility to protect rights, responsibility defaults to the second level consisting of the remainder of humanity, organized under the other national governments and constituting the remainder of the international community. This is essentially the model or picture underlying, for instance, what has come to be called the "responsibility to protect" (or R2P).[15]

One good question is why primary responsibility should be assigned to national states; often that question turns into the question of ideal theory: Is a system of states really the best way to organize the globe? This is a question that, as I have said, I leave aside for consideration in some less desperate time, in order to pursue another good question: Why, in the

---

[11] C. R. Beitz, *The Idea of Human Rights* (New York: Oxford University Press, 2009), 109.   [12] Ibid.
[13] Ibid., 108.   [14] Shue, *Basic Rights*, 2nd edn., 171 ff.
[15] International Commission on Intervention and State Sovereignty, *The Responsibility to Protect* (Ottawa: International Development Research Centre, 2001).

international system we now have, default duties or secondary responsibility should fall to "the international community"? This is so far a very general and rather grand question, and we want to focus here specifically on the pressing issue of accelerating climate change.[16] An initial part of the answer to the grand, general question is that it is very difficult to imagine what other agents could effectively step into the breach when a national state is failing to protect its own people other than some agency of the international community (perhaps, as in the case of climate change, an agency of a kind yet to be fully conceived and constructed). Because national states are so extremely jealous of their state sovereignty and their other prerogatives and tend to be armed and dangerous, very few effective options are available for bringing about inside their claimed jurisdiction any change that they view as unnecessary or oppose for any other reason. Action genuinely representative of the international community, even if implemented primarily by one or a few other national states, is one of the only options likely ever to have both the legitimacy and the power to accomplish inside a sovereign state anything not welcomed, much less seriously opposed, by the state in question. Otherwise, when national governments were hostile to rights, the individuals they control would simply have to be abandoned to their fates.

If it is difficult to conceive any effective agency other than some agency of the international community in most cases in which a state has failed to protect the most fundamental rights, it is truly inconceivable that a national state could protect the rights threatened specifically by climate change while acting alone. Every state is a "failed state" as far as climate is concerned. Climate change is such a deeply global problem that only coordinated international action could possibly deal with it. Climate is itself a set of literally global phenomena driven by transnational systems at various levels like the jet stream, the Atlantic thermohaline circulation, and El Niño and La Niña. And the greenhouse gases (GHGs) that are disturbing the historical climate equilibria, whatever their national origins, themselves mix into a single global atmospheric pool blanketing the entire planet and producing climate disturbances for emitters and non-emitters alike. Climate is one case

---

[16] The rate of climate change is itself increasing and appears to be feeding upon itself, that is, generating dominantly positive feedbacks – see P. Forster, V. Ramaswamy, P. Artaxo *et al.*, "Changes in Atmospheric Constituents and in Radiative Forcing," in S. Solomon, D. Qin, M. Manning *et al.* (eds.), *Climate Change 2007: The Physical Science Basis*, Working Group I Contribution to the Fourth Assessment Report of the Intergovernmental Panel on Climate Change (Cambridge University Press, 2007), 129–234; and H. Le Treut, R. Somerville, U. Cubasch *et al.*, "Historical Overview of Climate Change Science," in Solomon *et al.* (eds.), *Climate Change 2007: The Physical Science Basis*, 93–127.

in which effective national protective measures are simply not possible. Anything China does can be undone by the USA, and vice versa. The futility of uncoordinated national efforts at protection against the effects of climate change is certain. The only conceivable protection of any rights threatened by climate change is protection through concerted action by the international community as a whole.

It is worth noting, on the other hand, that international institutions need not be centralized. Given the failure so far of the Conference of the Parties to the Framework Convention on Climate Change to reach any general agreement about a multilateral treaty that will effectively eliminate carbon emissions in any reasonable period of time, it may well be that more hope lies with international coalitions, and coalitions of coalitions, of agents operating at various different levels than through standard international treaties.[17] If, say, the State of California, the Chinese Ministry of Environmental Protection, Wal-Mart, and the EU could agree on aggressive action, it might not be necessary to wait on ill-informed laggards like the majority of the US Senate. It would be outrageously tragic if the fate of the whole Earth depended on the US Senate. This is a largely non-normative issue at a different level of analysis that could not be pursued here even if I were the right one to pursue it. The important negative caution is simply that action can be international without needing to be based on the kind of treaty that the US Senate so rarely deigns to ratify.

2 NECESSARY FEATURES OF RIGHTS-PROTECTING INSTITUTIONS: II, INTERGENERATIONAL

It is equally certain that any effective protections for the people of later generations must be begun in prior generations, making climate change an inherently intertemporal/transgenerational, as well as an inherently international, challenge. It is quite simply impossible for the social arrangements necessary to control the severity of climate change to be put into place by the generation who need protection. The overall climate system has extraordinary momentum; it is difficult to divert it from whatever path it is on, but once it has been diverted, as it now has been since the fossil-fuel-fed Industrial Revolution, it is then equally difficult to divert it yet again from that new path over any period of time less than a century. Above all, this is because of the atmospheric residence time of the most dangerous GHG, carbon dioxide (CO_2). Only a few years ago scientists believed that the

---

[17] I am grateful to Scott Moore for raising this issue.

average atmospheric residence time of a molecule of CO_2 was roughly a century.[18] They have now realized that it is so much longer than a century that it is currently incalculable: "A[n atmospheric] lifetime for CO_2 cannot be defined . . . The behaviour of CO_2 is completely different from the trace gases with well-defined lifetimes. Stabilisation of CO_2 emissions at current levels would result in a continuous increase of atmospheric CO_2 over the 21st century and beyond . . . In fact, only in the case of essentially complete elimination of emissions can the atmospheric concentration of CO_2 ultimately be stabilised at a constant level . . . More specifically, the rate of emission of CO_2 currently greatly exceeds its rate of removal, and the slow and incomplete removal implies that small to moderate reductions in its emissions would not result in the stabilisation of CO_2 concentrations, but rather would only reduce the rate of its growth in coming decades."[19]

The practical implication of this rather mind-boggling conclusion is this: Once CO_2 is emitted into the atmosphere, it stays there over any period of time of interest to humans; and once more CO_2 is emitted, more stays there. Or in other words, at whatever level of CO_2 the atmospheric concentration peaks, that concentration will stay for a long, long time – multiple generations at a bare minimum. This makes the duration of climate change like few other problems, except perhaps the generation of nuclear waste, which is also extraordinarily persistent, and the manufacture of the most persistent toxic chemicals. And of course the pervasiveness of climate change is incomparably greater than nuclear waste or any toxics about which we so far know. While it is always good for rights-protecting institutions to be enduring, for them to deal specifically with the dangers of climate change no other option is possible.

When I was a small boy in rural Virginia in the 1940s, traveling evangelists would pitch their tent for a week in our county and attempt to convert us (although most of us thought we had signed up locally). One of their standard ploys was to try to terrorize us by preaching on the final evening about the "unforgivable sin": It was essential to convert before you committed it – later would be too late because on this one there was no going back. I used to lie awake after returning home from the tent meeting, worrying that I might have committed the "unforgivable sin" already without having realized it at the time, since the evangelist's account of it was as

---

[18] This was already a great oversimplification – for a more sophisticated approach, see F. Joos, M. Bruno, R. Fink *et al.*, "An Efficient and Accurate Representation of Complex Oceanic and Biospheric Models of Anthropogenic Carbon Uptake," *Tellus*, 48B (1996), 397–417.
[19] G. A. Meehl, T. F. Stocker, W. D. Collins *et al.*, "Global Climate Projections," in Solomon *et al.* (eds.), *Climate Change 2007: The Physical Science Basis*, 824–5.

vague as it was ominous, and so be eternally damned before I had even
gotten a good start on life (or had much fun). Adolescence, of course,
brought other worries, and I gave up on the "unforgivable sin," coming to
doubt that there was any such thing. Now, however, I sometimes think the
atmospheric scientists may have figured out what the "unforgivable sin" is
after all: emitting so much CO_2 that uncountable generations face a severely
disrupted and worsening climate that blights their lives from the beginning!
The penalty is not quite the promised eternal damnation, but bad enough,
and, worse, the penalty falls not on the unforgiven sinners/emitters but on
their innocent descendants, dooming them from the start.

3 NECESSARY FEATURES OF RIGHTS-PROTECTING INSTITUTIONS: III, IMMEDIATE

Philosophers and normative theorists have an unfortunate tendency often
to begin offering prescriptions for dealing with problems before appreciating in sufficient depth the specific features of the particular problem. We
ought to consider further the implications of the recent IPCC finding,
quoted in the third paragraph above, that in the case of CO_2, in effect, what
goes up does not come down any time soon. Common sense suggests that
although we may be constantly adding CO_2 to the atmosphere, what is
there already must somehow be drifting off somewhere else or otherwise be
being neutralized. It is correct that, on the scale of centuries, the molecules
of CO_2 decay, with the carbon and oxygen atoms entering new combinations. But the fact that the decomposition occurs only over such an
extended time frame means that for all practical purposes – that is, as far
as human interests are concerned – the concentration of CO_2 is simply
expanding indefinitely, now that we are ejecting so much more from fossil-
fuel burning. It is absolutely critical how large the concentration is when it
peaks – when it reaches its largest size – and it will not peak as long as we
keep adding to it, because all additions amount practically to net additions.

Therefore, we need an immediate exit strategy from fossil fuel consumption and institutions to begin implementing it. I quote the alarming crucial
finding again: "In fact, only in the case of essentially complete elimination
of emissions can the atmospheric concentration of CO_2 ultimately be
stabilised at a constant level." This is a stunning fact about the atmospheric
chemistry of our planet, which makes continued use of fossil fuel (by far the
primary source of the CO_2 emissions) seem much like indulgence in an
"unforgivable sin." And this is why the proposal by the Obama administration to cut annual GHG emissions flows by only 4 percent below 1990

flows is such a sad gesture – this would continue rapidly to increase the size of atmospheric stocks of GHG. As a physicist friend familiar with the US election campaign of 1992 recently put it, "It's the stocks, Stupid!"

Indeed, these recent empirical findings support three momentous judgments critical to the nature and extent of our responsibility to terminate fossil fuel usage. First, if we do not act to prevent future climate change from becoming more severe, we have no grounds for confidence that anyone else ever can – there can be no later reversal of additional deteriorations, for a long time, with any known technology.[20] We know of no acceptable way for subsequent generations to reverse whatever damage our actions will unleash upon them. In order to bring about any mitigation of climate change a century from now, vigorous action needs to be taken now. Second, we in any case ought to be the ones to act, in part, because we are the ones currently making matters (irreversibly) worse for everyone to come in the foreseeable future. We ought to take the vigorous action. Third, thanks to their hard-won understanding of the effects of reliance on carbon fuels, our scientists have opened the door to our bequeathing to those who follow us a priceless legacy: institutions that will nurture alternative energy sources that do not progressively undermine the environment to which the human species and other contemporary species are successfully adapted. Vigorous action now, by us, can produce invaluable results in the protection of threatened rights. What must we do?

4 NECESSARY TASKS FOR RIGHTS-PROTECTING INSTITUTIONS: I, "DO NO HARM"

Like increasing numbers of scientists, I personally believe that there is a strong case for choosing the minimum goal mentioned in the Copenhagen Outcome Document of limiting warming to 2°C above pre-industrial levels.[21] The World Bank accepts that even a rise of 2°C would include these results: "Between 100 million and 400 million more people could be at risk of hunger. And 1 billion to 2 billion more people may no longer have

---

[20] One can of course fantasize about unknown technological fixes. Some of the more popular ones, like geo-engineering, are morally highly dubious – see S. M. Gardiner, "Is 'Arming the Future' with Geoengineering Really the Lesser Evil?," in Gardiner *et al.* (eds.), *Climate Ethics*, 284–311.

[21] Indeed, much can be said for a more ambitious goal of a global temperature rise of no more than 1.5°C above pre-industrial levels, which would require constraining atmospheric carbon to 350 ppm – hence the name of the website: 350.org. The World Bank has endorsed the goal of preventing temperature rises from exceeding 2°C – see International Bank for Reconstruction and Development, *Development and Climate Change*, 3.

enough water to meet their needs."[22] Others who do not accept this particular goal may of course view the following as a hypothetical example with an arbitrary goal; the general logic of the position remains, whichever goal one selects from within the range of what could plausibly be genuinely believed to be likely to avoid dangerous interference with the climate system.[23] The same logic holds provided any firm limit on emissions is adopted.

If we want to limit global warming, for example, then, to 2°C above pre-industrial levels, we must avoid emitting the trillionth metric ton of carbon (Tt C) to be confident of having even a 50 percent chance of meeting our target.[24] We have already emitted 0.5 Tt C and are therefore already committed to 1°C of warming. "Having taken 250 years to burn the first half-trillion tonnes of carbon, we look set, on current trends, to burn the next half trillion in less than 40."[25] Recent research suggests that the most helpful way to conceive our challenge, if we want to avoid warming of more than 2°C, is as the challenge of humanity's remaining within a total cumulative carbon budget of 1 Tt C, although of course it may turn out that this cumulative cap needs to be revised as scientific understanding progresses.[26] Total cumulative emissions of carbon must not surpass 1 Tt C, or global average surface temperature will, as likely as not, rise more than 2°C above pre-industrial levels, due to CO_2 alone. As shorthand, then, we can view our challenge as staying within a cumulative carbon budget of 1 Tt C,

---

[22] Ibid., 5. Both projections are taken from the 2007 IPCC report.

[23] For well-informed speculations about what the world would look like after temperature rises of various amounts up to 6°C, see M. Lynas, *Six Degrees: Our Future on a Hotter Planet* (Washington, DC: National Geographic, 2008). For a more technical account, see G. W. Yohe, R. D. Lasco, Q. K. Ahmad et al., "Perspectives on Climate Change and Sustainability," in M. Parry, O. Canziani, J. Palutikof et al. (eds.), *Climate Change 2007: Impacts, Adaptation and Vulnerability*, Working Group II Contribution to the Fourth Assessment Report of the Intergovernmental Panel on Climate Change (Cambridge University Press, 2007), Tables 20.8 and 20.9, 828–9. It is obvious that even a 3°C rise would be intolerable, with greater rises being increasingly catastrophic.

[24] M. R. Allen, D. J. Frame, C. Huntingford et al., "Warming Caused by Cumulative Carbon Emissions Towards the Trillionth Tonne," *Nature*, 458 (April 30, 2009), 1163–6.

[25] M. Allen, D. Frame, K. Frieler et al., "The Exit Strategy," *Nature Reports Climate Change*, 3 (May 2009), 57. Atmospheric physicists at the University of Oxford maintain a website on which a ticking meter shows the blinding speed of the global release of carbon from fossil fuels at http://trillionthton. org/.

[26] Three factors might cause us to revise the cap: (1) we might make revisions regarding our conception of "dangerous anthropogenic interference in the climate system", and change the 2°C target; (2) we might choose a larger or smaller cap to reflect risk preferences regarding the costs of abatements and climate change damages; or (3), as scientific uncertainty resolves, we might find we need to revise the cap even while holding the target and risk preference fixed. I am grateful to my colleague David Frame for these points and other helpful suggestions.

or, in short, avoiding the trillionth metric ton.[27] We have only forty years – about half a normal affluent lifetime – to eliminate completely the use of fossil fuel if the temperature rise is not to exceed 2 °C.[28] Before today's college students retire, humanity must either have perfected carbon sequestration or stopped digging coal and pumping oil, the remaining reserves of which hang over us like a sledgehammer.

Accepting that the fundamental specific challenge has the shape of a need to stay within a cumulative budget of carbon emissions, which one must accept if one adopts any ceiling on temperature rises, has radical implications for how we ought to behave. Because we, together with all the generations who follow in our wake, must stay inside a single limit, carbon emissions are zero-sum across generations. That carbon emissions are *zero-sum across all emitters throughout foreseeable time* is profoundly important for the nature of our responsibilities for our actions regarding the rights of people to come. Every metric ton of carbon emissions for which one person is responsible is one less metric ton of carbon emissions available for all the other persons who will live during the foreseeable future. We are in direct competition for a scarce resource with our own great-grandchildren, and everyone else's great-grandchildren. Every time I fly across the Atlantic is one less time that anyone else can fly across the Atlantic in a plane burning fossil fuel. Our challenge is intergenerationally zero-sum.

This is obviously far from making the problem unique. The consumption of any non-renewable resource is intergenerationally zero-sum. Any unit of it that I consume is one less unit for everyone else across time to consume. We have simply learned recently that on the time scales that matter to humans, the planet's capacity to deal with carbon without rises in surface temperature is one of the non-renewables, even if over several centuries the atmospheric carbon will break down.

Business-as-usual consumption of carbon fuel is, then, doubly dangerous for the people to come. First, carbon emissions contribute to making climate change more severe. Second, carbon emissions contribute to using

---

[27] Obviously if our goal were to be more ambitious than remaining below 2 °C above pre-industrial levels, we would need to adopt an even tighter budget. At this point what is crucial is that we move aggressively downward in our carbon emissions; the ultimate target can be, and surely will be, adjusted as time goes by.

[28] I despaired of the capacity of US institutions of governance to rise to the challenge even before the January 2010 decision of the US Supreme Court in *Citizens United* v. *Federal Election Commission* (558 U.S. [2010]) opened the floodgates for the use of unlimited corporate funds against political candidates who support any legislation aimed at the necessary limitation of the use of fossil fuel. Now it is difficult to imagine effective political action in the USA against such wealthy entrenched power as the oil and coal industries, but legal action could conceivably work – see next note.

up whatever quota is set on behalf of humanity for total cumulative carbon emissions; e.g., the remaining 0.5 Tt C if we want to have a 50 percent chance of avoiding a temperature rise of more than 2°C. These two are each separately compelling reasons to institutionalize the initial exit strategy from fossil fuel as soon as humanly possible and assign responsibilities accordingly. The strategy will naturally need, as already indicated, to be adjusted as events develop and better understanding of climate dynamics grows. But these double dangers make vigorous action, far more robust than anything now contemplated by conventional politicians, urgent.

The main argument usually given for inaction is uncertainty. What if, in spite of the growing sophistication and increasing validation of climate science, the dangers of climate change have somehow been exaggerated? I have dealt with this complex issue at length elsewhere, where the upshot was what amounted to a modified formulation of the "precautionary principle": one ought to ignore entirely questions of probability beyond a certain minimal level of likelihood in cases with "three features: (1) *massive loss*: the magnitude of the possible losses is massive; (2) *threshold likelihood*: the likelihood of the losses is significant, even if no precise probability can be specified, because (a) the mechanism by which the losses would occur is well-understood, and (b) the conditions for the functioning of the mechanism are accumulating; and (3) *non-excessive costs*: the costs of prevention are not excessive (a) in light of the magnitude of the possible losses and (b) even considering the other important demands on our resources."[29] The danger of climate change clearly has these three features, so uncertainty ought not to prevent us from taking action that does not have excessive costs, with "excessive" measured only by the magnitude of the possible losses, not discounted by their probability.

The specification of the appropriate bearers of the duties to protect any particular right can often be the intellectually most challenging aspect of spelling out the form that protection for the right might best take, as Beitz

---

[29] H. Shue, "Deadly Delays, Saving Opportunities: Creating a More Dangerous World?," in Stephen M. Gardiner, Simon Caney, Dale Jamieson, and Henry Shue (eds.), *Climate Ethics: Essential Readings* (New York: Oxford University Press, 2010), 146–62, at 148. The fossil fuel industries continue to do their utmost to confuse the general public by the propagation of what has been aptly called "manufactured uncertainty" – see S. Vanderheiden, *Atmospheric Justice: A Political Theory of Climate Change* (New York: Oxford University Press, 2008), 192–202, for documentation of some of the notorious intentional deceptions during the Bush–Cheney Administration. However, lawsuits based on damage caused by climate change may be beginning to worry the fossil-fuel industry – see J. Schwartz, "Courts as Battlefields in Climate Fights," *New York Times*, January 26, 2010. In *Native Village of Kivalina* v. *ExxonMobil Corp.* it is alleged that "the industry conspired 'to suppress the awareness of the link' between emissions and climate change through 'front groups, fake citizens organizations and bogus scientific bodies.'"

rightly emphasizes.[30] The fact that continuing carbon emissions by us exacerbate both the double dangers is also the first of three powerful reasons why we clearly ought to begin to curtail carbon emissions sharply now. Business as usual is not inaction – it is the knowing infliction of more straitened circumstances of life on those to come. One of the most compelling principles for the assignment of responsibility is the principle that the persons inflicting a harm must stop – first, do no harm – and must, if possible, compensate for the damage they themselves have already done. Our generation cannot plead ignorance of the implications of our choice of energy source. The harm consists of the creation of circumstances in which the fulfillment of such rights as those to life, health, and subsistence comes under increasing threat.[31] Those circumstances will evidently arise no later than the emission of the trillionth metric ton.[32] The best way to begin to protect the rights of future generations is to stop threatening them – to cease contributing to conditions that may make enjoyment of those rights impossible. If we do not act vigorously now, we will inflict avoidable harm on people who live in our polluted wake. While I believe that this first reason is by itself entirely sufficient, an additional, although more complex, reason further strengthens the case for major responsibility for urgent action falling upon the generations now alive, the first to begin to understand climate dangers.

5 NECESSARY TASKS FOR RIGHTS-PROTECTING INSTITUTIONS: II, FAIRNESS

We have noted already that continuation of the consumption of fossil fuels not only contributes to greater severity of climate change but also continues to use up whatever the remaining quota of "tolerable" additional carbon emissions may be – "tolerable" in the extremely weak sense of not necessarily expected to undermine the conditions in which human rights can be fulfilled.[33] If the temperature ceiling is to be 2°C above pre-industrial levels, then the remaining quota is 0.5 Tt C. If someone can believe that higher

---

[30] See especially Beitz, *Human Rights*, ch. 7.
[31] Harm need not be done to an individual identifiable in advance.
[32] As indicated in note 21 above, harm undoubtedly begins at lower rises in temperature – indeed, it has already begun today. Many species will be driven to extinction by radical changes in habitat, for example. Setting the bar this low is merely a concession to the reality of the political power backing fossil fuel use. Strictly speaking, I am advocating merely "do no unavoidable harm," which is a pale copy of the Hippocratic principle.
[33] Tolerable for whom? The climate change that has already occurred has undermined the conditions necessary for the survival of many other species – most notoriously the polar bear, but various other animal and plant species as well. See *BBC News*, "Climate Focus 'Ignores Wildlife,'"

temperatures, and generally more severe climate change, are tolerable, her conception of the remaining quota will be correspondingly larger. But some ceiling – whether or not 2°C above pre-industrial levels – must be imposed unless climate change is to be allowed to grow indefinitely worse. Whatever the total emissions within the quota, questions of fairness arise about how those emissions are distributed. So the second major ethical issue, after harm, is fairness. The requirement to distribute fairly is no less fundamental than the requirement to do no harm (and to compensate for harm done), but its specification in this case is considerably more complicated.

As we have already seen, the central question is intergenerational at its core. Fairness – or "intergenerational equity," as the lawyers and economists like to call it – is then not an additional peripheral aspect of the question that we may optionally take up or not, as we choose. A single budget for carbon emissions, whatever its total size, is shared by us and every foreseeable generation to come. Consequently there simply is no such thing as doing what is fair "except for the intergenerational part"; in the concrete instance we face, what is fair is a pervasively intergenerational issue.[34]

Since all humans from now on share the same emissions budget and the budget is zero-sum, if one wants to be fair, one needs to leave for others their fair share and use only one's own fair share. But how can we think about this apparently novel issue of what are going to become fair shares of carbon emissions in the international institutional context we are about to create? One way to reduce carbon emissions is to tax them, but most of the plans currently discussed involve the creation and trading of permits to emit carbon – carbon trading, "cap-and-trade," which will be a transformative international and intergenerational institution. Under trading, the only way to create an incentive to reduce carbon emissions is to require the possession of a permit to emit carbon and to charge some people for carbon permits (and progressively to reduce the number of permits available – the "cap" – driving up the price of permits).[35] At present the world economy is

February 18, 2008. Other species are of great value – I am concentrating on humans in the text merely for simplicity. Few biological scientists would accept the idea that any level of additional human emissions is tolerable.

[34] I think most issues of justice are fundamentally intergenerational and international at bottom, but I do not seem to have convinced many people yet. For the central argument, see H. Shue, "The Burdens of Justice," *Journal of Philosophy*, 80, 10 (October 1983), 600–8. For one theorist who is convinced about the international aspect at least, see K.-C. Tan, *Justice Without Borders: Cosmopolitanism, Nationalism and Patriotism* (Cambridge University Press, 2004), 175.

[35] The political right wing in the USA is complaining mindlessly that "cap-and-trade" is "just another tax." The whole point, of course, is to make fossil-fuel consumption more expensive so that it will be reduced and – within 40 years, if possible – eliminated. Naturally the fossil-fuel sector and the politicians subservient to it oppose such measures which literally threaten their customary livelihood.

dependent upon fossil fuels and therefore dependent upon emitting carbon, but many people in the world are too poor to pay for not only what would otherwise be the price of fossil fuel but also the premium that will be added to the price by the cost of permits to release the amount of CO_2 generated by burning that much fossil fuel.

The basic requirement of human rights for any institution of carbon trading, then, is that it not make it impossible for people to survive by pricing those in the market for fossil fuels out of that market as long as so many people are still dependent on fossil fuel for lack of any affordable and sustainable alternative energy.[36] We are of course trying to reach a point at which none of us are dependent on fossil fuel, but we cannnot make the transition by simply pretending we are already there and ignoring the fact that most people are now dependent on fossil fuels.[37] So the fundamental issue of fairness that arises under carbon trading is: Who must pay for permits and who receives free permits? The distributive principle for free carbon emissions (emissions without purchase of a permit) needs to be a distributive principle appropriate to an intergenerationally zero-sum resource that is a necessity of life for as long as the current predominantly fossil-fuel energy regime retains its grip on us. Suppose for concreteness, as before, that the trading regime has a budget of 0.5 Tt C emissions remaining before the dominance of fossil fuels must be ended (if warming beyond 2°C

---

[36] The energy supplies of significant numbers of poor people would not be affected because they do not participate in markets for fossil fuel but instead, for example, gather sticks and dung to dry and burn in highly inefficient stoves. These practices, however, are unsustainable in their own way in that they also contribute substantially to climate change when, in particular, millions of primitive cookstoves generate massive clouds of black carbon ("soot") that settle in polar regions and reduce the albedo of the snow. "Black carbon might account for as much as half of Arctic warming" – see E. Rosenthal, "Third-World Stove Soot Is Target in Climate Fight," *New York Times*, April 15, 2009; and International Bank for Reconstruction and Development, *Development and Climate Change*, Box 7.10, 312. Higher market prices for coal, oil, and gas might tend to trap people in these other environmentally highly destructive technologies unless complementary action is taken to create opportunities for them to use energy in less damaging ways, like inexpensive but far more efficient cookstoves in the short term.

I will in the remainder, then, describe those who must purchase fossil fuel in the market as the "market-dependent poor." Strictly speaking, they are the fossil-fuel-market-dependent poor, but to avoid worse clumsiness I will generally use the shorter label. I am grateful to Denis G. Arnold for raising this important distinction.

[37] How the globe came to be dependent on an energy economy centered on fossil fuels is an important historical question. For a fascinating account of how it happened, see K. Pomeranz, *The Great Divergence: China, Europe, and the Making of the Modern World Economy* (Princeton University Press, 2000). For an introduction to the military implications of the dependency of the "superpower" on fossil fuel, see M. T. Klare, *Blood and Oil: The Dangers and Consequences of America's Growing Dependency on Imported Petroleum* (New York: Henry Holt, 2004); and for the role of the American car in particular, see I. Rutledge, *Addicted to Oil: America's Relentless Drive for Energy Security* (New York: I. B. Tauris, 2005).

above pre-industrial levels is to be avoided). So the principle must be appropriate to the distribution of those 0.5 Tt C. The logic is the same whatever the numbers are.

One point that is perfectly obvious – unless one believes that it is acceptable to create social institutions that cause widespread deaths – is that any acceptable distribution of the intergenerationally zero-sum quota of 0.5 Tt C must be compatible with every individual's benefiting from the minimal amount of carbon emissions made necessary for a decent life by the then-existing fossil-fuel energy regime until the regime's environmentally destructive dominance can be broken.

The institutional arrangements currently under discussion include no plan for a (hopelessly impractical) distribution of emissions permits directly to individuals around the world. Most proposed trading schemes involve the distribution (free or at auction) of permits to firms and/or nations; the additional cost created by the requirement to have permits will then of course be passed on to individuals by firms (just as it would if firms faced carbon taxes). But the logic of the situation is clearest if we simplify by describing the situation as if there were going to be permits for individuals, some sold and some provided free of charge. We would then need a priority list specifying who gets (free) emissions first from the remaining, but rapidly diminishing, budget of 0.5 Tt C. Obviously, unless some people are to be condemned to death for lack of benefiting from a minimal amount of carbon emissions, those who can least afford to pay for emissions ought to be at the top of the list of those who do not have to pay for emissions. For example, some people can grow adequate food only by using petroleum-based fertilizer or fossil-fuel-powered irrigation pumps. Unaided, people who now can barely afford the fertilizer or the fuel for the pump might not be able to afford any increase in prices driven by permit charges.

The next step in the argument will not be obvious, but it seems to me to be the only prudent approach, given how dangerous extreme climate change will be and how vital it therefore is to enforce a relatively low cap on total cumulative emissions (such as 1 Tt C) by the time fossil fuel use is eliminated completely (in order to avoid a temperature rise exceeding 2°C above pre-industrial levels). We do not know for how long the remaining budget consisting of the second 0.5 Tt C of possibly 'tolerable' emissions – 0.5 Tt C have already been emitted as of now[38] – will have to supply the

---

[38] The best estimate is that, as of February 20, 2010, we have emitted more than 535,157,625,000 metric tons of carbon into the atmosphere. See the results of the University of Oxford's calculations at http://trillionthton.org/.

for-the-meantime-unavoidable carbon-emission needs of many of the poor. As things are going now, the budget consisting of the second half of the total of 1 Tt C will likely be exhausted in less than 40 years – well before 2050.[39] The longer that many of the poor people on the planet must rely for survival on carbon emissions within a dominant fossil-fuel energy regime, the longer they will need to draw from whatever remains of this budget at any given time. If we are serious about not making the lives of the market-dependent poor impossible, and we accept the science, we must, in effect, reserve enough of the remaining budget of "tolerable" emissions for the fossil-fuel-market-dependent poor to use to maintain themselves at a decent level of existence for the duration of the period during which they must depend on the fossil-fuel regime. Obviously, the longer they are dependent on fossil fuels, the longer they will need to draw upon the budget and the more of it that will be needed strictly for them. On the one hand, the remaining budget of carbon emissions could be enlarged only by allowing warming beyond 2°C above pre-industrial levels, which is yet more dangerous. On the other hand, the time period of the dependence of the poor on carbon emissions can be shortened by making affordable alternative energy without carbon emissions available to them sooner, which is one of the actions most urgent to be taken, for this and other reasons.

It is vital not to confuse the number of Tt C withdrawn from the remaining budget of "tolerable" emissions by the poor without charge (without requiring purchase of a permit), a, with the total number of metric tons of carbon still able to be withdrawn from the budget altogether, N. Amount a will be only a fraction of total sum N, depending on all the variables affecting the purchase of permits, including how many are created, to whom they are distributed, what percentage are auctioned, etc., etc. The consumption of N, the pool of 0.5 Tt C in "tolerable" future emissions still remaining today, will consist of (a), the consumption by any market-dependent poor who are not required to have permits, and (b), the consumption by those who purchase, steal, or otherwise acquire permits. Any emissions regime less flagrant than business as usual in the past has been could extend the life of the total remaining budget of 0.5 Tt C beyond 2050 somewhat – the farther the arrangements depart from business as usual, the farther beyond 2050 the viability of this total remaining budget of

[39] See Allen *et al.*, "The Exit Strategy," 2. As I write this, the computer-based meter at http://trillionthton.org/ shows the trillionth metric ton probably being emitted very early in 2045 – 35 years from now.

emissions compatible with a 50 percent chance of the temperature rise not exceeding 2°C above pre-industrial levels could be extended.

Clearly it is an empirical question how long it will take to exhaust the emission budget consisting of the possibly "tolerable" 0.5 Tt C of emissions into the atmosphere,[40] given that both (a) free and (b) priced emissions will be coming out of the same budget. My suggestion – this is the non-obvious next step in the reasoning – is that *all* the free emissions should at least tentatively and temporarily be reserved entirely for the market-dependent poor.[41] Everyone else pays for permits. If and when further investigation provides solid grounds for a confident judgment about the total number of people across foreseeable generations who will both be unable to afford to purchase emission permits but will also need to benefit from carbon emissions because they do not yet have affordable alternative sources of energy, we might choose to provide free emissions to some larger group, if poor people are still subject to dependence on fossil fuel. How long the budget of possibly "tolerable" additional carbon emissions (0.5 Tt C) will last depends heavily upon (b) the emissions by those who purchase permits, but also upon the numbers of those people across the next generations who cannot afford permits for carbon emissions but nevertheless need to rely on carbon emissions and therefore must be allowed (a) emissions free of charge, unless they are to be condemned to death or desperation. The latter number, in turn, depends on how long it takes for the dependence of the poor on fossil fuels to be ended by the affordability of alternatives to fossil fuels. If and when the numbers become clearer, it might, depending upon what the actual numbers are, no longer be necessary to reserve all free emissions for the poor. Until evidence to the contrary appears, however, the priority list for free emissions from the budget should contain, I would suggest, absolutely no one other than the market-dependent poor. Otherwise, we risk pricing living people into desperation while trying to eliminate carbon

---

[40] Further consideration may show that fewer emissions can be treated as not constituting dangerous interference with the climate system – see note 21, above.

[41] A definite specification of who precisely count as the poor is clearly needed, but this is a familiar problem to which I can contribute nothing special here. It may be, depending on how the free permits would in effect reach the poor (continuing our simplification of permits for individuals), that the free permits ought also to be inalienable or non-tradable in order to make it pointless for "unreliable trustees," like rapacious governments, to steal them – for an interesting discussion, see Simon Caney, "Equality in the Greenhouse." If the mechanics of cap-and-trade seems to be becoming unworkably complex, I would be perfectly happy to see carbon taxes with compensating tax credits for the poor, except perhaps for the giant utilities and the other largest emitters, who could be allowed to trade permits among themselves. The EU has such a mixed system. I am grateful to Ed Page for discussion of this point.

emissions for the sake of future people by way of a trading mechanism to which some living people simply cannot afford to adjust.

What percentage of the permits should be distributed free to the poor? Whatever percentage they need to fulfill their basic rights. Otherwise, our institutional arrangements to protect the poor of the future would be arrangements for starving the poor of the present, which would be a crime against humanity. This will of course reduce the supply of permits to be purchased and drive up their price. The only way to reduce the price pressure is to bring affordable alternative energy on line more quickly.

Those familiar with the climate change debates will realize that my argument here has reached at least as strong a conclusion as the conclusion ordinarily reached by those who appeal to historical responsibility, but I have not relied upon their controversial premise that there is a universal right to equal per capita emissions. The premise playing the analogous role in this reformulation is merely that we could not in decency condemn those who need carbon emissions but cannot afford to purchase permits, by refusing to guarantee them within our new carbon trading institution the emissions they cannot do without, thereby violating their rights to life, health, and/or subsistence. I do not claim to have disproven anywhere the premise about equal per capita rights to emissions underlying the standard versions of the argument about historical responsibility; I have simply not needed that strong a premise because of what the nature of the challenge is empirically turning out to be, namely, the distribution of an intergenerational zero-sum carbon budget.[42] Basically, I have not needed the stronger premise about equal rights to emissions because our situation is so direly constrained by the necessity of remaining within a zero-sum emissions budget. Extreme situations are in some respects clearer – their very starkness simplifies our choices. All that we need to be committed to is the protection of the rights of the current, and soon-to-be-born, most vulnerable against the workings of the very permit system we may feel compelled to create for the sake of the most vulnerable of the more distant future. If we do not act vigorously now, we will be unfair to people who will live in our shadow.

The significance of the recommendation that all free emissions be reserved for the market-dependent poor can be brought out by contrast with another proposal. In his excellent chapter on the competitiveness

---

[42] Compare H. Shue, "Historical Responsibility," Technical Briefing for Ad Hoc Working Group on Long-Term Cooperative Action Under the Convention [AWG-LCA], SBSTA, UNFCC, Bonn, June 4, 2009.

effects of various climate policies, Richard D. Morgenstern (Chapter 10 in this volume) notes that one proposal is for free allowance allocations under the cap-and-trade system for "carbon-intensive" industries. In my terminology this would amount to granting some free emissions to these industries. An implicit appeal to fairness seems to me to be being made by these industries, who are, in effect, arguing as follows: "We understand that emissions must be reduced overall, and we are willing to do our fair share. But as luck would have it, our industry depends on much higher emissions than most. So a fair share for us would be, not an equal share, but a larger share more in line with our need for more 'carbon-intensive' emissions." But what, after all, is a "carbon-intensive" industry? It is an industry that emits an especially large quantity of carbon for the value of what it produces! If the solution to the competitiveness challenges for industries with high emissions of carbon is to concede them special privileges to emit, we simply continue the rush toward emission of the trillionth metric ton and use up portions of the quota of remaining emissions that could be protected for the market-dependent poor. It is the "carbon-intensive" industries whose emissions it is most important to challenge, not to indulge.

6 A SPECTACULAR LEGACY: TRANSCENDING
THE STANDARD CRUEL DILEMMA

Clearly, then, the third reason for urgent vigorous action is that for now, but not indefinitely, we face an opportunity to arrange for the protection of two sets of human rights that will become more and more difficult to protect simultaneously. On the one hand, we can protect against undermining by severe climate change the ability of people of the more distant future to enjoy their rights to life, subsistence, and health by avoiding the emission of the trillionth metric ton of carbon. On the other hand, we can protect against undermining, by means of the very cap-and-trade institution being created for the first purpose, the ability of the market-dependent poor of the present and the near future to enjoy their rights by guaranteeing them carbon emission permits without charge. As time goes by, we are liable to be told, as we often are, that we must choose between the "present poor" and the "future poor." As the remaining pool of carbon emissions possibly "tolerable" by the planetary climate system shrinks, we are likely to be told that everyone must, in order to drive down carbon emissions, pay more to emit carbon, which could price the then-current poor out of the energy market even for what have sometimes been called "subsistence emissions," carbon emissions essential to survival and subsistence.[43] This would sacrifice the present poor to the future poor. Or, we will be told, we must relax the ceiling on total cumulative carbon emissions and let them run on beyond 1 Tt C, which will likely produce more severe climate change and greater obstacles to the fulfillment of the rights of the future poor, sacrificing them to the present poor (and whoever else is emitting carbon!).

The most significant point is that we do not need to face any such dilemma between present rights and future rights if – and, as far as I can see, only if – we take robust action immediately that cuts carbon emissions sharply (so the future poor are not threatened by a deteriorating environment) and does it while protecting the urgent interests of the current poor, which are the substance of their same rights. The longer we continue to fiddle with our current casualness, the closer we will approach a dilemma in which a sudden crackdown on carbon emissions, designed to forestall the trillionth metric ton, which would threaten the subsistence emissions of the then-current poor, will seem to be the only alternative to an abandonment of the ceiling of 1 Tt C, which would threaten the future poor (and possibly everyone else as well, not to mention innumerable other species). But there is no need to put ourselves – or, rather, the current and future poor – into this box by continuing to delay facing reality.[44]

Instead, action is urgent on two converging fronts. First, carbon emissions need to be cut back sharply and aggressively. The atmospheric concentration of carbon will not stop growing until emissions are zero, as the language quoted twice above from the latest IPCC report indicates. Probably the maximum carbon concentration will determine the maximum climate change. Second, alternative energy technologies need to be developed as quickly as humanly possible, aiming at an early day when prices of the alternative technologies are competitive with the prices of fossil fuel and become affordable for the poorest. Fossil fuels are notoriously cheap, of course, which is the main reason we need the cap-and-trade (or carbon tax) institutions to drive up their price by political choice. We must aim for the point of crossover at which declines in the prices of alternative technologies and rises in the prices of fossil fuels mean that fossil fuels lose their

---

[43] The distinction between subsistence emissions and luxury emissions was introduced in the first influential discussion of the ethics of climate change: A. Agarwal and S. Narain, *Global Warming in an Unequal World* (New Delhi: Centre for Science and Environment, 1991), 5. I carried their suggestion forward in Shue, "Subsistence Emissions and Luxury Emissions." Unlike my usage, the two categories are treated as mutually exhaustive of all emissions in Vanderheiden, *Atmospheric Justice*, 67–73 and 242–3.

[44] For discussion of related issues, see S. Humphreys (ed.), *Human Rights and Climate Change* (Cambridge University Press, 2010), *passim.*

competitive price advantage. The farther we move on either front – making fossil fuels more expensive and making alternative energy technologies less expensive – the less far we need to move on the other front. Once the crossover occurs, even the purely selfish who care nothing for the environment and nothing for the rights of others will simply find it efficient to use alternative fuels. At that point, humanity might be out of the woods, provided that we have meanwhile not emitted the trillionth metric ton, or whatever the rapidly advancing science tells us is the outer boundary of environmentally "tolerable" carbon emissions. If we act vigorously and creatively now, we can invent institutions that will provide a priceless legacy of rights protection for multiple generations. Blinkered commission of the "unforgivable sin" of myopic self-indulgence or farsighted creation of invaluable institutions of rights protection – which choice will your generation make? To its undying shame, mine appears to have chosen.

Select bibliography

Adger, W. Neil. "Social Capital, Collective Action and Adaptation to Climate Change." *Economic Geography* 79 (2003): 387–404.
"The Right to Keep Cold." *Environment and Planning A* 36 (2004): 1711–1715.
Adger, W. Neil, Nigel W. Arnell and Emma L. Tompkins. "Successful Adaptation to Climate Change Across Scales." *Global Environmental Change* 15 (2005): 77–86.
Adger, W. Neil, Jon Barnett and Heidi Ellemor. "Unique and Valued Places." In *Climate Change Science and Policy*, ed. Steven H. Schneider *et al.*, 131–138. Washington, DC: Island Press, 2010.
Adger, W. Neil, Irene Lorenzoni and Karen L. O'Brien, eds. *Adapting to Climate Change: Thresholds, Values, Governance*. Cambridge University Press, 2009.
Arnold, Denis G. and Keith Bustos. "Business, Ethics, and Global Climate Change." *Business and Professional Ethics Journal* 22: nos. 2/3 (Summer/Fall 2005): 103–130.
Attfield, Robin. "Mediated Responsibilities, Global Warming and the Scope of Ethics." *Journal of Social Philosophy* 40, no. 2 (2009): 225–236.
Baer, Paul *et al.* "Equity and Greenhouse Gas Responsibility." *Science* 289 (5488), (2000): 2287.
Barnes, Peter. *Who Owns the Sky? Our Common Assets and the Future of Capitalism.* Washington, DC: Island Press, 2001.
Barnes, Peter *et al.* "Creating an Earth Atmospheric Trust." *Science* 319 (5864) (2008): 724.
Barnett, Jon. "Climate Change, Insecurity and Injustice." In *Fairness in Adaptation to Climate Change*, ed. W. Neil Adger *et al.*, 115–129. Cambridge, MA: MIT Press, 2006.
Bell, Derek. "Carbon Justice? The Case Against a Universal Right to Equal Carbon Emissions." In *Seeking Environmental Justice*, ed. Sarah Wilks, 239–257. Amsterdam: Rodolphi, 2008.
Bell, Derek and Simon Caney. *Global Justice and Climate Change*. Oxford University Press, forthcoming.
Bertram, Christopher. "Exploitation and Intergenerational Justice." In *Intergenerational Justice*, ed. Axel Gosseries and Lukas H. Meyer, 147–166. Oxford University Press, 2009.

Brown, Donald. *American Heat: Ethical Problems with the United States' Response to Global Warming.* Lanham, MD: Rowman & Littlefield, 2002.

Brown, Donald et al. *White Paper on the Ethical Dimensions of Climate Change.* Pennsylvania State University: Rock Ethics Institute, 2007.

Brundtland, Gro H. *Our Common Future: The World Commission on Environment and Development.* Oxford University Press, 1987.

Caney, Simon. "Climate Change, Justice and the Duties of the Advantaged." *Critical Review of International Social and Political Philosophy* 13, no. 1 (2010): 203–228.

"Climate Change and the Future: Time, Wealth and Risk." *Journal of Social Philosophy* 40, no. 2 (2009): 163–186.

"Climate Change, Human Rights and Moral Thresholds." In *Human Rights and Climate Change*, ed. Stephen Humphreys, 69–90. Cambridge University Press, 2009.

"Cosmopolitan Justice, Responsibility and Global Climate Change." *Leiden Journal of International Law* 18, no. 4 (2005): 747–775.

"Environmental Degradation, Reparations and the Moral Significance of History." *Journal of Social Philosophy* 73, no. 3 (2006): 464–482.

"Human Rights and Global Climate Change." In *Cosmopolitanism in Context: Perspectives from International Law and Political Theory*, ed. Roland Pierik and Wouter Werner, 43–44. Cambridge University Press, 2010.

"Human Rights, Climate Change and Discounting." *Environmental Politics* 17 (2008): 536–555.

"Justice and the Distribution of Greenhouse Gas Emissions." *Journal of Global Ethics* 5 (2009): 125–146.

Doran, Peter and Maggie K. Zimmerman. "Examining the Scientific Consensus on Climate Change." *EOS* 20, no. 3 (2009): 21–22.

Gardiner, Stephen M. "A Contract on Future Generations?" In *Intergenerational Justice*, ed. Axel Gosseries and Lukas H. Meyer, 77–118. Oxford University Press, 2009.

"A Perfect Moral Storm: Climate Change, Intergenerational Ethics and the Problem of Corruption." *Environmental Values* 15, no. 3 (2006): 397–413.

"Climate Change as a Global Test for Contemporary Political Institutions and Theories." In *Climate Change, Ethics and Human Security*, ed. Karen O'Brien, Asunción Lera St. Clair and Berit Kristoffersen, 131–153. Cambridge University Press, 2010.

"Ethics and Global Climate Change." *Ethics* 114 (2004): 555–600.

"Is 'Arming the Future' with Geoengineering Really the Lesser Evil? Some Doubts About the Ethics of Intentionally Manipulating the Climate System." In *Climate Ethics: Essential Readings*, ed. Stephen Gardiner et al., 284–312. Oxford University Press, 2010.

"Rawls and Climate Change: Does Rawlsian Political Philosophy Pass the Global Test?" *Critical Review of International Social and Political Philosophy*, forthcoming.

Gardiner, Stephen, Simon Caney, Dale Jamieson and Henry Shue, eds. *Climate Ethics: Essential Readings.* Oxford University Press, 2010.

Gosseries, Axel. "Cosmopolitan Luck Egalitarianism and the Greenhouse Effect." *Canadian Journal of Philosophy* 31 (2005): 279–310.

"Historical Emissions and Free-Riding." *Ethical Perspectives* 11 (2003): 36–60.

Gosseries, Axel and Lukas H. Meyer, eds. *Intergenerational Justice*. Oxford University Press, 2009.

Hayward, Tim. "Human Rights Versus Emissions Rights: Climate Justice and the Equitable Distribution of Ecological Space." *Ethics and International Affairs* 21, no. 4 (2007): 431–450.

Houghton, John. *Global Warming: The Complete Briefing*, 4th edn. Cambridge University Press, 2009.

Jamieson, Dale. "Adaptation, Mitigation, and Justice." In *Perspectives on Climate Change*, ed. Walter Sinnott-Armstrong, and Richard Howarth, 221–253. New York: Elsevier, 2005.

"Climate Change and Global Environmental Justice." In *Changing the Atmosphere: Expert Knowledge and Environmental Governance*, ed. Clark A. Miller and Paul N. Edwards, 287–307. Cambridge University Press, 2001.

"Climate Change, Responsibility, and Justice." *Science and Engineering Ethics*, forthcoming.

"Duties to the Distant: Humanitarian Aid, Development Assistance, and Humanitarian Intervention." *Journal of Ethics* 9, no. 1–2 (2005): 151–170.

"Ethics and Intentional Climate Change." *Climatic Change* 33, no. 3 (1996): 323–336.

"Ethics, Public Policy and Global Warming." *Science, Technology and Human Values* 17, no. 2 (1992): 139–153.

"The Epistemology of Climate Change: Some Morals for Managers." *Society and Natural Resources* 4, no. 4 (1991): 319–329.

"The Moral and Political Challenges of Climate Change." In *Creating a Climate for Change: Communicating Climate Change and Facilitating Social Change*, ed. Susanne C. Moser and Lisa Dilling, 475–482. New York: Cambridge University Press, 2007.

"The Post-Kyoto Climate: A Gloomy Forecast." *Georgetown Journal of International Environmental Law* 20, no. 4 (2008): 537–551.

"The Rights of Animals and the Demands of Nature." *Environmental Values* 17, no. 2 (2008): 181–199.

Johnson, Lawrence E. "Future Generations and Contemporary Ethics." *Environmental Values* 12 (2003): 471–487.

Lomborg, Bjørn. *Cool It: The Skeptical Environmentalist's Guide to Global Warming*. New York: Random House, 2007.

The Skeptical Environmentalist: Measuring the Real State of the World. Cambridge University Press, 2001.

Lynas, Mark. *Six Degrees: Our Future on a Hotter Planet*. Washington, DC: National Geographic, 2008.

Miller, David. "Global Justice and Climate Change: How Should Responsibilities Be Distributed?" The Tanner Lectures on Human Values. Tsinghua University, Beijing (March 24–25, 2008), www.tannerlectures.utah.edu/lectures/documents/ Miller_08.pdf.

Moellendorf, Darrel. "Justice and the Intergenerational Assignment of the Costs of Climate Change," *Journal of Social Philosophy* 40 (2009): 204–224.
  "Treaty Norms and Climate Change Mitigation." *Ethics and International Affairs* 23 (2009): 247–265.
Mulgan, Tim. *Future People: A Moderate Consequentialist Account of Our Obligations to Future Generations.* Oxford University Press, 2006.
Neumayer, Eric. "In Defense of Historical Accountability for Greenhouse Gas Emissions." *Ecological Economics* 33 (2000): 185–192.
  Weak Versus Strong Sustainability: Exploring the Limits of Two Opposing Paradigms, 2nd edn. Cheltenham: Edward Elgar, 2003.
Nolt, John. "Hope, Self-Transcendence and the Justification of Environmental Ethics." *Inquiry* 53, 2 (2010): 162–182.
Nordhaus, William. *A Question of Balance: Weighing the Options on Global Warming Policies.* New Haven, CT: Yale University Press, 2008.
Oreskes, Naomi. "Beyond the Ivory Tower: The Scientific Consensus on Climate Change." *Science* 306, no. 5702 (2004): 1686.
Oreskes, Naomi and Erik M. Conway. *Merchants of Doubt: How a Handful of Scientists Obscured the Truth on Issues from Tobacco Smoke to Global Warming.* New York: Bloomsbury Press, 2010.
Paavola, Jouni and W. Neil Adger. "Fair Adaptation to Climate Change." *Ecological Economics* 56 (2006): 594–609.
Pacala, Stephen and Robert H. Socolow. "Stabilization Wedges: Solving the Climate Problem for the Next 50 Years with Current Technologies." *Science* 305, no. 5686 (2004): 968–972.
Page, Edward. *Climate Change, Justice and Future Generations.* Cheltenham: Edward Elgar, 2006.
Roberts, J. Timmons and Bradley C. Parks. *A Climate of Injustice: Global Inequality, North–South Politics and Climate Policy.* Cambridge, MA: MIT Press, 2007.
Sandler, Ronald L. and Philip Cafaro, eds. *Environmental Virtue Ethics.* Oxford: Rowman & Littlefield, 2005.
Shue, Henry. "Bequeathing Hazards: Security Rights and Property Rights of Future Humans." In *Global Environmental Economics*, ed. Mohammed H. I. Dore, and Timothy D. Mount, 40–42. Malden, MA: Blackwell, 1999.
  "Global Environment and International Inequality." *International Affairs* 75 (1999), 531–545.
  "Subsistence Emissions and Luxury Emissions." *Law and Policy* 15, no. 1 (1993): 39–59.
Singer, Peter. *One World: The Ethics of Globalization.* New Haven, CT: Yale University Press, 2002.
Sinnott-Armstrong, Walter and Richard B. Howarth, eds. *Perspectives on Climate Change: Science, Economics, Politics and Ethics, Advances in the Economics of Environmental Resources*, vol. 5. Amsterdam: Elsevier, 2005.
Socolow, Robert H. and Sau-Hai Lam. "Good Enough Tools for Global Warming Policy Making." *Philosophical Transactions of the Royal Society* 365 (2007): 897–934.

Socolow, Robert H. and Stephen W. Pacala. "A Plan to Keep Carbon in Check." *Scientific American* 295, no. 3 (2006): 50–57.

Stern, Nicholas. *The Economics of Climate Change: The Stern Review.* Cambridge University Press, 2007.

Traxler, Martino. "Fair Chore Division for Climate Change." *Social Theory and Practice* 28 (2002): 101–134.

Vanderheiden, Steve. *Atmospheric Justice: A Political Theory of Climate Change.* New York: Oxford University Press, 2008.

Vanderheiden, Steve, ed. *Political Theory and Global Climate Change.* Cambridge, MA: MIT Press, 2008.

Weart, Spencer R. *The Discovery of Global Warming.* Cambridge University Press, 2003.

Index

CPSIA information can be obtained at www.ICGtesting.com
Printed in the USA
LVOW06s0339050815

448743LV00007B/96/P

9 781107 666016